Cracking the

AP®

CALCULUS AB EXAM

2020 Edition

David S. Kahn

PrincetonReview.com

Penguin
Random
House

The Princeton Review
110 East 42nd Street, 7th Floor
New York, NY 10017
Email: editorialsupport@review.com

Published in the United States by Penguin Random House LLC, New York, and in Canada by Random House of Canada, a division of Penguin Random House Ltd., Toronto.

ISBN: 978-0-525-56815-5
eBook ISBN: 978-0-525-56853-7
ISSN: 2334-2633

Editor: Aaron Riccio
Production Editors: Liz Dacey, Lyssa Mandel
Production Artist: Deborah Weber

Printed in the United States of America.

10 9 8 7 6 5 4 3 2 1

2020 Edition

Editorial
Rob Franek, Editor-in-Chief
David Soto, Director of Content Development
Stephen Koch, Survey Manager
Deborah Weber, Director of Production
Gabriel Berlin, Production Design Manager
Selena Coppock, Managing Editor
Aaron Riccio, Senior Editor
Meave Shelton, Senior Editor
Chris Chimera, Editor
Sarah Litt, Editor
Orion McBean, Editor
Brian Saladino, Editor
Eleanor Green, Editorial Assistant

Penguin Random House Publishing Team
Tom Russell, VP, Publisher
Alison Stoltzfus, Publishing Director
Amanda Yee, Associate Managing Editor
Ellen Reed, Production Manager
Suzanne Lee, Designer

Acknowledgments

First of all, I would like to thank Arnold Feingold and Peter B. Kahn for once again doing every problem, reading every word, and otherwise lending their invaluable assistance. I also want to thank Aaron Riccio for being a terrific editor on this edition of the book, Gary King for his first-rate reading, analysis, and contributions, and the production team of Liz Rutzel, Harmony Quiroz, and Deborah Weber. Thanks to Frank, without whose advice I never would have taken this path. Thanks to Jeffrey, Miriam, and Vicki for moral support. Thanks Mom.

Finally, I would like to thank the people who really made all of this effort worthwhile—my students. I hope that I haven't omitted anyone, but if I have, the fault is entirely mine.

Aaron and Sasha, Aaron, Aaron T., Abby B., Abby F., Abby H., Abby L., Abbye, Abigail H., Addie, Aidan, Alan M., Albert G., Alec G, Alec M., Alec R., Alex and Claire, Alex A., Alex B., Alex and Daniela B., Alex F., Alex D., Alex G., Alex H., Alex I., Alex S., Alex and Gabe, Alexa, Alexandra, Natalie and Jason, Alexes, Alexis and Brittany, Ali and Jon, Ali and Amy Z., Ali H., Alice, Alice C., Alicia, Alisha, Allegra, Allie and Lauren, Allison and Andrew, Allison H., Allison and Matt S., Allison R., Ally, Ally M., Ally T., Ally W., Aly and Lauren, Alyssa C., Alyssa TDC, Alyssa and Courtney, Alvin, Amanda, Brittany, and Nick A., Amanda and Pamela B., Amanda C., Amanda E., Amanda H., Amanda M., Amanda R., Amanda S., Amanda Y., Amber and Teal, Amparo, Andrea T., Andrea V., Andrew A., Andrei, Andrew B., Andrew C., Andrew D., Andrew E., Andrew H., Andrew M., Andrew S., Andy and Sarah R., Andy and Allison, Angela, Angela F., Angela L., Anisha, Ann, Anna C-S., Anna D., Anna and Jon, Anna and Max, Anna L., Anna M., Anna S., Anna W., Annabelle M., Annerose, Annie, Annie B., Annie W., Antigone, Antonio, Anu, April, Ares, Ariadne, Ariane, Ariel, Arielle and Gabrielle, Arthur and Annie, Arya, Asheley and Freddy, Ashley A., Ashley K., Ashley and Sarah, Ashley and Lauren, Aubrey, Avra, Becca A., Becky B., Becky S., Becky H., Ben S., Ben and Andrew Y., Ben D., Ben S., Benjamin D., Benjamin H., Benji, Beth and Sarah, Bethany and Lesley, Betsy and Jon, Blythe, Bianca and Isabella, Biz, Bonnie, Bonnie C., Boris, Braedan, Brendan, Brett, Brett A., Brette and Josh, Brian, Brian C., Brian N., Brian W., Brian Z., Brigid, Brin, Brittany E., Brittany S., Brooke, Devon, and Megan, Brooke and Lindsay E., Butch, Cailyn H., Caitlin, Caitlin F., Caitlin M., Caitlin and Anna S., Camilla and Eloise, Camryn, Candace, Cara, Cara G., Carinna, Caroline A., Caroline C., Caroline H., Caroline S., Caroline and Peter W., Carrie M., Catherine W., Cecilia, Celia, Chad, Channing, Charlie, Charlotte, Charlotte B., Charlotte M., Chase, Chelsea, Chloe, Chloe K., Chris B., Chris C., Chris K., Chris P., Chrissie, Christa S., Christen C., Christian, Christian C., Christina F., Christine, Christine W., CJ, Claire H., Clare, Claudia, Clio, Colette, Conor G., Corey, Corinne, Coryn, Courtney and Keith, Courtney B., Courtney F., Courtney S., Courtney W., Craig, Dan M., Dana J., Dani and Adam, Daniel, Daniel and Jen, Daniella, Daniella C., Danielle and Andreas, Danielle and Nikki D., Danielle G., Danielle H., Danny K., Dara and Stacey, Dara M., Darcy, Darya, David B., David C., David R., David S., Deborah and Matthew, Deniz and Destine, Deshaun, Deval, Devin, Devon and Jenna, Deyshaun, Diana H., Dilly, Disha, Dong Yi, Dora, Dorrien, Eairinn, Eddy, Elana, Eleanor K., Elexa and Nicky, Elisa, Eliza, Elizabeth, Elizabeth and Mary C., Elizabeth F., Ella, Ellie, Elly B., Elyssa T., Emily A., Emily B., Emily and Allison, Emily and Catie A., Emily C., Emily G., Emily H., Emily K., Emily L., Emily and Pete M., Emily R., Emily R-H., Emily S., Emily T., Emma, Emma and Sophie, Emma D., Emma S., Eric N., Erica F., Erica H., Erica R., Erica S., Eric and Lauren, Erica and Annie, Erika, Erin, Erin I., Ernesto, Esther, Ethan, Eugenie, Eva, Evan, Eve H., Eve M., Fifi, Francesca, Frank, Franki, Gabby, Geoffrey, George and Julian., George M., Gloria, Grace M., Grace P., Gracie, Graham and Will, Greg, Greg F., Griffin, Gussie, Gylianne, Haley, Halle R., Hallie, Hannah C., Hannah J., Hannah R., Hannah and Paige, Harrison, Annabel and Gillian, Harry, Hayley and Tim D., Hazel, Heather D., Heather E., Heather F., Heather and Gillian, Hernando and Vicki, Hilary and Lindsay, Hilary F., Holli, Holly G., Holly K., Honor, Ian, Ian P., India, Ingrid, Ira, Isa, Isabel, Isabella and Simone, Isabelle T., Ismini, Isobel, Ivy, Jabari, Jacob and Kara, Jack, Jack B., Jackie and Vicky,

Jackie, Jackie A., Jackie S., Jackson, Jaclyn and Adam, Jacob D., Jada, James S., Jamie, Jason P., Jason and Andrew, Jay, Jay K., Jae and Gideon, Jake T., Jarrod, Jason L., Jayne and Johnny, Jed D., Jed F., Jeffrey M., Jenna, Jen, Jen M., Jenn N., Jennifer B., Jennifer W., Jenny K, Jenny and Missy, Jeremy C., Jeremy and Caleb, Jess, Jess P., Jesse, Jessica, Jessica and Eric H., Jessica L., Jessica T., Jessie, Jessie C., Jessie and Perry N., Jill, Jillian, Daniel and Olivia, Jillian S., Jimin, Jimmy C., Jimmy P., Joanna, Joanna C., Joanna G., Joanna M., Joanna and Julia M., Joanna W., Jocelyn, Jody and Kim, Joe B., Johanna, John, John and Dan, Jonah and Zoe, Jonathan G., Jonathan P., Jonathan W., Jongwoo, Jordan C., Jordan and Blair F., Jordan G., Jordan P., Jordana, Jordyn, Jordyn K., Josh and Jesse, Josh and Noah, Judie and Rob, Julia and Caroline, Julia, Julia F., Julia G., Julia H., Julia and Charlotte P., Julia T., Julie, Julie H., Julie J., Julie and Dana, Julie P., Juliet, Juliette R., Justin L., Kara B., Kara O., Kasey, Kasia and Amy, Kat C., Kat R., Kate D., Kate F., Kate G., Kate L., Kate P., Kate S., Kate W., Katie C., Katie F., Katherine C., Kathryn, Leslie, and Travis, Katie, Katie M., Katrina, Keith M., Kelly C., Ken D., Kenzie, Keri, Kerri, Kim H., Kimberly, Kimberly H., Kirsten, Kitty and Alex, Kori, Kripali, Krista, Erika, and Karoline, Kristen, Kristen, Kristen Y., Magan, and Alexa, Laila and Olivia R., Laura B., Laura F., Laura G., Laura R., Laura S., Laura T., Laura Z., Lauren and Eric, Lauren R., Lauren S., Lauren T., Lauren and Allie, Lea K., Leah, Lee R., Leigh and Ruthie, Leigh, Lexi S., Lila, Lila M., Lilaj, Lili C., Lilla, Lily, Lily M-R., LilyHayes, Lisbeth and Charlotte, Lina, Lindsay F., Lindsay K., Lindsay L., Lindsay N., Lindsay R, Lindsay and Jessie S., Lindsey and Kari, Lisa, Liz D., Liz H., Liz M., Lizzi B., Lizzie A., Lizzie M., Lizzie W., Lizzy, Lizzy C., Lizzy R., Lizzy T., Louis, Lucas, Luke C., Lucinda, Lucy, Lucy D., Mackenzie, Maddie and JD, Maddie H., Maddie P., Maddy N., Maddy T., Maddy W., Madeline, Magnolia, Maggie L., Maggie T., Manuela, Mara and Steffie, Marcia, Margaret S., Maria B., Mariel C., Mariel L., Mariel S., Marielle K., Marielle S., Marietta, Marisa B., Marisa C., Marissa, Marlee, Marnie and Sam, Mary M., Matt, Matthew B., Matthew K., Matt F., Matt G., Matt and Caroline, Matt V., Matthew G., Mavis H., Max B., Max and Chloe K., Max M., Maxx, Ben, and Sam, Maya and Rohit, Maya N., Mckyla, Meesh, Megan G., Meliha, Melissa, Melissa and Ashley I., Meredith, Meredith and Gordie B., Meredith and Katie D., Meredith R., Meredith S-K., Michael and Dan, Michael H., Michael R., Michaela, Michal, Michele S., Mike C., Mike R., Miles, Milton, Miranda, Moira, Molly L., Molly R., Molly and Annie, Morgan, Morgan C., Morgan, Zoe, and Lila, Morgan K., Nadia and Alexis, NaEun, Nancy, Nanette, Naomi, Natalie and Andrew B., Natasha and Mikaela B., Natasha L., Nathania, Nico, Nick, Nick A., Nicky S., Nicole B., Nicole J., Nicole N., Nicole P., Nicole S., Nicolette, Nidhi, Nikki G., Nina R., Nina V., Noah, Nora, Nushien, Odean, Ogechi, Oli and Arni, Oliver, Oliver W., Olivia and Daphne, Olivia P., Omar, Oren, Pam, Paige, Pasan, Paul A., Peter, Philip and Peter, Phoebe, Pierce, Pooja, Priscilla, Quanquan, Quinn and Chris, Rachel A, Rachel B., Rachel and Adam, Rachel H., Rachel I., Rachel and Jake, Rachel K., Rachel L., Rachel and Mitchell M., Rachel S., Rachel and Eli, Rachel and Steven, Rachel T., Rachel W., Rae, Ramit, Randi and Samantha, Rasheeda, Ray B., Rayna, Rebecca A., Rebecca B., Rebecca G., Rebecca and Gaby K., Rebecca R., Rebecca W., Renee F., Resala, Richard, Tina, and Alice, Richie D., Ricky, Riley C., Rob L., Robin P., Romanah, Rose, Ryan, Ryan B., Ryan S., Saahil, Sabrina and Arianna, Sabrina O., Sally, Sam B., Sam C., Sam G., Sam L., Sam R., Sam S., Sam W., Sam Z., Samantha and Savannah, Samantha M., Samuel C., Samar, Sara G., Sara H., Sara L., Sara R., Sara S., Sara W., Sarah, Patty, and Kat, Sarah and Beth, Sarah A., Sarah C., Sarah D., Sarah F., Sarah G., Sarah and Michelle K., Sarah L., Sarah M-D., Sarah and Michael, Sarah P., Sarah S., Sascha, Samara, Saya, Sean, Selena G., Serena, Seung Woo, Shadae, Shannon, Shayne, Shayne, Shorty, Siegfried, Simon W., Simone, Sinead, Siobhan, Skye F., Skye L., Sofia G., Sofia M., Sonja, Sonja and Talya, Sophia A., Sophia B., Sophia J., Sophia S., Sophie and Tess, Sophie D., Sophie S., Sophie W., Stacey, Stacy, Steffi, Steph, Stephanie, Stephanie K., Stephanie L., Stephen C., Sumair, Sunaina, Suzie, Syd, Sydney, Sydney S., Tammy and Hayley, Tara and Max, Taryn, Tasnia, Tayler, Taylor, Tenley and Galen, Terrence, Tess, Theodora, Thomas B., Timothy H., Tina, Tita, Tom, Tori, Torri, Tracy, Tracy K., Tripp H., Tripp W., Tyler, Tyrik, Uma, Vana, Vanessa and John, Veda, Veronica M., Vicky B., Victor Z., Victoria, Victoria H., Victoria J., Victoria M., Vinny and Eric, Vivek, Vladimir, Waleed, Will A., Will, Teddy, and Ellie, William M., William Z., Wyatt and Ryan, Wyna, Xianyuan, Yakir, Yesha, Yujin, Yuou, Zach, Zach B., Zach S., Zach Y., Zachary N., Zaria, Zoë, Zoe L., and Zoey and Rachel.

Contents

Get More (Free) Content
at **PrincetonReview.com/cracking**

As easy as **1•2•3**

1 Go to PrincetonReview.com/cracking and enter the following ISBN for your book:

9780525568155

2 Answer a few simple questions to set up an exclusive Princeton Review account. *(If you already have one, you can just log in.)*

3 Enjoy access to your **FREE** content!

Once you've registered, you can...

- Get our take on any recent or pending updates to the AP Calculus AB Exam

- Take a full-length practice SAT and ACT

- Get valuable advice about the college application process, including tips for writing a great essay and where to apply for financial aid

- If you're still choosing between colleges, use our searchable rankings of *The Best 385 Colleges* to find out more information about your dream school.

- Access comprehensive study guides and a variety of printable resources, including bonus questions, an appendix on TI-84 use, additional bubble sheets, and important formulas

- Check to see if there have been any corrections or updates to this edition

Need to report a potential **content** issue?

Contact **EditorialSupport@review.com** and include:
- full title of the book
- ISBN
- page number

Need to report a **technical** issue?

Contact **TPRStudentTech@review.com** and provide:
- your full name
- email address used to register the book
- full book title and ISBN
- Operating system (Mac/PC) and browser (Firefox, Safari, etc.)

Look For These Icons Throughout The Book

 ONLINE ARTICLES

 PROVEN TECHNIQUES

 APPLIED STRATEGIES

 MORE GREAT BOOKS

 STUDY BREAK

 GOING DEEPER

Part I
Using This Book to Improve Your AP Score

- Preview: Your Knowledge, Your Expectations
- Your Guide to Using This Book
- How to Begin

PREVIEW: YOUR KNOWLEDGE, YOUR EXPECTATIONS

Ch-ch-ch-changes

The organizational structure of the AP Calculus AB Exam is changing for the 2019–2020 course. The material in this book covers all currently announced content for the course and exam, but please check your online student tools for any late-breaking news.

Your route to a high score on the AP Calculus AB Exam depends a lot on how you plan to use this book. Start thinking about your plan by responding to the following questions.

1. Rate your level of confidence about your knowledge of the content tested by the AP Calculus AB Exam:

    A. Very confident—I know it all
    B. I'm pretty confident, but there are topics for which I could use help
    C. Not confident—I need quite a bit of support
    D. I'm not sure

2. If you have a goal score in mind, circle your goal score for the AP Calculus AB Exam:

    5   4   3   2   1   I'm not sure yet

3. What do you expect to learn from this book? Circle all that apply to you.

    A. A general overview of the test and what to expect
    B. Strategies for how to approach the test
    C. The content tested by this exam
    D. I'm not sure yet

YOUR GUIDE TO USING THIS BOOK

This book is organized to provide as much—or as little—support as you need, so you can use this book in whatever way will be most helpful for improving your score on the AP Calculus AB Exam.

*   The remainder of **Part I** will provide guidance on how to use this book and help you determine your strengths and weaknesses.

*   **Part II** of this book contains Practice Test 1 and its answers and explanations. (Bubble sheets can be printed out from your free online student tools for easy use.) We strongly recommend that you take this test before going any further, in order to realistically determine:
    o   your starting point right now
    o   which question types you're ready for and which you might need to practice
    o   which content topics you are familiar with and which you will want to carefully review

Once you have nailed down your strengths and weaknesses with regard to this exam, you can focus your test preparation, build a study plan, and be efficient with your time.

- **Part III** of this book will
  - o provide information about the structure, scoring, and content of the AP Calculus AB Exam.
  - o help you to make a study plan.
  - o point you towards additional resources.

- **Part IV** of this book will explore the following strategies:
  - o how to attack multiple-choice questions
  - o how to write a high scoring free-response answer
  - o how to manage your time to maximize the number of points available to you

- **Part V** of this book covers the content you need for your exam.

- **Part VI** of this book contains two additional Practice Tests and their answers and explanations. If you skipped Practice Test 1, we recommend that you do all three (with at least a day or two between them) so that you can compare your progress between them. Additionally, this will help to identify any external issues: if you get a certain type of question wrong both times, you probably need to review it. If you only got it wrong once, you may have run out of time or been distracted by something. In either case, this will allow you to focus on the factors that caused the discrepancy and to be as prepared as possible on the day of the test.

Once you register your book online, you can download and print out the bubble sheets for all three practice tests. You can also find a handy Appendix that contains all the formulas you should know.

You may choose to use some parts of this book over others, or you may work through the entire book. This will depend on your needs and how much time you have. Let's now look at how to make this determination.

HOW TO BEGIN

1. Take a Test

   Before you can decide how to use this book, you need to take a practice test. Doing so will give you insight into your strengths and weaknesses, and the test will also help you make an effective study plan. If you're feeling test-phobic, remind yourself that a practice test is a tool for diagnosing yourself—it's not how well you do that matters but how you use information gleaned from your performance to guide your preparation.

   So, before you read further, take the AP Calculus AB Practice Test 1 starting at page 7 of this book. Be sure to do so in one sitting, following the instructions that appear before the test.

2. **Check Your Answers**

Using the answer key on page 38, count how many multiple-choice questions you got right and how many you missed. Don't worry about the explanations for now, and don't worry about why you missed questions. We'll get to that soon.

3. **Reflect on the Test**

After you take your first test, respond to the following questions:

- How much time did you spend on the multiple-choice questions?

- How much time did you spend on each free-response question?

- How many multiple-choice questions did you miss?

- Do you feel you had the knowledge to address the subject matter of the free-response questions?

- Do you feel your free responses were well organized and thoughtful?

- Circle the content areas that were most challenging for you and draw a line through the ones in which you felt confident/did well.

  o Functions, Graphs, and Limits

  o Differential Calculus

  o Integral Calculus

  o Applications of Derivatives

  o Applications of Integrals

4. **Read Part III and Complete the Self-Evaluation**

Part III will provide information on how the test is structured and scored. It will also set out areas of content that are tested.

As you read Part III, reevaluate your answers to the questions above. At the end of Part III, you will revisit the questions above and refine your answers to them. You will then be able to make a study plan, based on your needs and time available, that will allow you to use this book most effectively.

5. **Engage with Parts IV and V as Needed**

Notice the word *engage*. You'll get more out of this book if you use it intentionally than if you read it passively, hoping for an improved score through osmosis.

Strategy chapters will help you think about your approach to the question types on this exam. Part IV will open with a reminder to think about how you approach questions now and then close with a reflection section asking you to think about how/ whether you will change your approach in the future.

Content chapters in Part V are designed to provide a review of the content tested on the AP Calculus AB Exam, including the level of detail you need to know and how the content is tested. You will have the opportunity to assess your mastery of the content of each chapter through test-appropriate questions and a reflection section.

6. **Take Another Test and Assess Your Performance**

    Once you feel you have developed the strategies you need and gained the knowledge you lacked, you should take Test 2. You should do so in one sitting, following the instructions at the beginning of the test.

    When you are done, check your answers to the multiple-choice sections. See if a teacher will read your long-form calculus responses and provide feedback.

    Once you have taken the test, reflect on what areas you still need to work on, and revisit the chapters in this book that address those topics. Through this type of reflection and engagement, you will continue to improve.

7. **Keep Working**

    After you have revisited certain chapters in this book, continue the process of testing, reflecting, and engaging with Practice Test 3 on page 569. You want to be increasing your readiness by carefully considering the types of questions you are getting wrong and how you can change your strategic approach to different parts of the test.

    As we will discuss in Part III, there are other resources available to you, including a wealth of information on the College Board's AP Students website: https://apstudent.collegeboard.org/apcourse/ap-calculus-ab. You can continue to explore areas that can stand to improve and engage in those areas right up to the day of the test.

Another Course? Of Course!

If you can't get enough AP Calculus AB and want to review this material with an expert, we also offer an online Cram Course that you can sign up for here: https://www.princeton-review.com/college/ ap-test-prep.

Part II
Practice Test 1

Practice Test 1

AP® Calculus AB Exam

SECTION I: Multiple-Choice Questions

DO NOT OPEN THIS BOOKLET UNTIL YOU ARE TOLD TO DO SO.

At a Glance

Total Time
1 hour and 45 minutes
Number of Questions
45
Percent of Total Grade
50%
Writing Instrument
Pencil required

Instructions

Section I of this examination contains 45 multiple-choice questions. Fill in only the ovals for numbers 1 through 45 on your answer sheet.

CALCULATORS MAY NOT BE USED IN THIS PART OF THE EXAMINATION.

Indicate all of your answers to the multiple-choice questions on the answer sheet. No credit will be given for anything written in this exam booklet, but you may use the booklet for notes or scratch work. After you have decided which of the suggested answers is best, completely fill in the corresponding oval on the answer sheet. Give only one answer to each question. If you change an answer, be sure that the previous mark is erased completely. Here is a sample question and answer.

Sample Question   Sample Answer

Chicago is a
(A) state
(B) city
(C) country
(D) continent

Use your time effectively, working as quickly as you can without losing accuracy. Do not spend too much time on any one question. Go on to other questions and come back to the ones you have not answered if you have time. It is not expected that everyone will know the answers to all the multiple-choice questions.

About Guessing

Many candidates wonder whether or not to guess the answers to questions about which they are not certain. Multiple-choice scores are based on the number of questions answered correctly. Points are not deducted for incorrect answers, and no points are awarded for unanswered questions. Because points are not deducted for incorrect answers, you are encouraged to answer all multiple-choice questions. On any questions you do not know the answer to, you should eliminate as many choices as you can, and then select the best answer among the remaining choices.

CALCULUS AB

SECTION I, Part A

Time—60 Minutes

Number of questions—30

A CALCULATOR MAY NOT BE USED ON THIS PART OF THE EXAMINATION

Directions: Solve each of the following problems, using the available space for scratchwork. After examining the form of the choices, decide which is the best of the choices given and fill in the corresponding oval on the answer sheet. No credit will be given for anything written in the test book. Do not spend too much time on any one problem.

In this test: Unless otherwise specified, the domain of a function f is assumed to be the set of all real numbers x for which $f(x)$ is a real number.

1. If $f(x) = 3x^2 - 10x + \ln(x - 3)$, then $f'(4) =$

(A) 8
(B) 9
(C) 14
(D) 15

2. $\lim\limits_{x \to \infty} \dfrac{5x^2 - 3x + 1}{4x^2 + 2x + 5}$ is

(A) 0

(B) $\dfrac{4}{5}$

(C) $\dfrac{5}{4}$

(D) ∞

GO ON TO THE NEXT PAGE.

3. If $f(x) = \dfrac{3x^2 + x}{3x^2 - x}$, then $f'(x)$ is

(A) 1

(B) $\dfrac{6x^2 + 1}{6x^2 - 1}$

(C) $\dfrac{-6}{(3x - 1)^2}$

(D) $\dfrac{-2x^2}{(x^2 - x)^2}$

4. $\displaystyle\lim_{x \to 0} \dfrac{3\sin(2x)}{x} =$

(A) 0
(B) 3
(C) 6
(D) The limit does not exist.

GO ON TO THE NEXT PAGE.

5. If $y = \sin(xy)$, find $\dfrac{dy}{dx}$.

   (A) $\cos(xy)$

   (B) $\dfrac{y\cos(xy)}{1 - x\cos(xy)}$

   (C) $x\cos(xy)$

   (D) $\dfrac{y\cos(xy)}{1 - x}$

6. Using a right Riemann sum with four subintervals, approximate $\displaystyle\int_0^4 3x^2 + 1\,dx$.

   (A) 45
   (B) 46
   (C) 94
   (D) 95

7. If $h(x) = f(x)g(4x)$, then $h'(2) =$

   (A) $f'(2)g'(8)$
   (B) $f'(2)g(8) + 4f(2)g'(8)$
   (C) $f'(2)g(8) + f(2)g'(8)$
   (D) $4f'(2)g'(8)$

GO ON TO THE NEXT PAGE.

8. An equation of the line normal to the graph of $y = \sqrt{(3x^2 + 2x)}$ at (2, 4) is

(A) $4x + 7y = 20$
(B) $-7x + 4y = 2$
(C) $7x + 4y = 30$
(D) $4x + 7y = 36$

9. $4 \int \csc^2 x \, dx =$

(A) $-4\cot x + C$

(B) $-\dfrac{1}{4}\cot x + C$

(C) $\dfrac{4\csc^3 x}{3} + C$

(D) $-\dfrac{4\csc^3 x}{3} + C$

10. If $f(x) = \cos^2 x$, then $f''(\pi) =$

(A) -2
(B) 0
(C) 1
(D) 2

GO ON TO THE NEXT PAGE.

11. If $f(x) = \dfrac{5}{x^2+1}$ and $g(x) = 3x$, then $g(f(2)) =$

(A) $\dfrac{5}{37}$

(B) 3

(C) 5

(D) $\dfrac{37}{5}$

12. $\displaystyle\int x\sqrt{1+x^2}\,dx =$

(A) $\dfrac{1}{3}\left(1+x^2\right)^{\frac{3}{2}} + C$

(B) $\dfrac{4}{3}\left(1+x^2\right)^{\frac{3}{2}} + C$

(C) $-\left(1+x^2\right)^{-\frac{1}{2}} + C$

(D) $-\dfrac{1}{4}\left(1+x^2\right)^{-\frac{1}{2}} + C$

GO ON TO THE NEXT PAGE.

13. The slope of the line tangent to the graph of $3x^2 + 5 \ln y = 12$ at $(2, 1)$ is

(A) $-\dfrac{12}{5}$

(B) $\dfrac{12}{5}$

(C) $\dfrac{5}{12}$

(D) -7

---

14. The equation $y = 2 - 3 \sin \dfrac{\pi}{4} (x - 1)$ has a fundamental period of

(A) $\dfrac{1}{8}$

(B) $\dfrac{4}{\pi}$

(C) 8

(D) 2π

GO ON TO THE NEXT PAGE.

15. The graph of f' is given below. At which point is f continuous but not differentiable?

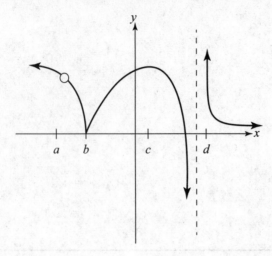

(A) a
(B) b
(C) c
(D) d

16. For what value of x does the function $f(x) = x^3 - 9x^2 - 120x + 6$ have a local minimum?

(A) 10
(B) 4
(C) −4
(D) −10

GO ON TO THE NEXT PAGE.

17. The acceleration of a particle moving along the x-axis at time t is given by $a(t) = 4t - 12$. If the velocity is 10 when $t = 0$ and the position is 4 when $t = 0$, then the particle is changing direction at

    (A) $t = 1$
    (B) $t = 3$
    (C) $t = 5$
    (D) $t = 1$ and $t = 5$

---

18. What is the area of the region between $y = 2x^2$ and $y = 12 - x^2$?

    (A)   0
    (B)   16
    (C)   32
    (D)   48

---

19. $\int (e^{3\ln x} + e^{3x})\, dx =$

    (A) $3 + \dfrac{e^{3x}}{3} + C$

    (B) $\dfrac{e^{x^4}}{4} + 3e^{3x} + C$

    (C) $\dfrac{e^{x^4}}{4} + \dfrac{e^{3x}}{3} + C$

    (D) $\dfrac{x^4}{4} + \dfrac{e^{3x}}{3} + C$

GO ON TO THE NEXT PAGE.

20. If $f(x) = (x^2 + x + 11)\sqrt{(x^3 + 5x + 121)}$, then $f'(0) =$

   (A) $\dfrac{5}{2}$

   (B) $\dfrac{27}{2}$

   (C) 22

   (D) $\dfrac{247}{2}$

21. If $f(x) = 5^{3x}$, then $f'(x) =$

   (A) $5^{3x}(\ln 125)$

   (B) $\dfrac{5^{3x}}{3\ln 5}$

   (C) $3(5^{2x})$

   (D) $3(5^{3x})$

GO ON TO THE NEXT PAGE.

22. A solid is generated when the region in the first quadrant enclosed by the graph of $y = (x^2 + 1)^3$, the line $x = 1$, the x-axis, and the y-axis is revolved about the x-axis. Its volume is found by evaluating which of the following integrals?

(A) $\pi \int_1^8 (x^2 + 1)^3 \, dx$

(B) $\pi \int_1^8 (x^2 + 1)^6 \, dx$

(C) $\pi \int_0^1 (x^2 + 1)^3 \, dx$

(D) $\pi \int_0^1 (x^2 + 1)^6 \, dx$

23. $\displaystyle\lim_{x \to 0} \dfrac{\sin\left(\dfrac{\pi}{2} + x\right) - 1}{x} =$

(A) -1
(B) 0
(C) 1
(D) The limit does not exist.

GO ON TO THE NEXT PAGE.

24. If $\dfrac{dy}{dx} = \dfrac{(3x^2 + 2)}{y}$ and $y = 4$ when $x = 2$, then when $x = 3$, $y =$

    (A) 18

    (B) 58

    (C) $\pm\sqrt{74}$

    (D) $\pm\sqrt{58}$

25. $\displaystyle\int \dfrac{dx}{9 + x^2} =$

    (A) $3\tan^{-1}\left(\dfrac{x}{3}\right) + C$

    (B) $\dfrac{1}{3}\tan^{-1}\left(\dfrac{x}{3}\right) + C$

    (C) $\dfrac{1}{3}\tan^{-1}(x) + C$

    (D) $\dfrac{1}{9}\tan^{-1}(x) + C$

26. If $f(x) = \cos^3(x + 1)$, then $f'(\pi) =$

    (A) $-3\cos^2(\pi + 1)\sin(\pi + 1)$
    (B) $3\cos^2(\pi + 1)$
    (C) $3\cos^2(\pi + 1)\sin(\pi + 1)$
    (D) 0

GO ON TO THE NEXT PAGE.

27. $\int x\sqrt{x+3}\,dx =$

(A) $\dfrac{2(x+3)^{\frac{3}{2}}}{3} + C$

(B) $\dfrac{2}{5}(x+3)^{\frac{5}{2}} - 2(x+3)^{\frac{3}{2}} + C$

(C) $\dfrac{3(x+3)^{\frac{3}{2}}}{2} + C$

(D) $\dfrac{4x^2(x+3)^{\frac{3}{2}}}{3} + C$

28. If $f(x) = \ln(\ln(1-x))$, then $f'(x) =$

(A) $-\dfrac{1}{\ln(1-x)}$

(B) $\dfrac{1}{(1-x)\ln(1-x)}$

(C) $\dfrac{1}{(1-x)^2}$

(D) $-\dfrac{1}{(1-x)\ln(1-x)}$

GO ON TO THE NEXT PAGE.

29. $\lim\limits_{x \to 0} \dfrac{xe^x}{1 - e^x} =$

   (A) $-\infty$
   (B) -1
   (C) 1
   (D) ∞

---

30. $\int \tan^6 x \ \sec^2 x \ dx =$

   (A) $\dfrac{\tan^7 x}{7} + C$

   (B) $\dfrac{\tan^7 x}{7} + \dfrac{\sec^3 x}{3} + C$

   (C) $\dfrac{\tan^7 x \sec^3 x}{21} + C$

   (D) $7 \tan^7 x + C$

END OF PART A, SECTION I
IF YOU FINISH BEFORE TIME IS CALLED, YOU MAY CHECK YOUR WORK ON PART A ONLY.
DO NOT GO ON TO PART B UNTIL YOU ARE TOLD TO DO SO.

CALCULUS AB

SECTION I, Part B

Time—45 Minutes

Number of questions—15

A GRAPHING CALCULATOR IS REQUIRED FOR SOME QUESTIONS ON THIS PART OF THE EXAMINATION

Directions: Solve each of the following problems, using the available space for scratchwork. After examining the form of the choices, decide which is the best of the choices given and fill in the corresponding oval on the answer sheet. No credit will be given for anything written in the test book. Do not spend too much time on any one problem.

In this test:

1. The **exact** numerical value of the correct answer does not always appear among the choices given. When this happens, select from among the choices the number that best approximates the exact numerical value.

2. Unless otherwise specified, the domain of a function f is assumed to be the set of all real numbers x for which $f(x)$ is a real number.

31. $\int_0^{\frac{\pi}{4}} \sin x \, dx + \int_{-\frac{\pi}{4}}^0 \cos x \, dx =$

    (A) -1

    (B) 0

    (C) 1

    (D) $\sqrt{2}$

GO ON TO THE NEXT PAGE.

32. A particle is moving along the y-axis. The position of the particle at time $t \geq 0$ is given by $y(t) = y^3 - 4y^2 + 7$. Which of the following expressions gives the total distance that the particle traveled in the first four seconds?

(A) $\displaystyle\int_0^{\frac{8}{3}} y^3 - 4y^2 + 7 \, dy - \int_{\frac{8}{3}}^4 y^3 - 4y^2 + 7 \, dy$

(B) $\displaystyle -\int_0^{\frac{8}{3}} y^3 - 4y^2 + 7 \, dy + \int_{\frac{8}{3}}^4 y^3 - 4y^2 + 7 \, dy$

(C) $\displaystyle\int_0^{\frac{8}{3}} 3y^2 - 8y \, dy - \int_{\frac{8}{3}}^4 3y^2 - 8y \, dy$

(D) $\displaystyle -\int_0^{\frac{8}{3}} 3y^2 - 8y \, dy + \int_{\frac{8}{3}}^4 3y^2 - 8y \, dy$

33. If $\displaystyle\int_0^4 f(x) \, dx = 10$ and $\displaystyle\int_8^0 f(x) \, dx = -23$, what is $\displaystyle\int_4^8 f(x) \, dx$?

(A) −33
(B) −13
(C) 13
(D) 33

34. If $f'(x) = 3\sin x + 2\cos x$ and $f(0) = 5$, find $f(x)$.

(A) $-3\cos x + 2\sin x + 5$
(B) $-3\cos x + 2\sin x + 8$
(C) $3\cos x - 2\sin x + 2$
(D) $3\cos x - 2\sin x + 5$

GO ON TO THE NEXT PAGE.

35. The graph of *f* is shown below. Which of the following is true about the function at $x = 5$?

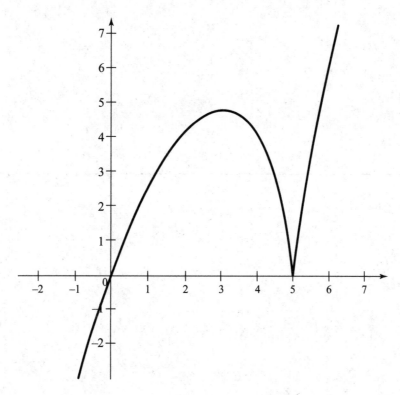

(A) $f'(5)$ is undefined.
(B) $f(5)$ is undefined.
(C) $f'(5) = 0$
(D) None of the above is true.

---

36. The function *f* is defined by $f(x) = e^{-\frac{x^2}{4}}$. What is the area between *f* and the *x*-axis from $x = -1.65$ to $x = 1.65$?

(A) 0
(B) 1.341
(C) 2.682
(D) 3.950

GO ON TO THE NEXT PAGE.

37. The average value of the function $f(x) = \ln^2 x$ on the interval $[2, 4]$ is

    (A) 1.204
    (B) 2.159
    (C) 2.408
    (D) 8.636

---

38. $\dfrac{d}{dx}\displaystyle\int_0^{3x} \cos(t)\, dt =$

    (A) $\sin 3x$
    (B) $\cos 3x$
    (C) $3 \sin 3x$
    (D) $3 \cos 3x$

GO ON TO THE NEXT PAGE.

39. The graph of the function *f* shown in the figure below. For which of the following values of *x* is $f'(x)$ positive and decreasing?

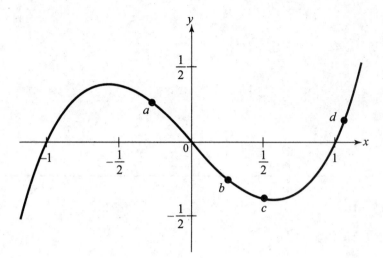

(A) *a*
(B) *b*
(C) *c*
(D) *d*

40. The radius of a sphere is increasing at a rate proportional to itself. If the radius is 4 initially, and the radius is 10 after two seconds, what will the radius be after three seconds?

(A) 15.81
(B) 16.00
(C) 25.00
(D) 62.50

GO ON TO THE NEXT PAGE.

41. Suppose $f(x) = \int_0^x (t^3 + t)\,dt$. Find $f'(5)$.

    (A) 130
    (B) 120
    (C) 76
    (D) 74

42. $\int \ln 2x\,dx =$

    (A) $\dfrac{\ln 2x}{2x} + C$

    (B) $x\ln x - x + C$

    (C) $x\ln 2x - x + C$

    (D) $2x\ln 2x - 2x + C$

43. Given $f(x) = \begin{cases} ax^2 + 3bx + 14; x \le 2 \\ 3ax + 5b; x > 2 \end{cases}$, find the values of a and b that make f differentiable for all x.

    (A) $a = -6, b = 2$
    (B) $a = 6, b = -2$
    (C) $a = 2, b = 6$
    (D) $a = 2, b = -6$

GO ON TO THE NEXT PAGE.

44. A circular shock wave is spreading from its point of origin at the rate of $640\ \dfrac{\text{m}^2}{\text{s}}$. How fast is a point on the boundary of the shock wave moving outward (in $\dfrac{\text{m}}{\text{s}}$) when the boundary is 100 m from the origin?

   (A) 1.019
   (B) 2.037
   (C) 3.201
   (D) 6.400

---

45. Find the distance traveled (to three decimal places) in the first four seconds, for a particle whose velocity is given by $v(t) = 7e^{-t^2}$, where t stands for time.

   (A)   0.976
   (B)   6.204
   (C)   6.359
   (D) 12.720

STOP
END OF PART B, SECTION I
IF YOU FINISH BEFORE TIME IS CALLED, YOU MAY CHECK YOUR WORK ON PART B ONLY.
DO NOT GO ON TO SECTION II UNTIL YOU ARE TOLD TO DO SO.

SECTION II
GENERAL INSTRUCTIONS

You may wish to look over the problems before starting to work on them, since it is not expected that everyone will be able to complete all parts of all problems. All problems are given equal weight, but the parts of a particular problem are not necessarily given equal weight.

A GRAPHING CALCULATOR IS REQUIRED FOR SOME PROBLEMS OR PARTS OF PROBLEMS ON THIS SECTION OF THE EXAMINATION.

- You should write all work for each part of each problem in the space provided for that part in the booklet. Be sure to write clearly and legibly. If you make an error, you may save time by crossing it out rather than trying to erase it. Erased or crossed-out work will not be graded.

- Show all your work. You will be graded on the correctness and completeness of your methods as well as your answers. Correct answers without supporting work may not receive credit.

- Justifications require that you give mathematical (noncalculator) reasons and that you clearly identify functions, graphs, tables, or other objects you use.

- You are permitted to use your calculator to solve an equation, find the derivative of a function at a point, or calculate the value of a definite integral. However, you must clearly indicate the setup of your problem, namely the equation, function, or integral you are using. If you use other built-in features or programs, you must show the mathematical steps necessary to produce your results.

- Your work must be expressed in standard mathematical notation rather than calculator syntax. For example, $\int_{1}^{5} x^2 \, dx$ may not be written as fnInt $(X^2, X, 1, 5)$.

- Unless otherwise specified, answers (numeric or algebraic) need not be simplified. If your answer is given as a decimal approximation, it should be correct to three places after the decimal point.

- Unless otherwise specified, the domain of a function f is assumed to be the set of all real numbers x for which $f(x)$ is a real number.

GO ON TO THE NEXT PAGE.

SECTION II, PART A
Time—30 minutes
Number of problems—2

A graphing calculator is required for some problems or parts of problems.

During the timed portion for Part A, you may work only on the problems in Part A.

On Part A, you are permitted to use your calculator to solve an equation, find the derivative of a function at a point, or calculate the value of a definite integral. However, you must clearly indicate the setup of your problem, namely the equation, function, or integral you are using. If you use other built-in features or programs, you must show the mathematical steps necessary to produce your results.

1. A particle moves along the x-axis so that its acceleration at any time $t > 0$ is given by $a(t) = 12t - 18$. At time $t = 1$, the velocity of the particle is $v(1) = 0$ and the position is $x(1) = 9$.

   (a) Write an expression for the velocity of the particle $v(t)$.

   (b) At what values of t does the particle change direction?

   (c) Write an expression for the position $x(t)$ of the particle.

   (d) Find the total distance traveled by the particle from $t = \dfrac{3}{2}$ to $t = 6$.

_____

GO ON TO THE NEXT PAGE.

2. Let R be the region in the first quadrant bounded from above by $g(x) = 19 - x^2$ and from below by $f(x) = x^2 + 1$.

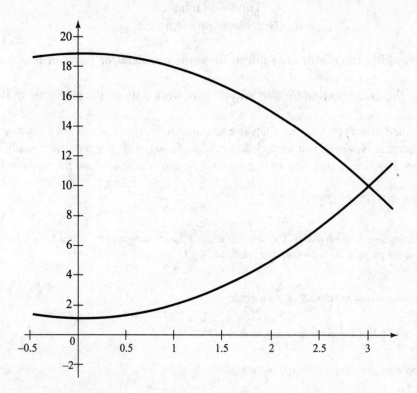

(a) Find the area of R.

(b) A solid is formed by revolving R around the x-axis. Find the volume of the solid.

(c) A solid has its base as the region R, whose cross-sections perpendicular to the x-axis are squares. Find the volume of the solid.

GO ON TO THE NEXT PAGE.

SECTION II, PART B
Time—1 hour
Number of problems—4

No calculator is allowed for these problems.

During the timed portion for Part B, you may continue to work on the problems in Part A without the use of any calculator.

---

3. Consider the equation $x^2 - 2xy + 4y^2 = 84$.

   (a) Write an expression for the slope of the curve at any point (x, y).

   (b) Find the equation of the tangent lines to the curve at the point $x = 2$.

   (c) Find $\dfrac{d^2y}{dx^2}$ at $(0, \sqrt{21})$.

---

4. Water is draining at the rate of 48π ft³/second from the vertex at the bottom of a conical tank whose diameter at its base is 40 feet and whose height is 60 feet.

   (a) Find an expression for the volume of water in the tank, in terms of its radius, at the surface of the water.

   (b) At what rate is the radius of the water in the tank shrinking when the radius is 16 feet?

   (c) How fast is the height of the water in the tank dropping at the instant that the radius is 16 feet?

GO ON TO THE NEXT PAGE.

5. Let f be the function given by $f(x) = 2x^4 - 4x^2 + 1$.

   (a)   Find an equation of the line tangent to the graph at $(-2, 17)$.

   (b)   Find the x- and y-coordinates of the relative maxima and relative minima. Verify your answer.

   (c)   Find the x- and y-coordinates of the points of inflection. Verify your answer.

---

6. Grain is being loaded into a silo at the rate of $G(t) = 400e^{\frac{-t^2}{4}}$ ft^3/hr, where t is the number of hours that it is being loaded, $0 \le t \le 8$. At time $t = 0$, there is 100 ft^3 of grain in the silo. Grain is also being removed through the base of the silo at the following rates, where $R(t)$ is the amount of grain being removed in ft^3/hour, $0 \le t \le 8$:

t	0	2	5	7	8
$R(t)$	60	90	110	120	125

   (a)   Estimate the total amount of grain removed from the silo at $t = 8$ hrs, using a left-hand Riemann sum and 4 subintervals.

   (b)   Estimate the amount of grain in the silo at the end of 8 hours, using your answer from part (a).

   (c)   Estimate $R'(5)$, showing your work. Indicate the units of measure.

STOP

END OF EXAM

Practice Test 1:
Answers and
Explanations

PRACTICE TEST 1 MULTIPLE CHOICE ANSWER KEY

The following answer key has been specifically formatted for diagnostic purposes. Please use the "Correct" column to check off the questions you not only answered correctly, but feel you fully understood. Use a squiggle to mark questions you used POE to guess on, or weren't sure about. Once you've marked up this table use either just the wrong answers OR the wrong and the unsure answers to tally up a percentage for each chapter.

If you are pressed for time, begin your review (in sequential order) with the chapters in which you scored the lowest percentage.

Question Number	Question Answer	Correct? ✎	See Chapter(s) #
1	D		6
2	C		3
3	C		6
4	C		3, 12
5	B		7
6	C		14
7	B		6
8	D		8
9	A		13
10	A		16
11	B		N/A*
12	A		13
13	A		8
14	C		N/A*
15	B		4, 5
16	A		9
17	D		10
18	C		16
19	D		13
20	B		6
21	A		11, 15
22	D		17
23	B		3, 12

Question Number	Question Answer	Correct? ✎	See Chapter(s) #
24	D		18
25	B		13
26	A		6
27	B		13
28	D		6
29	B		3, 12
30	A		13, 15
31	C		13
32	D		13
33	C		13
34	B		6
35	A		7
36	C		16
37	A		14
38	D		13
39	A		9
40	A		10
41	A		14
42	C		13, 15
43	B		6
44	D		10
45	B		10

*A small percentage of questions on the AP Calculus AB Exam pull from precalculus topics. This book does not re-cover that material, so there is no chapter reference. Do, however, refer to the explanation for more details.

PRACTICE TEST 1 FREE-RESPONSE ANSWER KEY

Question Number	Question Answer	Correct? ✎	See Chapter(s) #	Question Number	Question Answer	Correct? ✎	See Chapter(s) #
1a	See explanation		10	4a	See explanation		10
1b	See explanation		10	4b	See explanation		10
1c	See explanation		10	4c	See explanation		10
1d	See explanation		10	4d	See explanation		10
2a	See explanation		16	5a	See explanation		8
2b	See explanation		17	5b	See explanation		9
2c	See explanation		17	5c	See explanation		9
3a	See explanation		7	6a	See explanation		14
3b	See explanation		8	6b	See explanation		14
3c	See explanation		7	6c	See explanation		5

CHAPTER 3 TEST SCORE SELF-EVALUATION

CHAPTER 6 TEST SCORE SELF-EVALUATION

CHAPTER 4 TEST SCORE SELF-EVALUATION

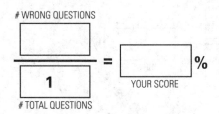

CHAPTER 7 TEST SCORE SELF-EVALUATION

CHAPTER 5 TEST SCORE SELF-EVALUATION

CHAPTER 8 TEST SCORE SELF-EVALUATION

CHAPTER 9 TEST SCORE SELF-EVALUATION

CHAPTER 10 TEST SCORE SELF-EVALUATION

CHAPTER 11 TEST SCORE SELF-EVALUATION

CHAPTER 12 TEST SCORE SELF-EVALUATION

CHAPTER 13 TEST SCORE SELF-EVALUATION

CHAPTER 14 TEST SCORE SELF-EVALUATION

CHAPTER 15 TEST SCORE SELF-EVALUATION

CHAPTER 16 TEST SCORE SELF-EVALUATION

CHAPTER 17 TEST SCORE SELF-EVALUATION

CHAPTER 18 TEST SCORE SELF-EVALUATION

ANSWERS AND EXPLANATIONS TO SECTION I

1. **D** Take the derivative: $f'(x) = 6x - 10 + \dfrac{1}{x-3}$ and plug in $x = 4$.

$$f'(4) = 6(4) - 10 + \frac{1}{(4)-3} = 15$$

2. **C** **Step 1:** To solve this problem, you need to remember how to evaluate limits. Always do limit problems on the first pass. Whenever we have a limit of a polynomial fraction where $x \to \infty$, divide the numerator and the denominator, separately, by the highest power of x in the fraction.

$$\lim_{x\to\infty} \frac{5x^2 - 3x + 1}{4x^2 + 2x + 5} = \lim_{x\to\infty} \frac{\dfrac{5x^2}{x^2} - \dfrac{3x}{x^2} + \dfrac{1}{x^2}}{\dfrac{4x^2}{x^2} + \dfrac{2x}{x^2} + \dfrac{5}{x^2}}$$

Step 2: Simplify $\displaystyle\lim_{x\to\infty} \frac{5 - \dfrac{3}{x} + \dfrac{1}{x^2}}{4 + \dfrac{2}{x} + \dfrac{5}{x^2}}$.

Step 3: Now take the limit. Remember that the $\displaystyle\lim_{x\to\infty} \frac{k}{x^n} = 0$, if $n > 0$, where k is a constant. Thus, we get

$$\lim_{x\to\infty} \frac{5 - \dfrac{3}{x} + \dfrac{1}{x^2}}{4 + \dfrac{2}{x} + \dfrac{5}{x^2}} = \lim_{x\to\infty} \frac{5 - 0 + 0}{4 + 0 + 0} = \frac{5}{4}$$

3. **C** **Step 1:** We need to use the Quotient Rule to evaluate this derivative. Remember, the derivative of $\dfrac{u}{v} = \dfrac{v\dfrac{du}{dx} - u\dfrac{dv}{dx}}{v^2}$. But, before we take the derivative, we should factor an x out of the top and bottom and cancel, simplifying the quotient.

$$f(x) = \frac{3x^2 + x}{3x^2 - x} = \frac{x(3x+1)}{x(3x-1)} = \frac{3x+1}{3x-1}$$

Step 2: Now take the derivative.

$$f'(x) = \frac{(3x-1)(3) - (3x+1)(3)}{(3x-1)^2}$$

Step 3: Simplify.

$$\frac{9x - 3 - 9x - 3}{(3x-1)^2} = \frac{-6}{(3x-1)^2}$$

4.   **C**   We know that $\lim\limits_{x\to 0}\dfrac{\sin(ax)}{x} = a$, so we get $3(2) = 6$. Or we could get the answer using L'Hôpital's

Rule, because the limit is of an indeterminate form: $\dfrac{0}{0}$. We take the derivative of the top and bottom

and get $\lim\limits_{x\to 0}\dfrac{3\sin(2x)}{x} = \lim\limits_{x\to 0}\dfrac{3(2)\cos(2x)}{1} = 3(2) = 6$.

5.   **B**   We need to use implicit differentiation to find $\dfrac{dy}{dx}$. First, take the derivative with respect to x of

both sides: $\dfrac{dy}{dx} = \cos(xy)\left(x\dfrac{dy}{dx} + y \right)$. Remember that you need to use the Product Rule to find the

derivative of xy.

Now distribute on the right side: $\dfrac{dy}{dx} = x\dfrac{dy}{dx}\cos(xy) + y\cos(xy)$.

Next, group the terms containing $\dfrac{dy}{dx}$ on the left side: $\dfrac{dy}{dx} - x\dfrac{dy}{dx}\cos(xy) = y\cos(xy)$.

Factor out $\dfrac{dy}{dx}$: $x\dfrac{dy}{dx}(1 - x\cos(xy)) = y\cos(xy)$.

And divide to isolate $\dfrac{dy}{dx}$: $\dfrac{dy}{dx} = \dfrac{y\cos(xy)}{1 - x\cos(xy)}$.

6.   **C**   The right Riemann sum will use the values on the right-hand sides of the intervals [0, 1], [1, 2], [2, 3], and [3, 4]. We get $f(1) + f(2) + f(3) + f(4) = 4 + 13 + 28 + 49 = 94$.

7.   **B**   First, differentiate h using the Product Rule: $h'(x) = f'(x)g(4x) + 4f(x)g'(4x)$. Now substitute $x = 2$ and get $h'(2) = f'(2)g(8) + 4f(2)g'(8)$.

8.   **D**   Here we do everything that we normally do for finding the equations of tangent lines, except that we use the negative reciprocal of the slope to find the normal line. This is because the normal line is perpendicular to the tangent line.

Step 1: First, find the slope of the tangent line.

$$\frac{dy}{dx} = \frac{1}{2}(3x^2 + 2x)^{-\frac{1}{2}}(6x + 2)$$

Step 2: DON'T SIMPLIFY. Immediately plug in $x = 2$. We get

$$\frac{dy}{dx} = \frac{1}{2}(3x^2 + 2x)^{-\frac{1}{2}}(6x + 2) = \frac{1}{2}(3(2)^2 + 2(2))^{-\frac{1}{2}}(6(2) + 2) = \frac{1}{2}(16)^{-\frac{1}{2}}(14) = \frac{7}{4}$$

This means that the slope of the tangent line at $x = 2$ is $\dfrac{7}{4}$, so the slope of the normal

line is $-\dfrac{4}{7}$.

Step 3: Then the equation of the normal line is $(y - 4) = -\dfrac{4}{7}(x - 2)$.

Step 4: Multiply through by 7 and simplify.

$$7y - 28 = -4x + 8$$
$$4x + 7y = 36$$

9. **A** The derivative of $\cot x$ is $-\csc^2 x$, so $4\int \csc^2 x \, dx = -4\cot x + C$.

10. **A** This problem is just asking us to find a higher order derivative of a trigonometric function.

    Step 1: The first derivative requires the Chain Rule.

    $f(x) = \cos^2 x$
    $f'(x) = 2(\cos x)(-\sin x) = -2\cos x \sin x$

    Step 2: The second derivative requires the Product Rule.

    $f'(x) = -2\cos x \sin x$
    $f''(x) = -2(\cos x \cos x - \sin x \sin x) = -2(\cos^2 x - \sin^2 x)$

    Step 3: Now plug in π for x and simplify.

    $-2(\cos^2 (\pi) - \sin^2 (\pi)) = -2(1 - 0) = -2$

11. **B** **Step 1:** To find $g(f(x))$, all you need to do is to replace all of the x's in $g(x)$ with $f(x)$'s.

    $$g(f(x)) = 3f(x) = 3\left(\frac{5}{x^2 + 1}\right) = \frac{15}{x^2 + 1}$$

    Step 2: Now all we have to do is plug in 2 for x.

    $$g(f(2)) = \frac{15}{2^2 + 1} = 3$$

12. **A** We can evaluate this integral using u-substitution. Let $u = 1 + x^2$ and $du = 2x dx$, $\frac{1}{2} du = x dx$.

    Substituting into the integrand, we get $\frac{1}{2}\int \sqrt{u} \, du$. Evaluate the integral:

    $\frac{1}{2}\int \sqrt{u} \, du = \frac{1}{2}\frac{u^{\frac{3}{2}}}{\frac{3}{2}} + C = \frac{1}{3}u^{\frac{3}{2}} + C$. And substitute back: $\frac{1}{3}\left(1 + x^2\right)^{\frac{3}{2}} + C$.

13. **A** This is another equation of a tangent line problem, combined with implicit differentiation. Often, the AP Exam has more than one tangent line problem, so make sure that you can do these well!

By the way, do you remember the derivative of $\ln(f(x))$? It is $\dfrac{f'(x)}{f(x)}$.

Step 1: First, take the derivative of the equation.

$$6x\frac{dx}{dx} + \frac{5}{y}\frac{dy}{dx} = 0$$

Step 2: Next, we simplify and solve for $\dfrac{dy}{dx}$.

$$6x + \frac{5}{y}\frac{dy}{dx} = 0$$

$$\frac{dy}{dx} = \frac{-6xy}{5}$$

Step 3: Now, we plug in 2 for x and 1 for y to get the slope of the tangent line.

$$\frac{dy}{dx} = \frac{-6(2)(1)}{5} = \frac{-12}{5}$$

14. **C** The AP people expect you to remember a lot of your trigonometry, so if you're rusty, review the unit in the Appendix.

Step 1: In an equation of the form $f(x) = A \sin B\,(x \pm C) \pm D$, you should know four components. The amplitude of the equation is $|A|$, the horizontal or phase shift is $\pm C$, the vertical shift is $\pm D$, and the fundamental period is $\dfrac{2\pi}{B}$.

The same is true for $f(x) = A \cos B\,(x \pm C) \pm D$.

Step 2: All we have to do is plug into the formula for the period.

$$\frac{2\pi}{B} = \frac{2\pi}{\dfrac{\pi}{4}} = 8$$

15. **B** f is not continuous at a and d. At b, the function is continuous but has a cusp, which will have a vertical tangent line and so not be differentiable.

16. **A** This problem requires you to know how to find maxima/minima. This is a part of curve sketching and is one of the most important parts of differential calculus. A function has *critical points* where the derivative is zero or undefined (which is never a problem when the function is an ordinary polynomial). After finding the critical points, test them to determine whether they are maxima or minima or something else.

Step 1: First, as usual, take the derivative and set it equal to zero.

$$f'(x) = 3x^2 - 18x - 120$$
$$3x^2 - 18x - 120 = 0$$

Step 2: Find the values of x that make the derivative equal to zero. These are the critical points.

$$3x^2 - 18x - 120 = 0$$
$$x^2 - 6x - 40 = 0$$
$$(x - 10)(x + 4) = 0$$
$$x = \{10, -4\}$$

Step 3: In order to determine whether a critical point is a maximum or a minimum, we need to take the second derivative.

$$f''(x) = 6x - 18$$

Step 4: Now, we plug the critical points from Step 2 into the second derivative. If it yields a negative value, then the point is a maximum. If it yields a positive value, then the point is a minimum. If it yields zero, it is neither, and is most likely a point of inflection.

$$6(10) - 18 = 42$$
$$6(-4) - 18 = -42$$

Therefore, 10 is a minimum.

17. **D**   **Step 1:** Because acceleration is the derivative of velocity, if we know the acceleration of a particle, we can find the velocity by integrating the acceleration with respect to t.

$$v = \int a\, dt = \int (4t - 12)\, dt = 2t^2 - 12t + C$$

Next, because the velocity is 10 at $t = 0$, we can plug in 0 for t and solve for the constant.

$$2(0)^2 - 12(0) + C = 10$$

Therefore, $C = 10$ and the velocity, $v(t)$, is $2t^2 - 12t + 10$.

Step 2: In order to find when the particle is changing direction, we need to know when the velocity is equal to zero, so we set $v(t) = 0$ and solve for t.

$$2t^2 - 12t + 10 = 0$$
$$t^2 - 6t + 5 = 0$$
$$(t - 5)(t - 1) = 0$$
$$t = \{1, 5\}$$

Now, provided that the acceleration is not also zero at $t = \{1, 5\}$, the particle will be changing direction at those times. The acceleration is found by differentiating the equation for velocity with respect to time: $a(t) = 4t - 12$. This is not zero at either $t = 1$ or $t = 5$. Therefore, the particle is changing direction when $t = 1$ and $t = 5$.

18.   **C**   First, find where the two curves intersect. Set them equal to each other and solve:

$$12 - x^2 = 2x^2$$
$$12 = 3x^2$$
$$x^2 = 4$$
$$x = \pm 2$$

Therefore, in order to find the area of the region, we need to evaluate the integral:

$$\int_{-2}^{2} (12 - x^2) - (2x^2)\, dx$$

Simplify the integrand: $\int_{-2}^{2} (12 - x^2) - (2x^2)\, dx = \int_{-2}^{2} (12 - 3x^2)\, dx$.

Use the Power Rule: $\int_{-2}^{2} (12 - 3x^2)\, dx = (12x - x^3)\Big|_{-2}^{2}$.

And evaluate: $(12x - x^3)\Big|_{-2}^{2} = (24 - 8) - (-24 + 8) = 32$.

19.   **D**   This problem requires that you know your rules of exponential functions.

Step 1: First of all, $e^{3\ln x} = e^{\ln x^3} = x^3$. So we can rewrite the integral as

$$\int (e^{3\ln x} + e^{3x})\, dx = \int (x^3 + e^{3x})\, dx$$

Step 2: The rule for the integral of an exponential function is $\int e^k\, dx = \frac{1}{k}e^{kx} + C$.

Now we can do the integral: $\int (x^3 + e^{3x})\, dx = \frac{x^4}{4} + \frac{1}{3}e^{3x} + C$.

20.   **B**   This problem is just a complicated derivative, requiring you to be familiar with the Chain Rule and the Product Rule.

Step 1: $f'(x) = \frac{1}{2}(x^3 + 5x + 121)^{-\frac{1}{2}}(3x^2 + 5)(x^2 + x + 11) + (x^3 + 5x + 121)^{\frac{1}{2}}(2x + 1)$.

Step 2: Whenever a problem asks you to find the value of a complicated derivative at a particular point, NEVER simplify the derivative. Immediately plug in the value for x and do arithmetic instead of algebra.

$$f'(0) = \frac{1}{2}(0^3 + 5(0) + 121)^{-\frac{1}{2}}(3(0)^2 + 5)((0)^2 + (0) + 11) + ((0)^3 + 5(0) + 121)^{\frac{1}{2}}(2(0) + 1)$$

$$= \frac{1}{2}(121)^{-\frac{1}{2}}(5)(11) + (121)^{\frac{1}{2}}(1) = \frac{5}{2} + 11 = \frac{27}{2}$$

21.   **A**   This problem requires you to know how to find the derivative of an exponential function. According to the rule, if a function is of the form $a^{f(x)}$, its derivative is $a^{f(x)}(\ln a)\,f'(x)$. Now all we have to do is follow the rule!

Step 1: $f(x) = 5^{3x}$

$f'(x) = 5^{3x}(\ln 5)(3)$

Step 2: If you remember your rules of logarithms, $3\ln 5 = \ln(5^3) = \ln 125$.

So we can rewrite the answer to $f'(x) = 5^{3x}(\ln 5)(3) = 5^{3x}(\ln 125)$.

22. **D** This problem requires you to know how to find the volume of a solid of revolution.

If you have a region between two curves, from $x = a$ to $x = b$, then the volume generated when the region is revolved around the x-axis is $\pi\int_a^b\left[f(x)^2 - g(x)^2\right]dx$, if $f(x)$ is above $g(x)$ throughout the region.

Step 1: First, we have to determine what the region looks like. The curve looks like the following:

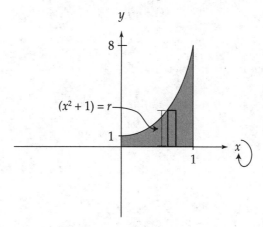

The shaded region is the part that we are interested in. Notice that the curve is always above the x-axis (which is $g(x)$). Now we just follow the formula.

$$\pi\int_0^1\left[\left[(x^2 + 1)^3\right]^2 - [0]^2\right]dx = \pi\int_0^1(x^2 + 1)^6\,dx$$

23. **B** We can get the answer using L'Hôpital's Rule, because the limit is of an indeterminate form: $\dfrac{0}{0}$. Take the derivative of the top and bottom and get $\lim_{x\to0}\dfrac{\sin\left(\dfrac{\pi}{2} + x\right) - 1}{x} = \lim_{x\to0}\dfrac{\cos\left(\dfrac{\pi}{2} + x\right)}{1} = \cos\dfrac{\pi}{2} = 0$.

Now break this into limits that we can easily evaluate.

24. **D** This is a very basic differential equation. See page 362 for a discussion of *separation of variables*.

Step 1: First, separate the variables. Then, we get

$$y\,dy = (3x^2 + 2)\,dx$$

Step 2: Now integrate both sides.

$$\int y \, dy = \int (3x^2 + 2) \, dx$$

$$\frac{y^2}{2} = x^3 + 2x + C$$

Notice how we use only one constant. All we have to do now is solve for C. We do this by plugging in 2 for x and 4 for y.

$$\frac{16}{2} = 2^3 + 4 + C$$

$$C = -4$$

So we can rewrite the equation as $\frac{y^2}{2} = x^3 + 2x - 4$.

Step 3: Now, if we plug in 3 for x, we can find y.

$$\frac{y^2}{2} = 27 + 6 - 4$$

$$y^2 = 58$$

$$y = \pm\sqrt{58}$$

25. **B** This is another inverse trigonometric integral.

Step 1: We know that $\int \frac{dx}{1 + x^2} = \tan^{-1}(x) + C$.

(See problem 9 if you're not sure of this.) The trick here is to get the denominator of the fraction in the integrand to be of the correct form. If we factor 9 out of the denominator, we get

$$\int \frac{dx}{9 + x^2} = \int \frac{dx}{9\left(1 + \frac{x^2}{9}\right)} = \frac{1}{9}\int \frac{dx}{1 + \frac{x^2}{9}} = \frac{1}{9}\int \frac{dx}{1 + \left(\frac{x}{3}\right)^2}$$

Step 2: Now use u-substitution to evaluate this integral.

Let $u = \frac{x}{3}$ and $du = \frac{1}{3} \, dx$ or $3 \, du = dx$. Then we have

$$\frac{1}{9}\int \frac{dx}{1 + \left(\frac{x}{3}\right)^2} = \frac{1}{9}\int \frac{3 \, du}{1 + u^2} = \frac{1}{3}\int \frac{du}{1 + u^2} = \frac{1}{3}\tan^{-1}(u) + C$$

Step 3: Now all we have to do is reverse the u-substitution and we're done.

$$\frac{1}{3}\tan^{-1}(u) + C = \frac{1}{3}\tan^{-1}\left(\frac{x}{3}\right) + C$$

26. **A** Think of $\cos^3(x + 1)$ as $[\cos(x + 1)]^3$.

Step 1: First, we take the derivative of the outside function and ignore the inside functions. The derivative of u^3 is $3u^2$. We get $\frac{d}{dx}[u]^3 = 3[u]^2$.

Step 2: Next, take the derivative of the cosine term and multiply. The derivative of $\cos u$ is $-\sin u$.

$$\frac{d}{dx}[\cos(u)]^3 = -3[\cos(u)]^2\sin(u)$$

Step 3: Finally, we take the derivative of $x + 1$ and multiply. The derivative of $x + 1$ is 1.

$$\frac{d}{dx}[\cos(x + 1)]^3 = -3[\cos(x + 1)]^2\sin(x + 1)$$

27. **B** We can do this integral with u-substitution.

Step 1: Let $u = x + 3$. Then $du = dx$ and $u - 3 = x$.

Step 2: Substituting, we get

$$\int x\sqrt{x + 3}\ dx = \int (u - 3)u^{\frac{1}{2}}\ du$$

Why is this better than the original integral, you might ask? Because now we can distribute and the integral becomes easy.

Step 3: When we distribute, we get

$$\int (u - 3)u^{\frac{1}{2}}\ du = \int \left(u^{\frac{3}{2}} - 3u^{\frac{1}{2}} \right) du$$

Step 4: Now we can integrate.

$$\int \left(u^{\frac{3}{2}} - 3u^{\frac{1}{2}} \right) du = \frac{2}{5}u^{\frac{5}{2}} - 3\left(\frac{2}{3}u^{\frac{3}{2}}\right) + C$$

Step 5: Substituting back, we get

$$\frac{2}{5}u^{\frac{5}{2}} - 3\left(\frac{2}{3}u^{\frac{3}{2}}\right) + C = \frac{2}{5}(x + 3)^{\frac{5}{2}} - 2(x + 3)^{\frac{3}{2}} + C$$

28. **D** Here, we use the Chain Rule.

Step 1: First, take the derivative of the outside function.

The derivative of $\ln u$ is $\dfrac{du}{u}$.

We get

$$\frac{d}{dx} \ln(\ln(u)) = \frac{1}{\ln(u)}$$

Step 2: Now we take the derivative of the function in the denominator. Once again, the function is $\ln u$.

We get

$$\frac{d}{dx} \ln(\ln(1-x)) = \frac{1}{\ln(1-x)} \left(\frac{-1}{1-x}\right) = \frac{-1}{(1-x)\ln(1-x)}$$

29. **B** Note that if we plug in zero, we get a limit of the indeterminate form $\dfrac{0}{0}$, so we can use L'Hôpital's Rule to find the limit. Take the derivative of the numerator and the denominator:

$$\lim_{x\to 0} \frac{xe^x}{1-e^x} = \lim_{x\to 0} \frac{xe^x + e^x}{-e^x}. \text{ Now, if we plug in 0 we get } \frac{0+1}{-1} = -1.$$

30. **A** We can do this integral with u-substitution.

Step 1: Let $u = \tan x$. Then $du = \sec^2 x\, dx$.

Step 2: Substituting, we get $\int \tan^6 x \sec^2 x\, dx = \int u^6\, du$.

Step 3: This is an easy integral: $\int u^6\, du = \dfrac{u^7}{7} + C$.

Step 4: Substituting back, we get $\dfrac{\tan^7 x}{7} + C$.

31. **C** These are a pair of basic trigonometric integrals. You should have memorized several trigonometric integrals, particularly $\int \sin x\, dx = -\cos x + C$ and $\int \cos x\, dx = \sin x + C$.

Step 1: $\displaystyle\int_0^{\frac{\pi}{4}} \sin x\, dx + \int_{-\frac{\pi}{4}}^0 \cos x\, dx = -\cos x \Big|_0^{\frac{\pi}{4}} + \sin x \Big|_{-\frac{\pi}{4}}^0$

Step 2: Now we evaluate the limits of integration, and we're done.

$$-\cos x \Big|_0^{\frac{\pi}{4}} + \sin x \Big|_{-\frac{\pi}{4}}^0 = \left[\left(-\cos\frac{\pi}{4}\right) - (-\cos(0))\right] + \left[(\sin(0)) - \left(\sin\left(-\frac{\pi}{4}\right)\right)\right] = -\frac{1}{\sqrt{2}} + 1 + 0 + \frac{1}{\sqrt{2}} = 1$$

32.   **D**   First, we will need to find when the particle is moving up (velocity is positive) or down (velocity is negative). We can find the velocity by taking the derivative of the position: $v(t) = y'(t) = 3y^2 - 8y$. Next, set it equal to zero: $3y^2 - 8y = 0$.

Factor: $y(3y - 8) = 0$, so the velocity is zero at $y = 0$ and $y = \dfrac{8}{3}$. We can determine whether the

velocity is positive or negative by using text values and plugging them into the derivative and checking

the sign: $v(1) = 3(1)^2 - 8(1) < 0$ (moving down) and $v(3) = 3(3)^2 - 8(3) > 0$ (moving up). If we want

to find the distance traveled, we now integrate the velocity function on the intervals, making the

integral negative for when the particle is moving down. We get $-\int_0^{\frac{8}{3}} 3y^2 - 8y\,dy + \int_{\frac{8}{3}}^4 3y^2 - 8y\,dy$.

33.   **C**   We can use the fundamental theorem of calculus for integrals to find this. First, note that

$\int_0^4 f(x)\,dx + \int_4^8 f(x)\,dx = \int_0^8 f(x)\,dx$. Next, note that if $\int_8^0 f(x)\,dx = -23$, then if we switch the

limits of integration $\int_0^8 f(x)\,dx = 23$. Now we plug in: $10 + \int_4^8 f(x)\,dx = 23$, so $\int_4^8 f(x)\,dx = 13$.

34.   **B**   First, we need to integrate: $f(x) = \int 3\sin x + 2\cos x\,dx = -3\cos x + 2\sin x + C$. Next, plug in the initial condition to solve for the constant.

$$5 = -3\cos(0) + 2\sin(0) + C$$

$$5 = -3(1) + C$$

$$C = 8$$

Therefore, $f(x) = -3\cos x + 2\sin x + 8$.

35.   **A**   The curve has a cusp at $x = 5$, so the curve is continuous there. But the slope is infinite at a cusp, so there is no derivative there. This means that $f'(5)$ is undefined.

36.   **C**   We need to integrate $f(x) = e^{-\frac{x^2}{4}}$ from $x = -1.65$ to $x = 1.65$. We can do this with the calculator.

We get $\displaystyle\int_{-1.65}^{1.65} e^{-\frac{x^2}{4}}\,dx = 2.682$.

37.   **A**   This problem requires you to be familiar with the Mean Value Theorem for Integrals, which we use to find the average value of a function.

Step 1: If you want to find the average value of $f(x)$ on an interval $[a, b]$, you need to evaluate the

integral $\dfrac{1}{b-a}\int_a^b f(x)\,dx$. So here we evaluate the integral $\dfrac{1}{2}\int_2^4 \ln^2 x\,dx$.

You have to do this integral on your calculator because you do not know how to evaluate this integral analytically unless you are very good with integration by parts!

Use fnInt. Divide this by 2, and you will get 1.204.

38. **D**   This problem is testing your knowledge of the Second Fundamental Theorem of Calculus. The theorem states that $\dfrac{d}{dx}\displaystyle\int_{a}^{u} f(t)\,dt = f(u)\dfrac{du}{dx}$, where a is a constant and u is a function of x. So all we have to do is follow the theorem: $\dfrac{d}{dx}\displaystyle\int_{0}^{3x} \cos(t)\,dt = 3\cos 3x$.

39. **A**   If $f'(x)$ is positive, then the curve will have an upward (positive) slope. If $f'(x)$ is decreasing, it means that the second derivative is negative, so the curve will be concave down. Point a has both of those attributes.

40. **A**   This is not a related-rate problem; this is a differential equation! It just happens to involve a rate.

Step 1: If we translate the first sentence into an equation, we get $\dfrac{dR}{dt} = kR$.

Put all of the terms that contain an R on the left of the equals sign and all of the terms that contain a t on the right-hand side.

$$\frac{dR}{R} = k\,dt$$

Step 2: Integrate both sides.

$$\int \frac{dR}{R} = k \int dt$$

Step 3: If we solve this for R, we get $R = Ce^{kt}$ (see the unit on Differential Equations).

Now we need to solve for C and k. First, we solve for C by plugging in the information that the radius is 4 initially. This means that $R = 4$ when $t = 0$.

If $4 = Ce^{0}$, then $C = 4$

Next, we solve for k by plugging in the information that $R = 10$ when $t = 2$.

$$10 = 4e^{2k}$$

$$\frac{5}{2} = e^{2k}$$

$$\ln\frac{5}{2} = 2k$$

$$\frac{1}{2}\ln\frac{5}{2} = k$$

Step 4: Now we have our final equation: $R = 4e^{\left(\frac{1}{2}\ln\frac{5}{2}\right)t}$.

If we plug in $t = 3$, we get $R = 4e^{\left(\frac{1}{2}\ln\frac{5}{2}\right)(3)} \approx 15.81$.

41. **A** We use the Second Fundamental Theorem of Calculus to find the derivative of the integral. In this case, we simply replace t in the integrand with x: $f'(x) = x^3 + x$. Now, all we have to do is plug in $x = 5$: $f'(5) = 5^3 + 5 = 130$.

42. **C** This is a simple integral that we do using integration by parts. You should memorize that $\int \ln(ax)\,dx = x\ln(ax) - x + C$, which makes this integral easy.

    Step 1: The formula for integration by parts is $\int u\,dv = uv - \int v\,du$.

    The trick is that we have to let $dv = dx$.

    $$\text{Let } u = \ln 2x \text{ and } dv = dx$$

    $$du = \frac{2}{2x}\,dx = \frac{1}{x}\,dx \text{ and } v = x$$

    Plugging in to the formula, we get

    $$\int \ln 2x\,dx = x\ln 2x - \int dx = x\ln 2x - x + C$$

43. **B** First, in order to be differentiable, f has to be continuous. Here, this means that if we plug 2 into both pieces of f, we need to get the same value: $a(2)^2 + 3b(2) + 14 = 3a(2) + 5b$, which simplifies to

    $$4a + 6b + 14 = 6a + 5b$$
    $$2a - b = 14$$

    Next, in order to be differentiable, if we plug 2 into both pieces of f', we need to get the same value.

    First, differentiate both pieces of f: $f'(x) = \begin{cases} 2ax + 3b; & x < 2 \\ 3a; & x > 2 \end{cases}$. Now we plug in 2 and set the two

    pieces equal to each other: $4a + 3b = 3a$, so $a = -3b$. Plug this into the first equation: $2(-3b) - b = 14$.

    Thus, $b = -2$ and $a = 6$.

44. **D** This is a related rates problem. We need to find out how fast the circumference of a circle, C,

    is increasing given how fast its area, A, is increasing. We are given that $\frac{dA}{dt} = 640$ and that

    $r = 100$. Now, we need to find the relationship between the area of a circle and its circumfer-

    ence. The area of a circle is $A = \pi r^2$ and the circumference is $C = 2\pi r$, where r is the radius.

Although we could isolate r in the second equation and substitute it into the first equation to get a direct relationship between A and C, it is easier to differentiate them separately and use $\dfrac{dr}{dt}$.

That is, we first differentiate the area equation with respect to time: $\dfrac{dA}{dt} = 2\pi r \dfrac{dr}{dt}$ and plug in. We get $640 = 2\pi(100)\dfrac{dr}{dt}$, so $\dfrac{dr}{dt} = \dfrac{640}{200\pi}$. Next, we differentiate the circumference equation with respect to time: $\dfrac{dC}{dt} = 2\pi \dfrac{dr}{dt}$. Now if we plug in $\dfrac{dr}{dt} = \dfrac{640}{200\pi}$, we get $\dfrac{dC}{dt} = 2\pi\left(\dfrac{640}{200\pi}\right) = 6.400$.

45. **B** **Step 1:** If we want to find the distance traveled, we take the integral of velocity from the starting time to the finishing time. Therefore, we need to evaluate $\displaystyle\int_0^4 7e^{-t^2}\,dt$.

Step 2: But we have a problem! We can't take the integral of e^{-t^2}. This means that the AP Exam wants you to find the answer using your calculator.

Rounded to three decimal places, the answer is 6.204.

ANSWERS AND EXPLANATIONS TO SECTION II

1. A particle moves along the x-axis so that its acceleration at any time $t > 0$ is given by $a(t) = 12t - 18$. At time $t = 1$, the velocity of the particle is $v(1) = 0$ and the position is $x(1) = 9$.

(a) Write an expression for the velocity of the particle $v(t)$.

Step 1: We know that the derivative of velocity with respect to time is acceleration, so the integral of acceleration with respect to time is velocity.

$$\int a(t)\,dt = v(t)$$
$$\int 12t - 18\,dt = 6t^2 - 18t + C = v(t)$$

If we plug in the information that at time $t = 1$, $v(1) = 0$, we can solve for C.

$$6(1)^2 - 18(1) + C = 0$$
$$-12 + C = 0$$
$$C = 12$$

This means that the velocity of the particle is $6t^2 - 18t + 12$.

(b) At what values of t does the particle change direction?

When a particle is in motion, it changes direction at the time when its velocity is zero. (As long as acceleration is not also zero.) So all we have to do is set velocity equal to zero and solve for t.

$$6t^2 - 18t + 12 = 0$$
$$t^2 - 3t + 2 = 0$$
$$(t - 2)(t - 1) = 0$$
$$t = 1, 2$$

(c) Write an expression for the position $x(t)$ of the particle.

We know that the derivative of position with respect to time is velocity, so the integral of velocity with respect to time is position.

$$\int v(t)\, dt = x(t)$$
$$\int (6t^2 - 18t + 12)\, dt = 2t^3 - 9t^2 + 12t + C = x(t)$$

If we plug in the information that at time $t = 1$, $x(1) = 9$, we can solve for C.

$$2(1)^3 - 9(1)^2 + 12(1) + C = 9$$
$$5 + C = 9$$
$$C = 4$$

Therefore, $x(t) = 2t^3 - 9t^2 + 12t + 4$.

(d) Find the total distance traveled by the particle from $t = \dfrac{3}{2}$ to $t = 6$.

Step 1: Normally, all that we have to do to find the distance traveled is to integrate the velocity equation from the starting time to the ending time. But we have to watch out for whether the particle changes direction. If so, we have to break the integration into two parts—a positive integral for when it is traveling to the right, and a negative integral for when it is traveling to the left.

One way to solve this is to find two integrals and add them together. Because you can use a calculator, it is simpler to use the fnInt calculation of the absolute value for $t = \dfrac{3}{2}$ and $t = 6$, using the function of velocity $6x^2 - 18x + 12$.

Graphing Calculator (TI-83 and TI-84)

Press MATH and select 9: fnInt from the list.

Press MATH then select the NUM menu, and choose 1: abs(

Enter the function $6x^2 - 18x + 12$ and follow with the closing parentheses. List the variable and low and high values for t, separated by commas, and follow with final closing parentheses so your expression looks like the following:

fnInt(abs($6x^2 - 18x + 12$), x, 3/2, 6)

Press ENTER

The result is 176.500.

2.   Let R be the region in the first quadrant bounded from above by $g(x) = 19 - x^2$ and from below by $f(x) = x^2 + 1$.

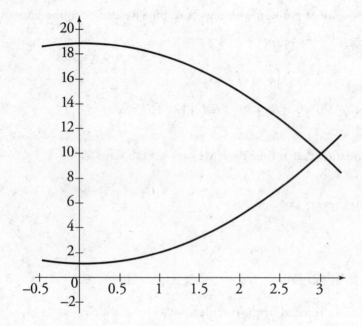

(a)   Find the area of R.

First, let's find where the curves intersect. Set them equal to each other:

$$19 - x^2 = x^2 + 1$$

$$2x^2 = 18$$

$$x^2 = 9$$

$$x = \pm 3$$

We are only interested in the first quadrant, so in order to find the area, we need to evaluate the integral $\int_0^3 (19 - x^2) - (x^2 + 1)dx$. We get

$$\int_0^3 (19 - x^2) - (x^2 + 1)dx = \int_0^3 (18 - 2x^2)dx = \left(18x - \frac{2x^3}{3}\right)\Big|_0^3 = (54 - 18) - (0 - 0) = 36$$

(b)   A solid is formed by revolving R around the x-axis. Find the volume of the solid.

In order to find the volume, we need to evaluate the integral $\pi\int_0^3 (19 - x^2)^2 - (x^2 + 1)^2 dx$. We get

$$\pi\int_0^3 (361 - 38x^2 + x^4) - (x^4 + 2x^2 + 1)dx =$$

$$\pi\int_0^3 (360 - 40x^2)dx =$$

$$\pi\int_0^3 (360 - 40x^2)dx = \pi\left(360x - \frac{40x^3}{3}\right)\Big|_0^3 = \pi(1080 - 360) - \pi(0 - 0) = 720\pi$$

(c)   A solid has its base as the region R, whose cross-sections perpendicular to the x-axis are squares. Find the volume of the solid.

In order to find the volume, we need to find the area of each cross-section, which is $side^2$ (because it is a square), where the side is the difference between the upper and lower curve. Thus, we need to evaluate the integral $\int_0^3 \left[(19 - x^2) - (x^2 + 1)\right]^2 dx$. We get

$$\int_0^3 \left[(19 - x^2) - (x^2 + 1)\right]^2 dx =$$

$$\int_0^3 (18 - 2x^2)^2 dx =$$

$$\int_0^3 (324 - 72x^2 + 4x^4)dx =$$

$$\left(324x - 24x^3 + \frac{4x^5}{5}\right)\Big|_0^3 = \left(972 - 648 + \frac{972}{5}\right) - (0) = 324 + \frac{972}{5} = \frac{2592}{5}$$

3.   Consider the equation $x^2 - 2xy + 4y^2 = 84$.

(a)   Write an expression for the slope of the curve at any point (x, y).

Step 1: The slope of the curve is just the derivative. But here we have to use implicit differentiation to find the derivative. If we take the derivative of each term with respect to x, we get

$$2x\frac{dx}{dx} - 2\left(x\frac{dy}{dx} + y\frac{dx}{dx}\right) + 8y\frac{dy}{dx} = 0$$

Remember that $\dfrac{dx}{dx} = 1$, which gives us

$$2x - 2\left(x\dfrac{dy}{dx} + y \right) + 8y\dfrac{dy}{dx} = 0$$

Step 2: Now just simplify and solve for $\dfrac{dy}{dx}$.

$$2x - 2x\dfrac{dy}{dx} - 2y + 8y\dfrac{dy}{dx} = 0$$

$$x - x\dfrac{dy}{dx} - y + 4y\dfrac{dy}{dx} = 0$$

$$-x\dfrac{dy}{dx} + 4y\dfrac{dy}{dx} = y - x$$

$$\left(4y - x\right)\dfrac{dy}{dx} = y - x$$

$$\dfrac{dy}{dx} = \dfrac{y - x}{4y - x}$$

(b) Find the equation of the tangent lines to the curve at the point $x = 2$.

Step 1: We are going to use the point-slope form of a line, $y - y_1 = m(x - x_1)$, where (x_1, y_1) is a point on the curve, and the derivative at that point is the slope m. First, we need to know the value of y when $x = 2$. If we plug 2 for x into the original equation, we get

$$4 - 4y + 4y^2 = 84$$

$$4y^2 - 4y - 80 = 0$$

$$y^2 - y - 20 = 0$$

This factors to $(y - 5)(y + 4) = 0$, so $y = 5$ or $y = -4$.

Notice that there are two values of y when $x = 2$, which is why there are two tangent lines.

Step 2: Now that we have our points, we need the slope of the tangent line at $x = 2$.

$$\dfrac{dy}{dx} = \dfrac{y - x}{4y - x}$$

$$\text{At } (2, 5), \quad \dfrac{dy}{dx} = \dfrac{5 - 2}{4(5) - 2} = \dfrac{3}{18} = \dfrac{1}{6}$$

$$\text{At } (2, -4), \quad \dfrac{dy}{dx} = \dfrac{-4 - 2}{4(-4) - 2} = \dfrac{-6}{-18} = \dfrac{1}{3}$$

Step 3: Plugging into our equation for the tangent line, we get

$$y - 5 = \frac{1}{6}(x - 2)$$

$$y + 4 = \frac{1}{3}(x - 2)$$

It is not necessary to simplify these equations.

(c)   Find $\dfrac{d^2 y}{dx^2}$ at $(0, \sqrt{21})$.

Step 1: Once we have the first derivative, we have to differentiate again to find $\dfrac{d^2 y}{dx^2}$.

We have to use implicit differentiation again.

$$\frac{dy}{dx} = \frac{y - x}{4y - x}$$

Use the Quotient Rule.

$$\frac{d^2 y}{dx^2} = \frac{(4y - x)\left(\dfrac{dy}{dx} - \dfrac{dx}{dx}\right) - (y - x)\left(4\dfrac{dy}{dx} - \dfrac{dx}{dx}\right)}{(4y - x)^2}$$

Simplifying, we get

$$\frac{d^2 y}{dx^2} = \frac{(4y - x)\left(\dfrac{dy}{dx} - 1\right) - (y - x)\left(4\dfrac{dy}{dx} - 1\right)}{(4y - x)^2}$$

Now we plug in $\dfrac{y - x}{4y - x}$ for $\dfrac{dy}{dx}$, which gives us

$$\frac{d^2 y}{dx^2} = \frac{(4y - x)\left(\dfrac{y - x}{4y - x} - 1\right) - (y - x)\left(4\dfrac{y - x}{4y - x} - 1\right)}{(4y - x)^2}$$

Now we would have to use a lot of algebra to simplify this but, fortunately, we can just plug in $(0, \sqrt{21})$ for x and y and solve from there.

$$\frac{d^2 y}{dx^2} = \frac{(4\sqrt{21})\left(\dfrac{\sqrt{21}}{4\sqrt{21}} - 1\right) - (\sqrt{21})\left(4\dfrac{\sqrt{21}}{4\sqrt{21}} - 1\right)}{(4\sqrt{21})^2} = \frac{-3\sqrt{21}}{336}$$

4.   Water is draining at the rate of 48π ft³/sec from the vertex at the bottom of a conical tank whose diameter at its base is 40 feet and whose height is 60 feet.

(a)   Find an expression for the volume of water (in ft³/sec) in the tank, in terms of its radius, at the surface of the water.

The formula for the volume of a cone is $V = \dfrac{1}{3}\pi r^2 H$, where r is the radius of the cone and h is the height. The ratio of the height of a cone to its radius is constant at any point on the edge of the cone, so we also know that $\dfrac{h}{r} = \dfrac{60}{20} = 3$. (Remember that the radius is half the diameter.) If we solve this for h and substitute, we get

$$h = 3r$$

$$V = \frac{1}{3}\pi r^2 (3r) = \pi r^3$$

(b)   At what rate (in ft/sec) is the radius of the water in the tank shrinking when the radius is 16 feet?

Step 1: This is a related rates question. We now have a formula for the volume of the cone in terms of its radius, so if we differentiate it in terms of t, we should be able to solve for the rate of change of the radius $\dfrac{dr}{dt}$.

We are given that the rate of change of the volume and the radius are, respectively,

$$\frac{dV}{dt} = -48\pi \text{ and } r = 16$$

Differentiating the formula for the volume, we get $\dfrac{dV}{dt} = 3\pi r^2 \dfrac{dr}{dt}$.

Now, we plug in and get $-48\pi = 3\pi(16)^2 \dfrac{dr}{dt}$. Finally, if we solve for $\dfrac{dr}{dt}$, we get

$$\frac{dr}{dt} = -\frac{1}{16} \text{ ft/sec}$$

(c)   How fast (in ft/sec) is the height of the water in the tank dropping at the instant that the radius is 16 feet?

Step 1: This is the same idea as the previous problem, except that we want to solve for $\dfrac{dh}{dt}$. In order to do this, we need to go back to our ratio of height to radius and solve it for the radius.

$$\frac{h}{r} = 3 \qquad \text{or} \qquad \frac{h}{3} = r$$

Substituting for r in the original equation, we get $V = \dfrac{1}{3}\pi\left(\dfrac{h}{3}\right)^2 h = \dfrac{\pi h^3}{27}$.

Step 2: Now we need to know what h is when r is 16. Using our ratio,

$$h = 3(16) = 48.$$

Step 3: Now if we differentiate, we get

$$\frac{dV}{dt} = \frac{\pi h^2}{9}\frac{dh}{dt}$$

Now we plug in and solve.

$$48\pi = \frac{\pi(48)^2}{9}\frac{dh}{dt}$$

$$\frac{dh}{dt} = -\frac{3}{16}$$

One should also note that, because $h = 3r$, $\dfrac{dh}{dt} = 3\dfrac{dr}{dt}$. Thus, after we found $\dfrac{dr}{dt}$ in part (b), we merely had to multiply it by 3 to find the answer for part (c).

5.   Let f be the function given by $y = f(x) = 2x^4 - 4x^2 + 1$.

(a)   Find an equation of the line tangent to the graph at $(-2, 17)$.

In order to find the equation of a tangent line at a particular point, we need to take the derivative of the function and plug in the x- and y-values at that point to give us the slope of the line.

Step 1: The derivative is $f'(x) = 8x^3 - 8x$. If we plug in $x = -2$, we get

$$f'(-2) = 8(-2)^3 - 8(-2) = -48$$

This is the slope m.

Step 2: Now we use the slope-intercept form of the equation of a line, $y - y_1 = m(x - x_1)$, and plug in the appropriate values of x, y, and m.

$$y - 17 = -48(x + 2)$$

If we simplify this, we get $y = -48x - 79$.

(b)   Find the x- and y-coordinates of the relative maxima and relative minima. Verify your answer.

If we want to find the maxima/minima, we need to take the derivative and set it equal to zero. The values that we get are called critical points. We will then test each point to see if it is a maximum or a minimum.

Step 1: We already have the first derivative from part (a), so we can just set it equal to zero.

$$8x^3 - 8x = 0$$

If we now solve this for x, we get

$$8x(x^2 - 1) = 0$$
$$8x(x + 1)(x - 1) = 0$$
$$x = 0, 1, -1$$

These are our critical points. In order to test if a point is a maximum or a minimum, we usually use the *second derivative test*. We plug each of the critical points into the second derivative. If we get a positive value, the point is a relative minimum. If we get a negative value, the point is a relative maximum. If we get zero, the point is a point of inflection.

Step 2: The second derivative is $f''(x) = 24x^2 - 8$. If we plug in the critical points, we get

$$f''(0) = 24(0)^2 - 8 = -8$$

$$f''(1) = 24(1)^2 - 8 = 16$$

$$f''(-1) = 24(-1)^2 - 8 = 16$$

So $x = 0$ is a relative maximum, and $x = 1, -1$ are relative minima.

Step 3: In order to find the y-coordinates, we plug the x-values back into the original equation and solve.

$$f(0) = 1$$
$$f(1) = -1$$
$$f(-1) = -1$$

And our points are

$(0, 1)$ is a relative maximum

$(1, -1)$ is a relative minimum

$(-1, -1)$ is a relative minimum

(c)   Find the *x*- and *y*- coordinates of the points of inflection. Verify your answer.

If we want to find the points of inflection, set the second derivative equal to zero. The values that we get are the *x*-coordinates of the points of inflection.

Step 1: We already have the second derivative from part (b), so all we have to do is set it equal to zero and solve for *x*.

$$24x^2 - 8 = 0$$

$$x^2 = \frac{1}{3}$$

$$x = \pm\sqrt{\frac{1}{3}}$$

Step 2: In order to find the *y*-coordinates, plug the *x*-values back into the original equation and solve.

$$f\left(\sqrt{\frac{1}{3}}\right) = 2\left(\sqrt{\frac{1}{3}}\right)^4 - 4\left(\sqrt{\frac{1}{3}}\right)^2 + 1 = \frac{2}{9} - \frac{4}{3} + 1 = -\frac{1}{9}$$

$$f\left(-\sqrt{\frac{1}{3}}\right) = 2\left(-\sqrt{\frac{1}{3}}\right)^4 - 4\left(-\sqrt{\frac{1}{3}}\right)^2 + 1 = \frac{2}{9} - \frac{4}{3} + 1 = -\frac{1}{9}$$

So the points of inflection are $\left(\sqrt{\frac{1}{3}}, -\frac{1}{9}\right)$ and $\left(-\sqrt{\frac{1}{3}}, -\frac{1}{9}\right)$.

6.   Grain is being loaded into a silo at the rate of $G(t) = 400e^{\frac{-t^2}{4}}$ ft^3 / hr, where *t* is the number of hours that it is being loaded, $0 \le t \le 8$. At time $t = 0$, there is 100 ft³ of grain in the silo. Grain is also being removed through the base of the silo at the following rates, where $R(t)$ is the amount of grain being removed in ft³/hr, $0 \le t \le 8$:

t	0	2	5	7	8
R(t)	60	90	110	120	125

(a)   Estimate the total amount of grain removed from the silo at *t* = 8 hrs, using a left-hand Riemann sum and 4 subintervals.

The Riemann sum is found by multiplying the width of each interval by *R* at each left endpoint of an interval. We get

$$(2 - 0)(60) + (5 - 2)(90) + (7 - 5)(110) + (8 - 7)(120)$$

$$(2)(60) + (3)(90) + (2)(110) + (1)(120) = 120 + 270 + 220 + 120 = 730$$

(b) Estimate the amount of grain in the silo at the end of 8 hours, using your answer from part (a).

We can estimate the amount of grain in the silo by taking the starting amount, 100, adding the amount that comes in, and subtracting the amount that is being removed (which we found in part a). We can find the amount that comes in by $\int_0^8 400e^{\frac{-t^2}{4}}\, dt$. If you plug this in your calculator, you get 708.982. (If you are unsure how to evaluate the integral on your calculator, check the online Appendix for some tips on using a TI-84 calculator.) Therefore, the amount in the silo at the end of 8 hours is $100 + 708.982 - 730 = 78.982$.

(c) Estimate $R'(5)$, showing your work. Indicate the units of measure.

We can estimate $R'(5)$ by finding the slope of the secant line from $t = 2$ to $t = 5$, and from $t = 5$ to $t = 7$ and averaging the two. The slope of the secant line from $t = 2$ to $t = 5$ is $\dfrac{110 - 90}{5 - 2} = \dfrac{20}{3}$, and the slope of the secant line from $t = 5$ to $t = 7$ is $\dfrac{120 - 110}{7 - 5} = 5$, so a good estimate for $R'(5)$ is $\dfrac{5 + \dfrac{20}{3}}{2} = \dfrac{35}{6}$.

Part III
About the AP Calculus AB Exam

- AB Calculus vs. BC Calculus
- The Structure of the Calculus Exam
- How the AP Calculus Exam is Scored
- Overview of Content Topics
- General Overview of This Book
- How AP Exams Are Used
- Other Resources
- Designing Your Study Plan

AB CALCULUS VS. BC CALCULUS

AP Calculus is divided into two types: AB and BC. The former is supposed to be the equivalent of a semester of college calculus; the latter, a year. In truth, AB calculus covers closer to three-quarters of a year of college calculus. In fact, the main difference between the two is that BC calculus tests some more theoretical aspects of calculus and covers a few additional topics. In addition, BC calculus is harder than AB calculus. The AB exam usually tests straightforward problems in each topic. They're not too tricky, and they don't vary very much. The BC exam asks harder questions. But neither exam is tricky in the sense that the SAT is, nor do they test esoteric aspects of calculus. Rather, both tests tend to focus on testing whether you've learned the basics of differential and integral calculus. The tests are difficult because of the breadth of topics that they cover, not the depth. You will probably find that many of the problems in this book seem easier than the problems you've had in school. This is because your teacher is giving you problems that are harder than those on the AP.

> Please note that this book will focus only on the topics that will appear on the AP Calculus AB Exam. If you are looking to prepare for the BC Calculus Exam, please purchase *Cracking the AP Calculus BC Exam,* in stores now!

THE STRUCTURE OF THE CALCULUS EXAM

Section/Type	Time	Score	More Score Info
Section I, Part A: 30 multiple-choice questions	60 minutes	Along with I-B, 50% of complete AP Exam Score	No guessing penalty, calculator NOT permitted
Section I, Part B: 15 multiple-choice questions	45 minutes	Along with I-A, 50% of complete AP Exam Score	No guessing penalty, calculator allowed
Section II, Part A: 2 open-ended questions	30 minutes	Along with II-B, 50% of complete AP Exam Score	calculator allowed
Section II, Part B: 4 open-ended questions	60 minutes	Along with II-A, 50% of complete AP Exam Score	calculator NOT permitted

There is no guessing penalty on the test, so don't leave anything blank in Section I. Section II, however, requires you to write out not only your solutions, but to show your steps. You can get partial credit, so make sure you put everything down on paper. You'll note, too, that some parts of the test allow you to use a graphing calculator and its programs. But here's the truth about calculus: most of the time, you don't need a calculator, so don't get stuck trying to find the right program. Go with what you know.

> Sign in to your online student tools to download and print a handy appendix of tips aimed at getting the most out of your TI-84.

HOW THE AP CALCULUS AB EXAM IS SCORED

A numeral score of 1 to 5 is going to be assigned to your test, based on the number of questions you've answered correctly.

5 = Extremely Well Qualified
4 = Well Qualified
3 = Qualified
2 = Possibly Qualified
1 = No Recommendation

Colleges decide for themselves the minimum score they will accept for college credit and/or advanced placement. The American Council on Education recommends the acceptance of grades 3 or above, and many colleges adhere to these standards. About 60 percent of students who take the AP Calculus AB Exam receive a score of 3 or higher. Check the website for each college you plan to apply to so that you know their policy on granting credit or advanced placement.

AP Calculus—2018 Score Distributions		
Score	**Number of Test Takers**	**Percentage**
5	59,733	19.4%
4	53,255	17.3%
3	64,768	21.0%
2	68,980	22.4%
1	61,802	20.0%
Total	308,538	100%

Data from The College Board, May 2018 AP Exam administrations

OVERVIEW OF CONTENT TOPICS

This list is drawn from the topical outline for AP Calculus furnished by the College Board. You might find that your teacher covers some additional topics, or omits some, in your course. Some of the topics are very broad, so we cannot guarantee that this book covers these topics exhaustively.

I. Functions, Graphs, and Limits
  A. Analysis of Graphs

- You should be able to analyze a graph based on "the interplay between geometric and analytic information." The preceding phrase comes directly from the College Board. Don't let it scare you. What the College Board really means is that you should have covered graphing in precalculus, and you should know (a) how to graph and (b) how to read a graph. **This is a precalculus topic, and we won't cover it in this book.**

  B. Limits

- You should be able to calculate limits algebraically or to estimate them from a graph or from a table of data.

Same Song, Slightly Different Verse
The 2019-2020 administration of the course will be modifying this outline. This book will still cover all the material, but in a potentially different order, one that's designed to help maximize your test-prep. To stay up-to-date with changes that might impact your test, register your book and check your online Student Tools. You can also go to the College Board's AP Central for this course (https://apcentral.collegeboard.org/courses/ap-calculus/course/updates-2019-20).

- You do **not** need to find limits using the Delta-Epsilon definition of a limit.

- You need to know how to use L'Hôpital's Rule to evaluate a limit of indeterminate form.

C. Asymptotes

- You should understand asymptotes graphically and be able to compare the growth rates of different types of functions (namely polynomial functions, logarithmic functions, and exponential functions). **This is a topic that should have been covered in precalculus, and we won't cover it in this book.**

- You should understand asymptotes in terms of limits involving infinity.

D. Continuity

- You should be able to test the continuity of a function in terms of limits, and you should understand continuous functions graphically.

- You should understand the Intermediate Value Theorem and the Extreme Value Theorem.

II. Differential Calculus

A. The Definition of the Derivative

- You should be able to find a derivative by finding the limit of the difference quotient.

- You should also know the relationship between differentiability and continuity. That is, if a function is differentiable at a point, it's continuous there. But if a function is continuous at a point, it's not necessarily differentiable there.

B. Derivative at a Point

- You should know the Power Rule, the Product Rule, the Quotient Rule, and the Chain Rule.

- You should be able to find the slope of a curve at a point, and the tangent and normal lines to a curve at a point.

- You should also be able to use local linear approximation and differentials to estimate the tangent line to a curve at a point.

- You should be able to find the instantaneous rate of change of a function using the derivative or the limit of the average rate of change of a function.

- You should be able to approximate the rate of change of a function from a graph or from a table of values.

- You should be able to find Higher-Order Derivatives and to use implicit differentiation.

C. Derivative of a Function

- You should be able to relate the graph of a function to the graph of its derivative, and vice-versa.

- You should know the relationship between the sign of a derivative and whether the function is increasing or decreasing (positive derivative means increasing; negative means decreasing).

- You should know how to find relative and absolute maxima and minima.

- You should know the Mean Value Theorem for derivatives and Rolle's Theorem.

D. Second Derivative

- You should be able to relate the graph of a function to the graph of its derivative and its second derivative, and vice-versa. This is tricky.

- You should know the relationship between concavity and the sign of the second derivative (positive means concave up; negative means concave down).

- You should know how to find points of inflection.

E. Applications of Derivatives

- You should be able to sketch a curve using first and second derivatives and be able to analyze the critical points.

- You should be able to solve Optimization problems (Max/Min problems) and Related Rates problems.

- You should be able to find the derivative of the inverse of a function.

- You should be able to solve Rectilinear Motion problems.

F. Computation of Derivatives

- You should be able to find the derivatives of Trig functions, Logarithmic functions, Exponential functions, and Inverse Trig functions.

III. Integral Calculus

A. Riemann Sums

- You should be able to find the area under a curve using left, right, and midpoint evaluations and the Trapezoid Rule.

- You should know the fundamental theorem of calculus:

$$\int_a^b f(x)dx = F(b) - F(a)$$

B. Applications of Integrals

- You should be able to find the area of a region, the volume of a solid of known cross-section, the volume of a solid of revolution, and the average value of a function.

- You should be able to solve acceleration, velocity, and position problems.

C. Fundamental Theorem of Calculus

- You should know the first and second fundamental theorems of calculus and be able to use them both to find the derivative of an integral and for the analytical and graphical analysis of functions.

D. Techniques of Antidifferentiation

- You should be able to integrate using the power rule and u-substitution.

E. Applications of Antidifferentiation

- You should be able to find specific antiderivatives using initial conditions.

- You should be able to solve separable differential equations and logistic differential equations.

- You should be able to interpret differential equations via slope fields. Don't be intimidated. These look harder than they are.

GENERAL OVERVIEW OF THIS BOOK

There's More Online
You can find a full list of all the formulas in this book in your online Student Tools, which also contains a wealth of other resources, like calculator shortcuts.

The key to doing well on the exam is to memorize a variety of techniques for solving calculus problems and to recognize when to use them. There's so much to learn in AP Calculus that it's difficult to remember everything. Instead, you should be able to derive or figure out how to do certain things based on your mastery of a few essential techniques. In addition, you'll be expected to remember a lot of the math that you did before calculus—particularly trigonometry. You should be able to graph functions, find zeros, derivatives, and integrals with the calculator.

Furthermore, if you can't derive certain formulas, you should memorize them! A lot of students don't bother to memorize the trigonometry special angles and formulas because they can do them on their calculators. This is a big mistake. You'll be expected to be very good with these in calculus, and if you can't recall them easily, you'll be slowed down and the problems will seem much harder. Make sure that you're also comfortable with analytic geometry. If you rely on your calculator to graph for you, you'll get a lot of questions wrong because you won't recognize the curves when you see them.

This advice is going to seem backward compared with what your teachers are telling you. In school you're often yelled at for memorizing things. Teachers tell you to understand the concepts, not just memorize the answers. Well, things are different here. The understanding will come later, after you're comfortable with the mechanics. In the meantime, you should learn techniques and practice them, and, through repetition, you will ingrain them in your memory.

Each chapter is divided into three types of problems: examples, solved problems, and practice problems. The first type is contained in the explanatory portion of the unit. The examples are designed to further your understanding of the subject and to show you how to get the problems right. Each step of the solution to the example is worked out, except for some simple algebraic and arithmetic steps that should come easily to you at this point.

The second type of problems is solved problems. The solutions are worked out in approximately the same detail as the examples. Before you start work on each of these, cover the solution with an index card, then check the solution afterward. And you should read through the solution, not just assume that you knew what you were doing because your answer was correct.

The third type is practice problems. Only the answer explanations to these are given. We hope you'll find that each chapter offers enough practice problems for you to be comfortable with the material. The topics that are emphasized on the exam have more problems; those that are de-emphasized have fewer. In other words, if a chapter has only a few practice problems, it's not an important topic on the AP Exam and you shouldn't worry too much about it.

These practice problems come in a variety of forms throughout the book (and in your online Student Tools). Practice Sets cover a set amount of material from a chapter in an open-ended format, while End-of-Chapter Drills present multiple-choice questions from topics across that whole chapter.

No, Really, There's Even More Online
Once you've registered your copy of the book, you can access drills on fundamental math skills, and also work on challenging End-of-Unit Drills that put everything to the test.

HOW AP EXAMS ARE USED

Different colleges use AP Exams in different ways, so it is important that you go to a particular college's website to determine how it uses AP Exams. The three items below represent the main ways in which AP Exam scores can be used.

- **College Credit.** Some colleges will give you college credit if you score well on an AP Exam. These credits count towards your graduation requirements, meaning that you can take fewer courses while in college. Given the cost of college, this could be quite a benefit, indeed.
- **Satisfy Requirements.** Some colleges will allow you to "place out" of certain requirements if you do well on an AP Exam, even if they do not give you actual college credits. For example, you might not need to take an introductory-level course, or perhaps you might not need to take a class in a certain discipline at all.
- **Admissions Plus.** Even if your AP Exam will not result in college credit or even allow you to place out of certain courses, most colleges will respect your decision to push yourself by taking an AP Course or even an AP Exam outside of a course. A high score on an AP Exam shows mastery of more difficult content than is taught in many high school courses, and colleges may take that into account during the admissions process.

More Great Books
Check out The Princeton Review's college guide books, including *The Best 385 Colleges, The Complete Book of Colleges, Paying for College Without Going Broke,* and many more!

OTHER RESOURCES

There are many resources available to help you improve your score on the AP Calculus AB Exam, not the least of which are your **teachers**. If you are taking an AP class, you may be able to get extra attention from your teacher, such as obtaining feedback on your free-response questions. If you are not in an AP course, reach out to a teacher who teaches calculus, and ask if the teacher will review your free-response questions or otherwise help you with content.

Another wonderful resource is **AP Students**, the official site of the AP Exams. The scope of the information at this site is quite broad and includes:

- a course description, which includes details on what content is covered and sample questions
- sample test questions
- free-response prompts from previous years

The AP Students home page address is: **https://apstudent.collegeboard.org**.

Finally, **The Princeton Review** offers tutoring and small group instruction. Our expert instructors can help you refine your strategic approach and add to your content knowledge. For more information, call 1-800-2REVIEW.

DESIGNING YOUR STUDY PLAN

As part of the Introduction, you identified some areas of potential improvement. Let's now delve further into your performance on Test 1, with the goal of developing a study plan appropriate to your needs and time commitment.

Read the answers and explanations associated with the multiple-choice questions (starting at page 41). After you have done so, respond to the following questions:

Break up your review into manageable portions. Download our helpful study guide for this book, once you register online.

- Review the Overview of Content Topics on pages 67–70. Next to each topic, indicate your rank of the topic as follows: "1" means "I need a lot of work on this," "2" means "I need to beef up my knowledge," and "3" means "I know this topic well."

- How many days/weeks/months away is your exam?

- What time of day is your best, most focused study time?

- How much time per day/week/month will you devote to preparing for your exam?

- When will you do this preparation? (Be as specific as possible: Mondays and Wednesdays from 3 to 4 p.m., for example.)

- Based on the answers above, will you focus on strategy (Part IV) or content (Part V) or both?

- What are your overall goals in using this book?

More AP Info Online!
We currently offer LiveOnline AP Cram Courses and Private tutoring for a handful of AP Exam subjects. For more information, head over to www.princeton-review.com/college/ap-test-prep.

Part IV
Test-Taking Strategies for the AP Calculus AB Exam

PREVIEW ACTIVITY

Review your responses to the first three questions on page 4 of Part I, and then respond to the following questions:

- How many multiple-choice questions did you miss even though you knew the answer?

- On how many multiple-choice questions did you guess blindly?

- How many multiple-choice questions did you miss after eliminating some answers and guessing based on the remaining answers?

- Did you find any of the free-response questions easier or harder than the others, and if so, why?

HOW TO USE THE CHAPTERS IN THIS PART

Before you read the following Strategy chapters, take a moment and think about what you are doing now. As you read and engage in the directed practice, be sure to appreciate the ways you can change your approach. At the end of each chapter in Part IV, you will have the opportunity to reflect on how you will change your approach.

Chapter 1
How to Approach
Multiple-Choice
Questions

CRACKING THE MULTIPLE-CHOICE QUESTIONS

Section I of the AP Calculus Exam consists of 45 multiple-choice questions, which you're given 105 minutes to complete. This section is worth 50 percent of your grade.

All the multiple-choice questions will have a similar format: each will be followed by four answer choices. At times, it may seem that there could be more than one possible correct answer. There is only one! Remember that the committee members who write these questions are calculus teachers. So, when it comes to calculus, they know how students think and what kind of mistakes they make. Answers resulting from common mistakes are often included in the four answer choices to trap you.

Use the Answer Sheet

For the multiple-choice section, you write the answers not in the test booklet but on a separate answer sheet (very similar to the ones we've supplied at the very end of this book). Four oval-shaped bubbles follow the question number, one for each possible answer. *Don't* forget to fill in all your answers on the answer sheet. Don't just mark them in the test booklet. Marks in the test booklet will not be graded. Also, make sure that your filled-in answers correspond to the correct question numbers! Check your answer sheet after every five answers to make sure you haven't skipped any bubbles by mistake.

Should You Guess?

Use Process of Elimination (POE) to rule out answer choices you know are wrong and increase your chances of guessing the right answer. Read all the answer choices carefully. Eliminate the ones that you know are wrong. If you only have one answer choice left, *choose it,* even if you're not completely sure why it's correct. Remember that questions in the multiple-choice section are graded by a computer, so it doesn't care *how* you arrived at the correct answer.

Proven Techniques
Use POE and the Two-Pass System to help boost your score.

Even if you can't eliminate answer choices, go ahead and guess. There is no guessing penalty for incorrect answers. You will be assessed only on the total number of correct answers, so be sure to fill in all the bubbles even if you have no idea what the correct answers are. When you get to questions that are too time-consuming, or that you don't know the answer to (and can't eliminate any options), don't just fill in any answer. Use what we call your "letter of the day" (LOTD). Selecting the same answer choice each time you guess will increase your odds of getting a few of those skipped questions right.

Use the Two-Pass System

Remember that you have about two and a quarter minutes per question on this section of the exam. Do not waste time by lingering too long over any single question. If you're having trouble, move on to the next question. After you finish all the questions, you can come back to the ones you skipped.

The best strategy is to go through the multiple-choice section twice. The first time, do all the questions that you can answer fairly quickly—the ones in which you feel confident about the correct answer. On this first pass, skip the questions that seem to require more thinking or the ones you need to read two or three times before you understand them. Circle the questions that you've skipped in the question booklet so that you can find them easily in the second pass. You must *be very careful* with the answer sheet by making sure the filled-in answers correspond correctly to the questions.

Once you have gone through all the questions, go back to the ones that you skipped in the first pass. But don't linger too long on any one question even in the second pass. Spending too much time wrestling over a hard question can cause two things to happen. One, you may run out of time and miss out on answering easier questions in the later part of the exam. Two, your anxiety might start building up, and this could prevent you from thinking clearly, which would make answering other questions even more difficult. If you simply don't know the answer, or can't eliminate any of them, just use your LOTD and move on.

REFLECT

Respond to the following questions:

- How long will you spend on multiple-choice questions?

- How will you change your approach to multiple-choice questions?

- What is your multiple-choice guessing strategy?

Chapter 2
How to Approach
Free-Response
Questions

CRACKING THE FREE-RESPONSE QUESTIONS

Section II is worth 50 percent of your grade on the AP Calculus Exam. This section is composed of two parts. Part A contains two free-response questions (you may use a calculator on this part); Part B contains four free-response questions where there are no calculators allowed. You're given a total of 90 minutes for this section.

Clearly Explain and Justify Your Answers

Remember that your answers to the free-response questions are graded by *readers* and not by computers. Communication is a very important part of AP Calculus. Compose your answers in precise sentences. Just getting the correct numerical answer is not enough. You should be able to *explain* your reasoning behind the technique that you selected and *communicate* your answer in the context of the problem. Even if the question does not explicitly say so, always explain and *justify* every step of your answer, including the final answer. Do not expect the graders to read between the lines. Explain everything as though somebody with no knowledge of calculus is going to read it. Be sure to present your solution in a systematic manner using solid logic and appropriate language. And remember, although you won't earn points for neatness, the graders can't give you a grade if they can't read and understand your solution!

Use Only the Space You Need

Do not try to fill up the space provided for each question. The space given is usually more than enough. The people who design the tests realize that some students write in big letters or make mistakes and need extra space for corrections. So if you have a complete solution, don't worry about the extra space. Writing more will not earn you extra credit. In fact, many students tend to go overboard and shoot themselves in the foot by making a mistake after they've already written the right answer.

Read the Whole Question!

Some questions might have several subparts. Try to answer them all, and don't give up on the question if one part is giving you trouble. For example, if the answer to part (b) depends on the answer to part (a), but you think you got the answer to part (a) wrong, you should still go ahead and do part (b) using your answer to part (a) as required. Chances are that the grader will not mark you wrong twice, unless it is obvious from your answer that you should have discovered your mistake.

REFLECT

Respond to the following questions:

- How much time will you spend on each free-response question?

- How will you change your approach to the free-response questions?

- Will you seek further help, outside of this book (such as a teacher, tutor, or AP Central), on how to approach the calculus exam?

Part V
Content Review for the AP Calculus AB Exam

HOW TO USE THE CHAPTERS IN THIS PART

The following chapters are organized into four units to help you break up your review into manageable portions. At the end of each unit, go to your free online student tools to download and print a comprehensive drill for all of the chapters covered in that unit. You may need to review some chapters more than once, as your goal is to obtain mastery of the content you may be missing.

Unit 1
Differential
Calculus Essentials

Chapter 3
Limits

WHAT IS A LIMIT?

In order to understand calculus, you need to know what a "limit" is. A limit is the value a function (which usually is written "$f(x)$" on the AP Exam) approaches as the variable within that function (usually "x") gets nearer and nearer to a particular value. In other words, when x is very close to a certain number, what is $f(x)$ very close to?

Let's look at an example of a limit: What is the limit of the function $f(x) = x^2$ as x approaches 2? In limit notation, the expression "the limit of $f(x)$ as x approaches 2" is written like this: $\lim_{x \to 2} f(x)$. In order to evaluate the limit, let's check out some values of $\lim_{x \to 2} f(x)$ as x increases and gets closer to 2 (without ever exactly getting there).

When $x = 1.9$, $f(x) = 3.61$.
When $x = 1.99$, $f(x) = 3.9601$.
When $x = 1.999$, $f(x) = 3.996001$.
When $x = 1.9999$, $f(x) = 3.99960001$.

As x increases and approaches 2, $f(x)$ gets closer and closer to 4. This is called the **left-hand limit** and is written: $\lim_{x \to 2^-} f(x)$. Notice the little minus sign!

What about when x is bigger than 2?

When $x = 2.1$, $f(x) = 4.41$.
When $x = 2.01$, $f(x) = 4.0401$.
When $x = 2.001$, $f(x) = 4.004001$.
When $x = 2.0001$, $f(x) = 4.00040001$.

As x decreases and approaches 2, $f(x)$ still approaches 4. This is called the **right-hand limit** and is written like this: $\lim_{x \to 2^+} f(x)$. Notice the little plus sign!

We got the same answer when evaluating both the left- and right-hand limits, because when x is 2, $f(x)$ is 4. You should always check both sides of the independent variable because, as you'll see shortly, sometimes you don't get the same answer. Therefore, we write that $\lim_{x \to 2} x^2 = 4$.

We didn't really need to look at all of these decimal values to know what was going to happen when x got really close to 2. But it's important to go through the exercise because, typically, the answers get a lot more complicated. Let's do a few examples.

Example 1: Find $\lim_{x \to 5} x^2$.

The approach is simple: Plug in 5 for x, and you get 25.

Example 2: Find $\lim_{x \to 3} x^3$.

Here, the answer is 27.

There are some simple algebraic rules of limits that you should know. These are:

$$\lim_{x \to a} kf(x) = k \lim_{x \to a} f(x)$$

Example: $\lim_{x \to 5} 3x^2 = 3 \lim_{x \to 5} x^2 = 75$

$$\text{If } \lim_{x \to a} f(x) = L_1 \text{ and } \lim_{x \to a} g(x) = L_2 \text{, then}$$

$$\lim_{x \to a} [f(x) + g(x)] = L_1 + L_2$$

Example: $\lim_{x \to 5} \left[x^2 + x^3 \right] = \lim_{x \to 5} x^2 + \lim_{x \to 5} x^3 = 150$

$$\text{If } \lim_{x \to a} f(x) = L_1 \text{ and } \lim_{x \to a} g(x) = L_2 \text{, then}$$

$$\lim_{x \to a} [f(x) \cdot g(x)] = L_1 \cdot L_2$$

Example: $\lim_{x \to 5} \left[\left(x^2 + 1 \right) \sqrt{x - 1} \right] = \lim_{x \to 5} \left(x^2 + 1 \right) \lim_{x \to 5} \sqrt{x - 1} = 52$

Example 3: Find $\lim_{x \to 0} \left(x^2 + 5x \right)$.

Plug in 0, and you get 0.

So far, so good. All you do to find the limit of a simple polynomial is plug in the number that the variable is approaching and see what the answer is. Naturally, the process can get messier— especially if x approaches zero.

Example 4: Find $\lim\limits_{x \to 0} \dfrac{1}{x^2}$.

If you plug in some very small values for x, you'll see that this function approaches ∞. And it doesn't matter whether x is positive or negative, you still get ∞. Look at the graph of $y = \dfrac{1}{x^2}$:

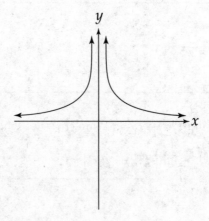

On either side of $x = 0$ (the y-axis), the curve approaches ∞.

Example 5: Find $\lim\limits_{x \to 0} \dfrac{1}{x}$.

Here you have a problem. If you plug in some very small positive values for x (0.1, 0.01, 0.001, and so on), you approach ∞. In other words, $\lim\limits_{x \to 0^+} \dfrac{1}{x} = \infty$. But, if you plug in some very small negative values for x (–0.1, –0.01, –0.001, and so on), you approach $-\infty$. That is, $\lim\limits_{x \to 0^-} \dfrac{1}{x} = -\infty$. Because the right-hand limit is not equal to the left-hand limit, the limit does not exist.

Look at the graph of $\dfrac{1}{x}$.

You can see that on the left side of $x = 0$, the curve approaches $-\infty$, and on the right side of $x = 0$, the curve approaches ∞. There are some very important points that we need to emphasize from the last two examples.

Why do we state the limit in Example 4 but not for Example 5? Because when we have $\dfrac{k}{x^2}$, the function is always positive no matter what the sign of x is and thus the function has the same limit from the left and the right. But when we have $\dfrac{k}{x}$, the function's sign depends on the sign of x, and you get a different limit from each side.

Let's look at a few examples in which the independent variable approaches infinity.

Example 6: Find $\lim\limits_{x\to\infty}\dfrac{1}{x}$.

As x gets bigger and bigger, the value of the function gets smaller and smaller. Therefore, $\lim\limits_{x\to\infty}\dfrac{1}{x}=0$.

Example 7: Find $\lim\limits_{x\to-\infty}\dfrac{1}{x}$.

It's the same situation as the one in Example 6; as x decreases (approaches negative infinity), the value of the function increases (approaches zero). We write the following:

$$\lim_{x\to-\infty}\frac{1}{x}=0$$

We don't have the same problem here that we did when x approached zero because "positive zero" is the same thing as "negative zero," whereas positive infinity is different from negative infinity.

Here's another rule:

(1) If the left-hand limit of a function is not equal to the right-hand limit of the function, then the limit does not exist.

(2) A limit equal to infinity is not the same as a limit that does not exist, but sometimes you will see the expression "no limit," which serves both purposes. If $\lim\limits_{x\to a}f(x)=\infty$, the limit, technically, does not exist.

(3) If k is a positive constant, then $\lim\limits_{x\to0^+}\dfrac{k}{x}=\infty$, $\lim\limits_{x\to0^-}\dfrac{k}{x}=-\infty$, and $\lim\limits_{x\to0}\dfrac{k}{x}$ does not exist.

(4) If k is a positive constant, then $\lim\limits_{x\to0^+}\dfrac{k}{x^2}=\infty$, $\lim\limits_{x\to0^-}\dfrac{k}{x^2}=\infty$, and $\lim\limits_{x\to0}\dfrac{k}{x^2}=\infty$.

If k and n are constants, $|x|>1$, and $n>0$, then $\lim\limits_{x\to\infty}\dfrac{k}{x^n}=0$, and $\lim\limits_{x\to-\infty}\dfrac{k}{x^n}=0$.

Example 8: Find $\lim\limits_{x \to \infty} \dfrac{3x+5}{7x-2}$.

When you have variables in both the top and the bottom, you can't just plug ∞ into the expression. You'll get $\dfrac{\infty}{\infty}$. We solve this by using the following technique:

When an expression consists of a polynomial divided by another polynomial, divide each term of the numerator and the denominator by the highest power of x that appears in the expression.

The highest power of x in this case is x^1, so we divide every term in the expression (both top and bottom) by x, like so:

$$\lim_{x \to \infty} \frac{3x+5}{7x-2} = \lim_{x \to \infty} \frac{\dfrac{3x}{x} + \dfrac{5}{x}}{\dfrac{7x}{x} - \dfrac{2}{x}} = \lim_{x \to \infty} \frac{3 + \dfrac{5}{x}}{7 - \dfrac{2}{x}}$$

Now when we take the limit, the two terms containing x approach zero. We're left with $\dfrac{3}{7}$.

Example 9: Find $\lim\limits_{x \to \infty} \dfrac{8x^2 - 4x + 1}{16x^2 + 7x - 2}$.

Divide each term by x^2. You get

$$\lim_{x \to \infty} \frac{8 - \dfrac{4}{x} + \dfrac{1}{x^2}}{16 + \dfrac{7}{x} - \dfrac{2}{x^2}} = \frac{8}{16} = \frac{1}{2}$$

Example 10: Find $\lim\limits_{x \to \infty} \dfrac{-3x^{10} - 70x^5 + x^3}{33x^{10} + 200x^8 - 1000x^4}$.

Divide each term by x^{10}.

$$\lim_{x \to \infty} \frac{-3x^{10} - 70x^5 + x^3}{33x^{10} + 200x^8 - 1000x^4} = \lim_{x \to \infty} \frac{-3 - \dfrac{70}{x^5} + \dfrac{1}{x^7}}{33 + \dfrac{200}{x^2} - \dfrac{1000}{x^6}} = -\frac{3}{33} = -\frac{1}{11}$$

Remember to focus your attention on the highest power of x.

The other powers don't matter because they're all going to disappear. Now we have three new rules for evaluating the limit of a rational expression as x approaches infinity.

> (1) If the highest power of x in a rational expression is in the numerator, then the limit as x approaches infinity is infinity.

Example: $\lim\limits_{x\to\infty}\dfrac{5x^7-3x}{16x^6-3x^2}=\infty$

> (2) If the highest power of x in a rational expression is in the denominator, then the limit as x approaches infinity is zero.

Example: $\lim\limits_{x\to\infty}\dfrac{5x^6-3x}{16x^7-3x^2}=0$

> (3) If the highest power of x in a rational expression is the same in both the numerator and denominator, then the limit as x approaches infinity is the coefficient of the highest term in the numerator divided by the coefficient of the highest term in the denominator.

Example: $\lim\limits_{x\to\infty}\dfrac{5x^7-3x}{16x^7-3x^2}=\dfrac{5}{16}$

LIMITS OF TRIGONOMETRIC FUNCTIONS

At some point during the exam, you'll have to find the limit of certain trig expressions, usually as x approaches either zero or infinity. There are four standard limits that you should memorize—with those, you can evaluate all of the trigonometric limits that appear on the test. As you'll see throughout this book, calculus requires that you remember all of your trig from previous years.

> Rule No. 1: $\lim\limits_{x\to 0}\dfrac{\sin x}{x}=1$ (x is in radians, *not* degrees)

This may seem strange, but if you look at the graphs of $f(x) = \sin x$ and $f(x) = x$, they have approximately the same slope near the origin (as x gets closer to zero). Because x and the sine of x are about the same as x approaches zero, their quotient will be very close to one. Furthermore, because $\lim_{x \to 0} \cos x = 1$ (review cosine values if you don't get this!), we know that

$$\lim_{x \to 0} \tan x = \lim_{x \to 0} \frac{\sin x}{\cos x} = 0.$$

Now we will find a second rule. Let's evaluate the limit $\lim_{x \to 0} \frac{\cos x - 1}{x}$. First, multiply the top and bottom by $\cos x + 1$. We get $\lim_{x \to 0} \left(\frac{\cos x - 1}{x} \right) \left(\frac{\cos x + 1}{\cos x + 1} \right)$. Now simplify the limit to $\lim_{x \to 0} \frac{\cos^2 x - 1}{x(\cos x + 1)}$. Next, we can use the trigonometric identity $\sin^2 x = 1 - \cos^2 x$ and rewrite the limit as $\lim_{x \to 0} \frac{-\sin^2 x}{x(\cos x + 1)}$. Now, break this into two limits: $\lim_{x \to 0} \frac{-\sin x}{x} \frac{\sin x}{(\cos x + 1)}$. The first limit is -1 (see Rule No. 1) and the second is 0 (why?), so the limit is 0.

$$\text{Rule No. 2: } \lim_{x \to 0} \frac{\cos x - 1}{x} = 0$$

Example 11: Find $\lim_{x \to 0} \frac{\sin 3x}{x}$.

Use a simple trick: Multiply the top and bottom of the expression by 3. This gives us $\lim_{x \to 0} \frac{3 \sin 3x}{3x}$.
Next, substitute a letter for $3x$; for example, a. Now, we get the following:

Remember that the $\lim_{x \to 0} \sin x = 0$.

$$\lim_{a \to 0} \frac{3 \sin a}{a} = 3 \lim_{a \to 0} \frac{\sin a}{a} = 3(1) = 3$$

Example 12: Find $\lim_{x \to 0} \frac{\sin 5x}{\sin 4x}$.

Now we get a bit more sophisticated. First, divide both the numerator and the denominator by x, like so:

$$\lim_{x \to 0} \frac{\dfrac{\sin 5x}{x}}{\dfrac{\sin 4x}{x}}$$

Next, multiply the top and bottom of the numerator by 5, and the top and bottom of the denominator by 4, which gives us

$$\lim_{x \to 0} \frac{\dfrac{5 \sin 5x}{5x}}{\dfrac{4 \sin 4x}{4x}}$$

From the work we did in Example 11, we can see that this limit is $\dfrac{5}{4}$.

Guess what! You have two more rules!

Rule No. 3: $\displaystyle\lim_{x \to 0} \frac{\sin ax}{x} = a$

Rule No. 4: $\displaystyle\lim_{x \to 0} \frac{\sin ax}{\sin bx} = \frac{a}{b}$

Example 13: Find $\displaystyle\lim_{x \to 0} \frac{x^2}{1 - \cos^2 x}$.

Using trigonometric identities, you can replace $(1 - \cos^2 x)$ with $\sin^2 x$.

$$\lim_{x \to 0} \frac{x^2}{1 - \cos^2 x} = \lim_{x \to 0} \frac{x^2}{\sin^2 x} = \lim_{x \to 0} \left(\frac{x}{\sin x} \cdot \frac{x}{\sin x} \right) = 1 \cdot 1 = 1$$

Here are other examples for you to try, with answers right beneath them. Give 'em a try, and check your work.

PROBLEM 1. Find $\lim\limits_{x \to 3} \dfrac{x-3}{x+2}$.

Answer: If you plug in 3 for x, you get $\lim\limits_{x \to 3} \dfrac{3-3}{3+2} = \dfrac{0}{5} = 0$.

PROBLEM 2. Find $\lim\limits_{x \to 3} \dfrac{x+2}{x-3}$.

Answer: The left-hand limit is $\lim\limits_{x \to 3^-} \dfrac{x+2}{x-3} = -\infty$.

The right-hand limit is $\lim\limits_{x \to 3^+} \dfrac{x+2}{x-3} = \infty$.

These two limits are not the same. Therefore, the limit does not exist.

PROBLEM 3. Find $\lim\limits_{x \to 3} \dfrac{x+2}{(x-3)^2}$.

Answer: The left-hand limit is $\lim\limits_{x \to 3^-} \dfrac{x+2}{(x-3)^2} = \infty$.

The right-hand limit is $\lim\limits_{x \to 3^+} \dfrac{x+2}{(x-3)^2} = \infty$.

If plugging in the value of x results in the denominator equaling zero and you cannot factor the quotient anymore, then check the left- and right-hand limits to find the limit of the expression.

These two limits are the same, so the limit is ∞.

PROBLEM 4. Find $\lim\limits_{x \to -4} \dfrac{x^2 + 6x + 8}{x+4}$.

If $\lim\limits_{x \to a} f(x) = \dfrac{0}{0}$, then $x - a$ is a factor of the top and the bottom.

Answer: If you plug -4 into the top and bottom, you get $\dfrac{0}{0}$. You have to factor the top into $(x+2)(x+4)$ to get this: $\lim\limits_{x \to -4} \dfrac{(x+2)(x+4)}{(x+4)}$.

Now it's time to cancel like terms: $\lim\limits_{x \to -4} \dfrac{(x+2)(x+4)}{(x+4)} = \lim\limits_{x \to -4}(x+2) = -2$.

PROBLEM 5. Find $\lim\limits_{x \to \infty} \dfrac{15x^2 - 11x}{22x^2 + 4x}$.

Answer: Divide each term by x^2.

$$\lim\limits_{x \to \infty} \dfrac{15x^2 - 11x}{22x^2 + 4x} = \lim\limits_{x \to \infty} \dfrac{15 - \dfrac{11}{x}}{22 + \dfrac{4}{x}} = \dfrac{15}{22}$$

PROBLEM 6. Find $\lim\limits_{x\to 0}\dfrac{4x}{\tan x}$.

Answer: Replace $\tan x$ with $\dfrac{\sin x}{\cos x}$, which changes the expression into

$$\lim_{x\to 0}\frac{4x}{\tan x}=\lim_{x\to 0}\frac{4x}{\dfrac{\sin x}{\cos x}}=\lim_{x\to 0}\frac{4x\cos x}{\sin x}$$

Because $\lim\limits_{x\to 0}\dfrac{\sin x}{x}=1$, the $\lim\limits_{x\to 0}\dfrac{x}{\sin x}=1$ as well. Thus, because $\lim\limits_{x\to 0}\dfrac{x}{\sin x}=1$ and $\lim\limits_{x\to 0}\cos x=1$,

the answer is 4.

PROBLEM 7. Find $\lim\limits_{h\to 0}\dfrac{(5+h)^2-25}{h}$.

Answer: First, expand and simplify the numerator.

$$\lim_{h\to 0}\frac{(5+h)^2-25}{h}=\lim_{h\to 0}\frac{25+10h+h^2-25}{h}=\lim_{h\to 0}\frac{10h+h^2}{h}$$

Next, factor h out of the numerator and the denominator.

$$\lim_{h\to 0}\frac{10h+h^2}{h}=\lim_{h\to 0}\frac{h(10+h)}{h}=\lim_{h\to 0}(10+h)$$

Taking the limit, you get $\lim\limits_{h\to 0}(10+h)=10$.

> Note: Pay careful attention to this problem. It will be very important when you work on problems in Chapter 6.

PRACTICE PROBLEM SET 1

Try these 22 problems to test your skill with limits. The answers are in Chapter 19.

1. $\lim\limits_{x\to 8}\left(x^2-5x-11\right)=$

2. $\lim\limits_{x\to 5}\left(\dfrac{x+3}{x^2-15}\right)=$

3. $\lim\limits_{x\to 3}\left(\dfrac{x^2-2x-3}{x-3}\right)=$

4. $\lim\limits_{x\to\infty}\left(\dfrac{x^4-8}{10x^2+25x+1}\right)=$

5. $\lim\limits_{x \to \infty} \left(\dfrac{x^4 - 8}{10x^4 + 25x + 1} \right) =$

6. $\lim\limits_{x \to 6^+} \left(\dfrac{x + 2}{x^2 - 4x - 12} \right) =$

7. $\lim\limits_{x \to 6^-} \left(\dfrac{x + 2}{x^2 - 4x - 12} \right) =$

8. $\lim\limits_{x \to 6} \left(\dfrac{x + 2}{x^2 - 4x - 12} \right) =$

9. $\lim\limits_{x \to 0^+} \left(\dfrac{x}{|x|} \right) =$

10. $\lim\limits_{x \to 7^+} \left(\dfrac{x}{x^2 - 49} \right) =$

11. $\lim\limits_{x \to 7} \left(\dfrac{x}{x^2 - 49} \right) =$

12. Let $f(x) = \begin{cases} x^2 - 5, & x \le 3 \\ x + 2, & x > 3 \end{cases}$

    Find: (a) $\lim\limits_{x \to 3^-} f(x)$; (b) $\lim\limits_{x \to 3^+} f(x)$; and (c) $\lim\limits_{x \to 3} f(x)$

13. Let $f(x) = \begin{cases} x^2 - 5, & x \le 3 \\ x + 1, & x > 3 \end{cases}$

    Find: (a) $\lim\limits_{x \to 3^-} f(x)$; (b) $\lim\limits_{x \to 3^+} f(x)$; and (c) $\lim\limits_{x \to 3} f(x)$

14. Find $\lim\limits_{x \to \frac{\pi}{4}} 3 \cos x$.

15. Find $\lim\limits_{x \to 0} 3 \dfrac{x}{\cos x}$.

16. Find $\lim\limits_{x \to 0} 3 \dfrac{x}{\sin x}$.

17. Find $\lim\limits_{x \to 0} \dfrac{\tan 7x}{\sin 5x}$.

18. Find $\lim\limits_{x \to \infty} \sin x$.

19. Find $\lim\limits_{x \to \infty} \sin \dfrac{1}{x}$.

20. Find $\lim\limits_{x \to 0} \dfrac{\sin^2 7x}{\sin^2 11x}$.

21. Find $\lim\limits_{h \to 0} \dfrac{(3+h)^2 - 9}{h}$.

22. Find $\lim\limits_{h \to 0} \dfrac{\dfrac{1}{x+h} - \dfrac{1}{x}}{h}$.

End of Chapter 3 Drill

The answers are in Chapter 20.

1. Evaluate $\lim_{x \to 2} \left(x^4 - 3x^2 + 5 \right)$.

    (A) 0
    (B) 2
    (C) 9
    (D) The limit does not exist.

2. Evaluate $\lim_{x \to 2} \dfrac{x^2 - x - 2}{x^2 - 8x + 15}$.

    (A) 0
    (B) $\dfrac{2}{3}$
    (C) 3
    (D) The limit does not exist.

3. Evaluate $\lim_{x \to 7} \dfrac{x^2 - 3x - 28}{x^2 - 5x - 14}$.

    (A) 0
    (B) $\dfrac{11}{9}$
    (C) 2
    (D) The limit does not exist.

4. Evaluate $\lim_{x \to \infty} \dfrac{4x^3 - 8x^2 + 11}{5x^3 + 3x - 9}$.

    (A) $-\dfrac{11}{9}$
    (B) 0
    (C) $\dfrac{4}{5}$
    (D) The limit does not exist.

5. Evaluate $\lim_{x \to 0} \dfrac{4 \sin 3x}{2 \tan 5x}$.

    (A) 0
    (B) $\dfrac{3}{5}$
    (C) 2
    (D) The limit does not exist.

6. Evaluate $\lim_{x \to 5} \dfrac{11}{x^2 - 25}$.

    (A) 0
    (B) $\dfrac{11}{25}$
    (C) ∞
    (D) The limit does not exist.

REFLECT

Respond to the following questions:

- For which topics discussed in this chapter do you feel you have achieved sufficient mastery to answer multiple-choice questions correctly?

- For which topics discussed in this chapter do you feel you have achieved sufficient mastery to answer open-ended questions correctly?

- For which topics discussed in this chapter do you feel you need more work before you can answer multiple-choice questions correctly?

- For which topics discussed in this chapter do you feel you need more work before you can answer open-ended questions correctly?

- What parts of this chapter are you going to re-review?

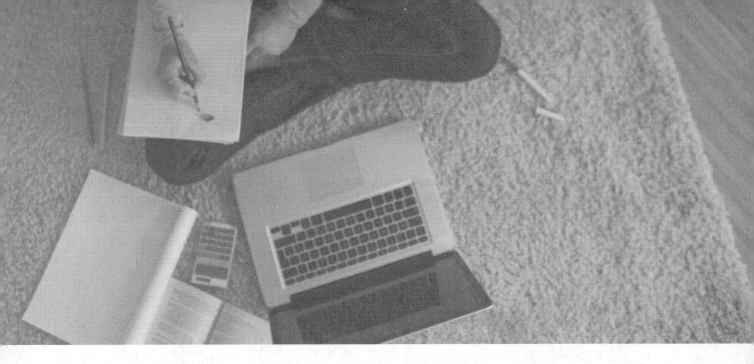

Chapter 4
Continuity

Every AP Exam has a few questions on continuity, so it's important to understand the basic idea of what it means for a function to be continuous. The concept is very simple: if the graph of the function doesn't have any breaks or holes in it within a certain interval, the function is continuous over that interval.

Simple polynomials are continuous everywhere; it's the other ones—trigonometric, rational, piecewise—that might have continuity problems. Most of the test questions concern these last types of functions. In order to learn how to test whether a function is continuous, you'll need some more mathematical terminology.

THE DEFINITION OF CONTINUITY

In order for a function $f(x)$ to be continuous at a point $x = c$, it must fulfill *all three* of the following conditions:

> **Condition 1:** $f(c)$ exists.
>
> **Condition 2:** $\lim_{x \to c} f(x)$ exists.
>
> **Condition 3:** $\lim_{x \to c} f(x) = f(c)$

Let's look at a simple example of a continuous function.

Example 1: Is the function $f(x) = \begin{cases} x+1, & x < 2 \\ 2x-1, & x \geq 2 \end{cases}$ continuous at the point $x = 2$?

Condition 1: Does $f(2)$ exist?

Yes. It's equal to $2(2) - 1 = 3$.

Condition 2: Does $\lim_{x \to 2} f(x)$ exist?

You need to look at the limit from both sides of 2. The left-hand limit is $\lim_{x \to 2^-} f(x) = 2 + 1 = 3$. The right-hand limit is $\lim_{x \to 2^+} f(x) = 2(2) - 1 = 3$.

Because the two limits are the same, the limit exists.

Condition 3: Does $\lim_{x \to 2} f(x) = f(2)$?

The two equal each other, so yes; the function is continuous at $x = 2$.

A simple and important way to check whether a function is continuous is to sketch the function. If you can't sketch the function without lifting your pencil from the paper at some point, then the function is not continuous.

Now let's look at some examples of functions that are not continuous.

> Polynomials are continuous for all reals in their domains.

Example 2: Is the function $f(x) = \begin{cases} x+1, x < 2 \\ 2x - 1, x > 2 \end{cases}$ continuous at $x = 2$?

Condition 1: Does $f(2)$ exist?

Nope. The function of x is defined if x is greater than or less than 2, but not if x is equal to 2. Therefore, the function is not continuous at $x = 2$. Notice that we don't have to bother with the other two conditions. Once you find a problem, the function is automatically not continuous, and you can stop.

Example 3: Is the function $f(x) = \begin{cases} x+1, \ x < 2 \\ 2x+1, \ x \geq 2 \end{cases}$ continuous at $x = 2$?

Condition 1: Does $f(x)$ exist?

Yes. It is equal to $2(2) + 1 = 5$.

Condition 2: Does $\lim\limits_{x \to 2} f(x)$ exist?

The left-hand limit is $\lim\limits_{x \to 2^-} f(x) = 2 + 1 = 3$.

The right-hand limit is $\lim\limits_{x \to 2^+} f(x) = 2(2) + 1 = 5$.

The two limits don't match, so the limit doesn't exist and the function is not continuous at $x = 2$.

Example 4: Is the function $f(x) = \begin{cases} x+1, \ x < 2 \\ x^2, \ x = 2 \\ 2x-1, \ x > 2 \end{cases}$ continuous at $x = 2$?

Condition 1: Does $f(2)$ exist?

Yes. It's equal to $2^2 = 4$.

Condition 2: Does $\lim\limits_{x \to 2} f(x)$ exist?

The left-hand limit is $\lim\limits_{x \to 2^-} f(x) = 2 + 1 = 3$.

The right-hand limit is $\lim\limits_{x \to 2^+} f(x) = 2(2) - 1 = 3$.

Because the two limits are the same, the limit exists.

Condition 3: Does $\lim\limits_{x \to 2} f(x) = f(2)$?

The $\lim\limits_{x \to 2} f(x) = 3$, but $f(2) = 4$. Because these aren't equal, the answer is "no" and the function is not continuous at $x = 2$.

Intermediate Learning

The intermediate value theorem says that if a function is continuous on the interval [a, b], then for any value c in the interval (a, b), there exists a value f(c) between f(a) and f(b) inclusive. In other words, if the function is continuous in some interval, then all of the possible values of f between the two end values of f exist in that interval.

TYPES OF DISCONTINUITIES

There are four types of discontinuities you have to know: jump, point, essential, and removable.

> A **jump** discontinuity occurs when the curve "breaks" at a particular place and starts somewhere else. In other words, $\lim\limits_{x \to a^-} f(x) \neq \lim\limits_{x \to a^+} f(x)$.

An example of jump discontinuity looks like this.

> A **point** discontinuity occurs when the curve has a "hole" in it from a missing point because the function has a value at that point that is "off the curve." In other words, $\lim\limits_{x \to a} f(x) \neq f(a)$.

Here's what a point discontinuity looks like.

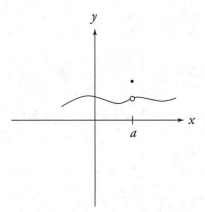

> An **essential** discontinuity occurs when the curve has a vertical asymptote.

This is an example of an essential discontinuity.

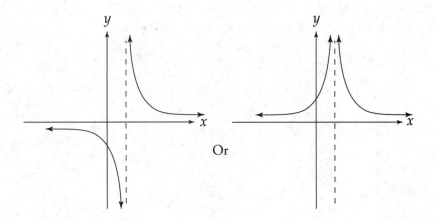

Or

> A **removable** discontinuity occurs when you have a rational expression with common factors in the numerator and denominator. Because these factors can be canceled, the discontinuity is "removable."

Here's an example of a removable discontinuity.

Now that you know what these four types of discontinuities look like, let's see what types of functions are not everywhere continuous.

Example 5: Consider the following function:

$$f(x) = \begin{cases} x+3, \ x \leq 2 \\ x^2, \ x > 2 \end{cases}$$

The left-hand limit is 5 as x approaches 2, and the right-hand limit is 4 as x approaches 2. Because the curve has different values on each side of 2, the curve is discontinuous at $x = 2$. We say that the curve "jumps" at $x = 2$ from the left-hand curve to the right-hand curve because the left and right-hand limits differ. It looks like the following:

This curve looks very similar to a point discontinuity, but notice that with a removable discontinuity, $f(x)$ is not defined at the point, whereas with a point discontinuity, $f(x)$ is defined there.

This is an example of a jump discontinuity.

Example 6: Consider the following function:

$$f(x) = \begin{cases} x^2, & x \neq 2 \\ 5, & x = 2 \end{cases}$$

Because $\lim\limits_{x \to 2} f(x) \neq f(2)$, the function is discontinuous at $x = 2$. The curve is continuous everywhere except at the point $x = 2$. It looks like the following:

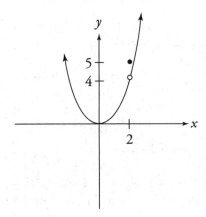

This is an example of a point discontinuity.

Example 7: Consider the following function: $f(x) = \dfrac{5}{x - 2}$.

The function is discontinuous because it's possible for the denominator to equal zero (at $x = 2$). This means that $f(2)$ doesn't exist, and the function has an asymptote at $x = 2$. In addition, $\lim\limits_{x \to 2^-} f(x) = -\infty$ and $\lim\limits_{x \to 2^+} f(x) = \infty$.

The graph looks like the following:

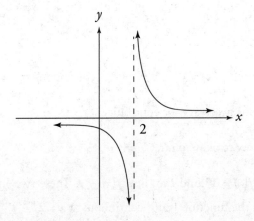

This is an example of an essential discontinuity.

Example 8: Consider the following function:

$$f(x) = \frac{x^2 - 8x + 15}{x^2 - 6x + 5}$$

If you factor the top and bottom, you can see where the discontinuities are.

$$f(x) = \frac{x^2 - 8x + 15}{x^2 - 6x + 5} = \frac{(x-3)(x-5)}{(x-1)(x-5)}$$

The function has a zero in the denominator when $x = 1$ or $x = 5$, so the function is discontinuous at those two points. But you can cancel the term $(x - 5)$ from both the numerator and the denominator, leaving you with

$$f(x) = \frac{x-3}{x-1}$$

Now the reduced function *is* continuous at $x = 5$. Thus, the original function has a removable discontinuity at $x = 5$. Furthermore, if you now plug $x = 5$ into the reduced function, you get

$$f(5) = \frac{2}{4} = \frac{1}{2}$$

The discontinuity is at $x = 5$, and there's a hole at $\left(5, \frac{1}{2}\right)$. In other words, if the original function were continuous at $x = 5$, it would have the value $\frac{1}{2}$. Notice that this is the same as $\lim\limits_{x \to 5} f(x)$.

These are the types of discontinuities that you can expect to encounter on the AP Exam. Here are some sample problems and their solutions. Cover the answers as you work, then check your results.

PROBLEM 1. Is the function $f(x) = \begin{cases} 2x^3 - 1, & x < 2 \\ 6x - 3, & x \geq 2 \end{cases}$ continuous at $x = 2$?

Answer: Test the conditions necessary for continuity.

Condition 1: $f(2) = 9$, so we're okay so far.

Condition 2: The $\lim\limits_{x \to 2^-} f(x) = 15$ and the $\lim\limits_{x \to 2^+} f(x) = 9$. These two limits don't agree, so the $\lim\limits_{x \to 2} f(x)$ doesn't exist and the function is not continuous at $x = 2$.

PROBLEM 2. Is the function $f(x) = \begin{cases} x^2 + 3x + 5, & x < 1 \\ 6x + 3, & x \geq 1 \end{cases}$ continuous at $x = 1$?

Answer: Condition 1: $f(1) = 9$.

Condition 2: The $\lim\limits_{x \to 1^-} f(x) = 9$ and the $\lim\limits_{x \to 1^+} f(x) = 9$.

Therefore, the $\lim\limits_{x \to 1} f(x)$ exists and is equal to 9.

Condition 3: $\lim\limits_{x \to 1} f(x) = f(1) = 9$.

The function satisfies all three conditions, so it is continuous at $x = 1$.

PROBLEM 3. For what value of a is the function $f(x) = \begin{cases} ax + 5, & x < 4 \\ x^2 - x, & x \geq 4 \end{cases}$ continuous at $x = 4$?

Answer: Because $f(4) = 12$, the function passes the first condition.

For Condition 2 to be satisfied, the $\lim\limits_{x \to 4^-} f(x) = 4a + 5$ must equal the $\lim\limits_{x \to 4^+} f(x) = 12$. So set

$4a + 5 = 12$. If $a = \dfrac{7}{4}$, the limit will exist at $x = 4$ and the other two conditions will also be

fulfilled. Therefore, the value $a = \dfrac{7}{4}$ makes the function continuous at $x = 4$.

PROBLEM 4. Where does the function $f(x) = \dfrac{2x^2 - 7x - 15}{x^2 - x - 20}$ have: (a) an essential discontinuity; and (b) a removable discontinuity?

Answer: If you factor the top and bottom of this fraction, you get

$$f(x) = \frac{2x^2 - 7x - 15}{x^2 - x - 20} = \frac{(2x + 3)(x - 5)}{(x + 4)(x - 5)}$$

Thus, the function has an essential discontinuity at $x = -4$. If we then cancel the term

$(x - 5)$, and substitute $x = 5$ into the reduced expression, we get $f(5) = \dfrac{13}{9}$. Therefore,

the function has a removable discontinuity at $\left(5, \dfrac{13}{9} \right)$.

> Note: Don't confuse coordinate parentheses with interval notation. In interval notation, square brackets include endpoints and parentheses do not. For example, the interval $2 \leq x \leq 4$ is written [2, 4] and the interval $2 < x < 4$ is written (2, 4).

PRACTICE PROBLEM SET 2

Now try these problems. The answers are in Chapter 19.

1. Is the function $f(x) = \begin{cases} x+7, & x<2 \\ 9, & x=2 \\ 3x+3, & x>2 \end{cases}$ continuous at $x = 2$?

2. Is the function $f(x) = \begin{cases} 4x^2 - 2x, & x<3 \\ 10x-1, & x=3 \\ 30, & x>3 \end{cases}$ continuous at $x = 3$?

3. Is the function $f(x) = \sec x$ continuous everywhere?

4. Is the function $f(x) = \sec x$ continuous on the interval $\left[-\dfrac{\pi}{2}, \dfrac{\pi}{2} \right]$?

5. Is the function $f(x) = \sec x$ continuous on the interval $\left(-\dfrac{\pi}{2}, \dfrac{\pi}{2} \right)$?

6. For what value(s) of k is the function $f(x) = \begin{cases} 3x^2 - 11x - 4, & x \le 4 \\ kx^2 - 2x - 1, & x > 4 \end{cases}$ continuous at $x = 4$?

7. At what point is the removable discontinuity for the function $f(x) = \dfrac{x^2 + 5x - 24}{x^2 - x - 6}$?

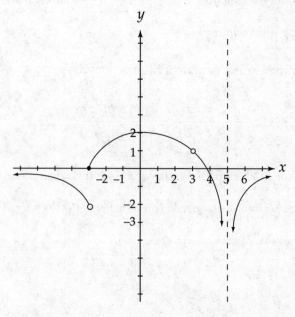

8.   Given the graph of $f(x)$ above, find

   (a) $\displaystyle\lim_{x \to -\infty} f(x)$

   (b) $\displaystyle\lim_{x \to \infty} f(x)$

   (c) $\displaystyle\lim_{x \to 3^-} f(x)$

   (d) $\displaystyle\lim_{x \to 3^+} f(x)$

   (e) $f(3)$

   (f)   Any discontinuities

End of Chapter 4 Drill

The answers are in Chapter 20.

1. Does the function $f(x) = \dfrac{x^2 - 8x + 15}{x^2 - 9}$ have any discontinuities? If so, where and what type?

   (A) An essential discontinuity at $x = 3$ and at $x = -3$
   (B) A removable discontinuity at $x = 3$ and at $x = -3$
   (C) A removable discontinuity at $x = 3$ and an essential discontinuity at $x = -3$.
   (D) The function is continuous for all real numbers.

2. Does the function $f(x) = \begin{cases} 5x - 11; & x \geq 4 \\ 2x + 1; & x < 4 \end{cases}$ have any discontinuities? If so, where and what type?

   (A) An essential discontinuity at $x = 4$
   (B) A removable discontinuity at $x = 4$
   (C) A jump discontinuity at $x = 4$
   (D) The function is continuous for all real numbers.

3. Does the function $f(x) = \begin{cases} 6x + 1; & x \geq 2 \\ 5x + 2; & x < 2 \end{cases}$ have any discontinuities? If so, where and what type?

   (A) An essential discontinuity at $x = 2$
   (B) A removable discontinuity at $x = 2$
   (C) A jump discontinuity at $x = 2$
   (D) The function is continuous for all real numbers.

4. For what value of k is $f(x) = \begin{cases} x^2 + kx - 4; & x \geq 1 \\ 6kx + 7; & x < 1 \end{cases}$ continuous for all real numbers?

   (A) $k = -2$

   (B) $k = -\dfrac{4}{5}$

   (C) $k = \dfrac{4}{5}$

   (D) $k = 2$

5. For what values of a and b is
   $$f(x) = \begin{cases} ax^2 + 2bx - 3; & x > 2 \\ 4bx^2 - 3ax + 3; & -2 \leq x \leq 2 \\ 4ax^2 + 5bx + 25; & x < -2 \end{cases}$$
   continuous for all real numbers?

   (A) $a = 2$ and $b = 3$
   (B) $a = 3$ and $b = 2$
   (C) $a = 1$ and $b = -1$
   (D) $a = -1$ and $b = 1$

REFLECT

Respond to the following questions:

- For which topics discussed in this chapter do you feel you have achieved sufficient mastery to answer multiple-choice questions correctly?

- For which topics discussed in this chapter do you feel you have achieved sufficient mastery to answer open-ended questions correctly?

- For which topics discussed in this chapter do you feel you need more work before you can answer multiple-choice questions correctly?

- For which topics discussed in this chapter do you feel you need more work before you can answer open-ended questions correctly?

- What parts of this chapter are you going to re-review?

Chapter 5
The Definition of
the Derivative

The main tool that you'll use in differential calculus is called the **derivative**. All of the problems that you'll encounter in differential calculus make use of the derivative, so your goal should be to become an expert at finding, or "taking," derivatives by the end of Chapter 6. However, before you learn a simple way to take a derivative, your teacher will probably make you learn how derivatives are calculated by teaching you something called the "Definition of the Derivative."

DERIVING THE FORMULA

The best way to understand the definition of the derivative is to start by looking at the simplest continuous function: a line. As you should recall, you can determine the slope of a line by taking two points on that line and plugging them into the slope formula.

$$m = \frac{y_2 - y_1}{x_2 - x_1} \qquad m \text{ stands for slope.}$$

For example, suppose a line goes through the points (3, 7) and (8, 22). First, you subtract the y-coordinates $(22 - 7) = 15$. Next, subtract the corresponding x-coordinates $(8 - 3) = 5$. Finally, divide the first number by the second: $\frac{15}{5} = 3$. The result is the slope of the line: $m = 3$.

Let's look at the graph of that line. The slope measures the steepness of the line, which looks like the following:

Notice that you can use the coordinates in reverse order and still get the same result. It doesn't matter in which order you do the subtraction as long as you're consistent.

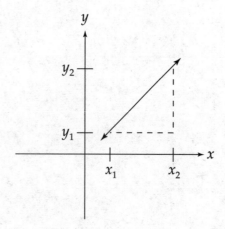

You probably remember your teachers referring to the slope as the "rise" over the "run." The rise is the difference between the y-coordinates, and the run is the difference between the x-coordinates. The slope is the ratio of the two.

Now for a few changes in notation. Instead of calling the x-coordinates x_1 and x_2, we're going to call them x_1 and $x_1 + h$, where h is the difference between the two x-coordinates. Second, instead of using y_1 and y_2, we use $f(x_1)$ and $f(x_1 + h)$. So now the graph looks like the following:

Occasionally, instead of h, some books use Δx.

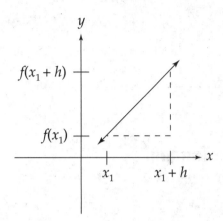

The picture is exactly the same—only the notation has changed.

The Slope of a Curve

Suppose that instead of finding the slope of a line, we wanted to find the slope of a curve. Here, the slope formula no longer works because the distance from one point to the other is along a curve, not a straight line. But we could find the approximate slope if we took the slope of the line between the two points. This is called the **secant line**.

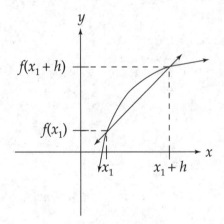

While at first the Difference Quotient may seem like another formula to memorize, keep in mind that it's just another way of writing the slope formula.

The formula for the slope of the secant line is

$$\frac{f(x_1 + h) - f(x_1)}{h}$$

Remember this formula! This is called the **Difference Quotient**.

The Secant and the Tangent

As you can see, the farther apart the two points are, the less the slope of the line corresponds to the slope of the curve.

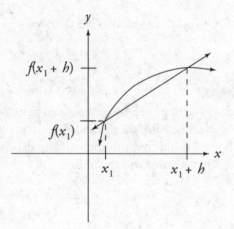

Conversely, the closer the two points are, the more accurate the approximation is.

In fact, there is one line, called the **tangent line**, that touches the curve at exactly one point. The slope of the tangent line is equal to the slope of the curve at exactly this point. The object of using the above formula, therefore, is to shrink h down to an infinitesimally small amount. If we could do that, then the difference between $(x_1 + h)$ and x_1 would be a point.

Graphically, it looks like the following:

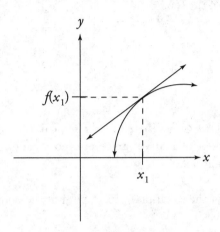

Keep in mind that there are infinitely many tangents for any curve because there are infinitely many points on the curve.

How do we perform this shrinking act? By using the limits we discussed in Chapter 3. We set up a limit during which h approaches zero, like the following:

$$\lim_{h \to 0} \frac{f(x_1 + h) - f(x_1)}{h}$$

This is the **definition of the derivative**, and we call it $f'(x)$.

Notice that the equation is just a slightly modified version of the difference quotient, with different notation. The only difference is that we're finding the slope between two points that are infinitesimally close to each other.

Example 1: Find the slope of the curve $f(x) = x^2$ at the point $(2, 4)$.

This means that $x_1 = 2$ and $f(2) = 2^2 = 4$. If we can figure out $f(x_1 + h)$, then we can find the slope. Well, how did we find the value of $f(x)$? We plugged x_1 into the equation $f(x) = x^2$. To find $f(x_1 + h)$ we plug $x_1 + h$ into the equation, which now looks like this

$$f(x_1 + h) = (2 + h)^2 = 4 + 4h + h^2$$

Now plug this into the slope formula.

$$\lim_{h \to 0} \frac{f(x_1 + h) - f(x_1)}{h} = \lim_{h \to 0} \frac{4 + 4h + h^2 - 4}{h} = \lim_{h \to 0} \frac{4h + h^2}{h}$$

Next, simplify by factoring h out of the top.

$$\lim_{h \to 0} \frac{4h + h^2}{h} = \lim_{h \to 0} \frac{h(4 + h)}{h} = \lim_{h \to 0}(4 + h)$$

Taking the limit as h approaches 0, we get 4. Therefore, the slope of the curve $y = x^2$ at the point (2, 4) is 4. Now we've found the slope of a curve at a certain point, and the notation looks like this: $f'(2) = 4$. Remember this notation!

Example 2: Find the derivative of the equation in Example 1 at the point (5, 25). This means that $x_1 = 5$ and $f(x) = 25$. This time,

$$(x_1 + h)^2 = (5 + h)^2 = 25 + 10h + h^2$$

Now plug this into the formula for the derivative.

$$\lim_{h \to 0} \frac{f(x_1 + h) - f(x_1)}{h} = \lim_{h \to 0} \frac{25 + 10h + h^2 - 25}{h} = \lim_{h \to 0} \frac{10h + h^2}{h}$$

Once again, simplify by factoring h out of the top.

$$\lim_{h \to 0} \frac{10h + h^2}{h} = \lim_{h \to 0} \frac{h(10 + h)}{h} = \lim_{h \to 0}(10 + h)$$

Taking the limit as h goes to 0, you get 10. Therefore, the slope of the curve $y = x^2$ at the point (5, 25) is 10, or: $f'(5) = 10$.

Using this pattern, let's forget about the arithmetic for a second and derive a formula.

Example 3: Find the slope of the equation $f(x) = x^2$ at the point $\left(x_1, x_1^2\right)$.

Follow the steps in the last two problems, but instead of using a number, use x_1. This means that $f(x_1) = x_1^2$ and $(x_1 + h)^2 = x_1^2 + 2x_1 h + h^2$. Then the derivative is

$$\lim_{h \to 0} \frac{x_1^2 + 2x_1 h + h^2 - x_1^2}{h} = \lim_{h \to 0} \frac{2x_1 h + h^2}{h}$$

Factor h out of the top.

$$\lim_{h \to 0} \frac{h(2x_1 + h)}{h} = \lim_{h \to 0}(2x_1 + h)$$

Now take the limit as h goes to 0: you get $2x_1$. Therefore, $f'(x_1) = 2x_1$.

This example gives us a general formula for the derivative of this curve. Now we can pick any point, plug it into the formula, and determine the slope at that point. For example, the derivative at the point $x = 7$ is 14. At the point $x = \dfrac{7}{3}$, the derivative is $\dfrac{14}{3}$.

DIFFERENTIABILITY

One of the important requirements for the differentiability of a function is that the function be continuous. But, even if a function is continuous at a point, the function is not necessarily differentiable there. Check out the graph below.

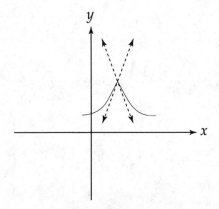

If a function has a "sharp corner," you can draw more than one tangent line at that point, and because the slopes of these tangent lines are not equal, the function is not differentiable there.

Another possible problem occurs when the tangent line is vertical (which can also occur at a cusp) because a vertical line has an infinite slope. For example, if the derivative of a function is $\dfrac{1}{x+1}$, it doesn't have a derivative at $x = -1$.

Try these problems on your own, then check your work against the answers immediately beneath each problem.

PROBLEM 1. Find the derivative of $f(x) = 3x^2$ at $(4, 48)$.

Answer: $f(4 + h) = 3(4 + h)^2 = 48 + 24h + 3h^2$. Use the definition of the derivative.

$$f'(4) = \lim_{h \to 0} \frac{48 + 24h + 3h^2 - 48}{h}$$

Simplify.

$$\lim_{h \to 0} \frac{24h + 3h^2}{h} = \lim_{h \to 0}(24 + 3h) = 24$$

The slope of the curve at the point $(4, 48)$ is 24.

PROBLEM 2. Find the derivative of $f(x) = 3x^2$.

Answer: $f(x + h) = 3(x + h)^2 = 3x^2 + 6xh + 3h^2$. Use the definition of the derivative.

$$f'(x) = \lim_{h \to 0} \frac{3x^2 + 6xh + 3h^2 - 3x^2}{h}$$

Simplify.

$$\lim_{h \to 0} \frac{6xh + 3h^2}{h} = \lim_{h \to 0}(6x + 3h) = 6x$$

The derivative is $6x$.

PROBLEM 3. Find the derivative of $f(x) = x^3$.

Answer: $f(x + h) = (x + h)^3 = x^3 + 3x^2h + 3xh^2 + h^3$. First, use the definition of the derivative.

$$f'(x) = \lim_{h \to 0} \frac{x^3 + 3x^2h + 3xh^2 + h^3 - x^3}{h}$$

And simplify.

$$\lim_{h \to 0} \frac{3x^2 h + 3xh^2 + h^3}{h} = \lim_{h \to 0} \left(3x^2 + 3xh + h^2\right) = 3x^2$$

The derivative is $3x^2$.

This next one will test your algebraic skills. Don't say we didn't warn you!

PROBLEM 4. Find the derivative of $f(x) = \sqrt{x}$.

Answer: $f(x + h) = \sqrt{x + h}$.

Use the definition of the derivative.

$$f'(x) = \lim_{h \to 0} \frac{\sqrt{x + h} - \sqrt{x}}{h}$$

Notice that this one doesn't cancel as conveniently as the other problems did. In order to simplify this expression, we have to multiply both the top and the bottom of the expression by $\sqrt{x + h} + \sqrt{x}$ (the conjugate of the numerator).

$$f'(x) = \lim_{h \to 0} \frac{\sqrt{x + h} - \sqrt{x}}{h} \left(\frac{\sqrt{x + h} + \sqrt{x}}{\sqrt{x + h} + \sqrt{x}} \right) = \lim_{h \to 0} \frac{x + h - x}{h\left(\sqrt{x + h} + \sqrt{x}\right)} = \lim_{h \to 0} \frac{h}{h\left(\sqrt{x + h} + \sqrt{x}\right)}$$

Simplify.

$$\lim_{h \to 0} \frac{1}{\left(\sqrt{x + h} + \sqrt{x}\right)} = \frac{1}{2\sqrt{x}}$$

The derivative is $\dfrac{1}{2\sqrt{x}}$.

PRACTICE PROBLEM SET 3

Now find the derivative of the following expressions. The answers are in Chapter 19.

1. $f(x) = 5x$ at $x = 3$

2. $f(x) = 4x$ at $x = -8$

3. $f(x) = 5x^2$ at $x = -1$

4. $f(x) = 8x^2$

5. $f(x) = -10x^2$

6. $f(x) = 20x^2$ at $x = a$

7. $f(x) = 2x^3$ at $x = -3$

8. $f(x) = -3x^3$

9. $f(x) = x^5$

10. $f(x) = 2\sqrt{x}$ at $x = 9$

11. $f(x) = 5\sqrt{2x}$ at $x = 8$

12. $f(x) = \sin x$ at $x = \dfrac{\pi}{3}$

13. $f(x) = x^2 + x$

14. $f(x) = x^3 + 3x + 2$

15. $f(x) = \dfrac{1}{x}$

16. $f(x) = ax^2 + bx + c$

End of Chapter 5 Drill

The answers are in Chapter 20.

1. Use the Definition of the Derivative to find $f'(x)$ for $f(x) = 8x - 11$.

    (A) -11
    (B) 0
    (C) 8
    (D) $8x$

2. Use the Definition of the Derivative to find $f'(x)$ for $f(x) = 3x^2 + 7x - 2$.

    (A) $6x$
    (B) $6x - 2$
    (C) $6x + 7$
    (D) $3x + 7$

3. Use the Definition of the Derivative to find $f'(x)$ for $f(x) = x^3 - 4x^2 + 11$.

    (A) $3x^2 - 4$
    (B) $3x^2 - 4x + 11$
    (C) $3x^2 - 8x + 11$
    (D) $3x^2 - 8x$

4. Use the Definition of the Derivative to find $f'(25)$, if $f(x) = 6\sqrt{x}$.

    (A) $\dfrac{3}{5}$

    (B) $\dfrac{3}{25}$

    (C) $\dfrac{6}{25}$

    (D) $\dfrac{6}{5}$

5. Use the Definition of the Derivative to find $f'(2)$, if $f(x) = \dfrac{1}{4x^2}$.

    (A) $-\dfrac{1}{16}$

    (B) $-\dfrac{1}{8}$

    (C) $\dfrac{1}{8}$

    (D) $\dfrac{1}{16}$

REFLECT

Respond to the following questions:

- For which topics discussed in this chapter do you feel you have achieved sufficient mastery to answer multiple-choice questions correctly?

- For which topics discussed in this chapter do you feel you have achieved sufficient mastery to answer open-ended questions correctly?

- For which topics discussed in this chapter do you feel you need more work before you can answer multiple-choice questions correctly?

- For which topics discussed in this chapter do you feel you need more work before you can answer open-ended questions correctly?

- What parts of this chapter are you going to re-review?

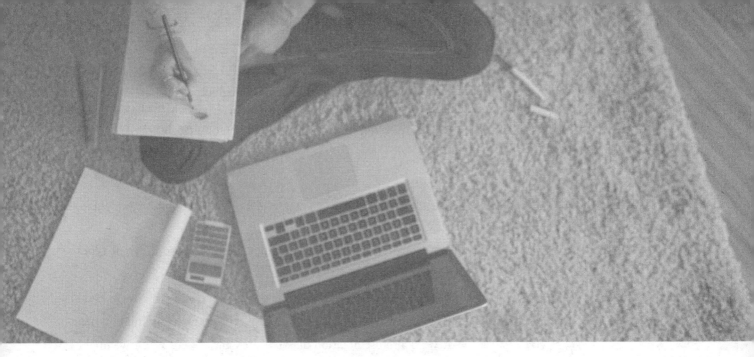

Chapter 6

Basic
Differentiation

In calculus, you'll be asked to do two things: differentiate and integrate. In this section, you're going to learn differentiation. Integration will come later, in the second half of this book. Before we go about the business of learning how to take derivatives, however, here's a brief note about notation. Read this!

NOTATION

There are several different notations for derivatives in calculus. We'll use two different types interchangeably throughout this book, so get used to them now.

We'll refer to functions three different ways: $f(x)$, u or v, and y. For example, we might write: $f(x) = x^3$, $g(x) = x^4$, $h(x) = x^5$. We'll also use notation like: $u = \sin x$ and $v = \cos x$. Or we might use: $y = \sqrt{x}$. Usually, we pick the notation that causes the least confusion.

The derivatives of the functions will use notation that depends on the function, as shown in the following table:

Function	First Derivative	Second Derivative
$f(x)$	$f'(x)$	$f''(x)$
$g(x)$	$g'(x)$	$g''(x)$
y	y' or $\dfrac{dy}{dx}$	y'' or $\dfrac{d^2y}{dx^2}$

In addition, if we refer to a derivative of a function in general (for example, $ax^2 + bx + c$), we might enclose the expression in parentheses and use either of the following notations:

$$\left(ax^2 + bx + c \right)', \text{ or } \frac{d}{dx}\left(ax^2 + bx + c \right)$$

Sometimes math books refer to a derivative using either D_x or f_x. We're not going to use either of them.

THE POWER RULE

In the last chapter, you learned how to find a derivative using the definition of the derivative, a process that is very time-consuming and sometimes involves a lot of complex algebra. Fortunately, there's a shortcut to taking derivatives, so you'll never have to use the definition again—except when it's a question on an exam!

The basic technique for taking a derivative is called the **Power Rule**.

Rule No. 1: If $y = x^n$, then $\dfrac{dy}{dx} = nx^{n-1}$

That's it. Wasn't that simple? Of course, this and all of the following rules can be derived easily from the definition of the derivative. Look at these next few examples of the Power Rule in action.

Example 1: If $y = x^5$, then $\dfrac{dy}{dx} = 5x^4$.

Example 2: If $y = x^{20}$, then $\dfrac{dy}{dx} = 20x^{19}$.

Example 3: If $f(x) = x^{-5}$, then $f'(x) = -5x^{-6}$.

Example 4: If $u = x^{\frac{1}{2}}$, then $\dfrac{du}{dx} = \dfrac{1}{2}x^{-\frac{1}{2}}$.

Example 5: If $y = x^1$, then $\dfrac{dy}{dx} = 1x^0 = 1$. (Because x^0 is 1!)

Example 6: If $y = x^0$, then $\dfrac{dy}{dx} = 0$.

Notice that when the power of the function is negative, the power of the derivative is more negative.

When the power is a fraction, you should be careful to get the subtraction right (you'll see the powers $\dfrac{1}{2}, \dfrac{1}{3}, \dfrac{3}{2}, -\dfrac{1}{2}$, and $-\dfrac{1}{3}$ often, so be comfortable with subtracting 1 from them).

When the power is 1, the derivative is just a constant. When the power is 0, the derivative is 0.

This leads to the next three rules.

Rule No. 2: If $y = x$, then $\dfrac{dy}{dx} = 1$

Rule No. 3: If $y = kx$, then $\dfrac{dy}{dx} = k$ (where k is a constant)

Rule No. 4: If $y = k$, then $\dfrac{dy}{dx} = 0$ (where k is a constant)

Note: For future reference, a, b, c, n, and k always stand for constants.

Example 7: If $y = 8x^4$, then $\dfrac{dy}{dx} = 32x^3$.

Example 8: If $y = 5x^{100}$, then $y' = 500x^{99}$.

Example 9: If $y = -3x^{-5}$, then $\dfrac{dy}{dx} = 15x^{-6}$.

Example 10: If $f(x) = 7x^{\frac{1}{2}}$, then $f'(x) = \dfrac{7}{2}x^{-\frac{1}{2}}$.

Example 11: If $y = x\sqrt{15}$, then $\dfrac{dy}{dx} = \sqrt{15}$.

Example 12: If $y = 12$, then $\dfrac{dy}{dx} = 0$.

If you have any questions about any of these 12 examples (especially the last two), review the rules. Now for one last rule.

THE ADDITION RULE

If $y = ax^n + bx^m$, where a and b are constants, then
$$\frac{dy}{dx} = a\left(nx^{n-1}\right) + b\left(mx^{m-1}\right)$$

This handy rule works for subtraction too.

Example 13: If $y = 3x^4 + 8x^{10}$, then $\frac{dy}{dx} = 12x^3 + 80x^9$.

Example 14: If $y = 7x^{-4} + 5x^{-\frac{1}{2}}$, then $\frac{dy}{dx} = -28x^{-5} - \frac{5}{2}x^{-\frac{3}{2}}$.

Example 15: If $y = 5x^4(2 - x^3)$, then $\frac{dy}{dx} = 40x^3 - 35x^6$.

Example 16: If $y = (3x^2 + 5)(x - 1)$, then

$$y = 3x^3 - 3x^2 + 5x - 5 \text{ and } \frac{dy}{dx} = 9x^2 - 6x + 5$$

Example 17: If $y = ax^3 + bx^2 + cx + d$, then $\frac{dy}{dx} = 3ax^2 + 2bx + c$.

After you've worked through all 17 of these examples, you should be able to take the derivative of any polynomial with ease.

As you may have noticed from the examples above, in calculus, you are often asked to convert from fractions and radicals to negative powers and fractional powers. In addition, don't freak out if your answer doesn't match any of the answer choices. Because answers to problems are often presented in simplified form, your answer may not be simplified enough.

There are two basic expressions that you'll often be asked to differentiate. You can make your life easier by memorizing the following derivatives:

If $y = \frac{k}{x}$, then $\frac{dy}{dx} = -\frac{k}{x^2}$.

If $y = k\sqrt{x}$, then $\frac{dy}{dx} = \frac{k}{2\sqrt{x}}$.

This first formula is also known as the Reciprocal Rule.

HIGHER ORDER DERIVATIVES

This may sound like a big deal, but it isn't. This term refers only to taking the derivative of a function more than once. You don't have to stop at the first derivative of a function; you can keep taking derivatives. The derivative of a first derivative is called the second derivative. The derivative of the second derivative is called the third derivative, and so on.

Generally, you'll have to take only first and second derivatives.

> Notice how we simplified the derivatives in the latter example? You should be able to do this mentally.

Function	First Derivative	Second Derivative
x^6	$6x^5$	$30x^4$
$8\sqrt{x}$	$\dfrac{4}{\sqrt{x}}$	$-2x^{-\frac{3}{2}}$

Here are some sample problems involving the rules we discussed above. As you work, cover the answers with an index card, and then check your work after you're done. By the time you finish them, you should know the rules by heart.

PROBLEM 1. If $y = 50x^5 + \dfrac{3}{x} - 7x^{-\frac{5}{3}}$, then $\dfrac{dy}{dx} =$

Answer: $\dfrac{dy}{dx} = 50\left(5x^4\right) + \left(-\dfrac{3}{x^2}\right) - 7\left(-\dfrac{5}{3}\right)x^{-\frac{8}{3}} = 250x^4 - \dfrac{3}{x^2} + \dfrac{35}{3}x^{-\frac{8}{3}}$

PROBLEM 2. If $y = 9x^4 + 6x^2 - 7x + 11$, then $\dfrac{dy}{dx} =$

Answer: $\dfrac{dy}{dx} = 9\left(4x^3\right) + 6\left(2x\right) - 7\left(1\right) + 0 = 36x^3 + 12x - 7$

PROBLEM 3. If $f(x) = 6x^{\frac{3}{2}} - 12\sqrt{x} - \dfrac{8}{\sqrt{x}} + 24x^{-\frac{3}{2}}$, then $f'(x) =$

Answer: $f'(x) = 6\left(\dfrac{3}{2}x^{\frac{1}{2}}\right) - \left(\dfrac{12}{2\sqrt{x}}\right) - 8\left(-\dfrac{1}{2}x^{-\frac{3}{2}}\right) + 24\left(-\dfrac{3}{2}x^{-\frac{5}{2}}\right) = 9\sqrt{x} - \dfrac{6}{\sqrt{x}} + 4x^{-\frac{3}{2}} - 36x^{-\frac{5}{2}}$

How'd you do? Did you notice the changes in notation? How about the fractional powers, radical signs, and x's in denominators? You should be able to switch back and forth between notations, between fractional powers and radical signs, and between negative powers in a numerator and positive powers in a denominator.

PRACTICE PROBLEM SET 4

Find the derivative of each expression and simplify. The answers are in Chapter 19.

1. $(4x^2 + 1)^2$

2. $(x^5 + 3x)^2$

3. $11x^7$

4. $18x^3 + 12x + 11$

5. $\dfrac{1}{2}(x^{12} + 17)$

6. $-\dfrac{1}{3}(x^9 + 2x^3 - 9)$

7. π^5

8. $-8x^{-8} + 12\sqrt{x}$

9. $6x^{-7} - 4\sqrt{x}$

10. $x^{-5} + \dfrac{1}{x^8}$

11. $(6x^2 + 3)(12x - 4)$

12. $(3 - x - 2x^3)(6 + x^4)$

13. $e^{10} + \pi^3 - 7$

14. $\left(\dfrac{1}{x} + \dfrac{1}{x^2}\right)\left(\dfrac{4}{x^3} - \dfrac{6}{x^4}\right)$

15. $\sqrt{x} + \dfrac{1}{\sqrt{3}}$

16. 0

17. $(x + 1)^3$

18. $\sqrt{x} + \sqrt[3]{x} + \sqrt[3]{x^2}$

19. $x(2x + 7)(x - 2)$

20. $\sqrt{x}\left(\sqrt[3]{x} + \sqrt[5]{x}\right)$

THE PRODUCT RULE

Now that you know how to find derivatives of simple polynomials, it's time to get more complicated. What if you had to find the derivative of this?

$$f(x) = (x^3 + 5x^2 - 4x + 1)(x^5 - 7x^4 + x)$$

You could multiply out the expression and take the derivative of each term, like

$$f(x) = x^8 - 2x^7 - 39x^6 + 29x^5 - 6x^4 + 5x^3 - 4x^2 + x$$

And the derivative is

$$f'(x) = 8x^7 - 14x^6 - 234x^5 + 145x^4 - 24x^3 + 15x^2 - 8x + 1$$

Needless to say, this process is messy. Naturally, there's an easier way. When a function involves two terms multiplied by each other, we use the **Product Rule**.

> The Product Rule: If $f(x) = uv$, then $f'(x) = u\dfrac{dv}{dx} + v\dfrac{du}{dx}$

With the Product Rule, the order of these two operations doesn't matter. It does matter with other rules, though, so it helps to use the same order each time.

To find the derivative of two things multiplied by each other, you multiply the first function by the derivative of the second, and add that to the second function multiplied by the derivative of the first.

Let's use the Product Rule to find the derivative of our example.

$$f'(x) = (x^3 + 5x^2 - 4x + 1)(5x^4 - 28x^3 + 1) + (x^5 - 7x^4 + x)(3x^2 + 10x - 4)$$

If we were to simplify this, we'd get the same answer as before. But here's the best part: we're not going to simplify it. One of the great things about the AP Exam is that when it's difficult to simplify an expression, you almost never have to. Nonetheless, you'll often need to simplify expressions when you're taking second derivatives, or when you use the derivative in some other equation. Practice simplifying whenever possible.

Example 1: $f(x) = (9x^2 + 4x)(x^3 - 5x^2)$

$$f'(x) = (9x^2 + 4x)(3x^2 - 10x) + (x^3 - 5x^2)(18x + 4)$$

Example 2: $y = \left(\sqrt{x} + 4\sqrt[3]{x}\right)\left(x^5 - 11x^8\right)$

$$y' = \left(\sqrt{x} + 4\sqrt[3]{x}\right)\left(5x^4 - 88x^7\right) + \left(x^5 - 11x^8\right)\left(\frac{1}{2\sqrt{x}} + \frac{4}{3\sqrt[3]{x^2}}\right)$$

Example 3: $y = \left(\frac{1}{x} + \frac{1}{x^2} - \frac{1}{x^3}\right)\left(\frac{1}{x} - \frac{1}{x^3} + \frac{1}{x^5}\right)$

$$y' = \left(\frac{1}{x} + \frac{1}{x^2} - \frac{1}{x^3}\right)\left(-\frac{1}{x^2} + \frac{3}{x^4} - \frac{5}{x^6}\right) + \left(\frac{1}{x} - \frac{1}{x^3} + \frac{1}{x^5}\right)\left(-\frac{1}{x^2} - \frac{2}{x^3} + \frac{3}{x^4}\right)$$

THE QUOTIENT RULE

What happens when you have to take the derivative of a function that is the quotient of two other functions? You guessed it: Use the **Quotient Rule**.

> The Quotient Rule: If $f(x) = \dfrac{u}{v}$, then $f'(x) = \dfrac{v\dfrac{du}{dx} - u\dfrac{dv}{dx}}{v^2}$

In this rule, as opposed to the Product Rule, the order in which you take the derivatives is very important, because you're subtracting instead of adding. It's always the bottom function times the derivative of the top minus the top function times the derivative of the bottom. Then divide the whole thing by the bottom function squared. A good way to remember this is to say the following:

$$\frac{"LoDeHi - HiDeLo"}{\left(Lo\right)^2}$$

You could also write

For the Quotient Rule, remember that order matters!

$$f(x) = \frac{u}{v} \text{ as } f(x) = u\frac{1}{v}$$

Then you could derive the Quotient Rule using the Product Rule.

$$f'(x) = u\left(-\frac{1}{v^2}\frac{dv}{dx}\right) + \frac{du}{dx}\frac{1}{v} = \frac{v\dfrac{du}{dx} - u\dfrac{dv}{dx}}{v^2}$$

Here are some more examples.

Example 4: $f(x) = \dfrac{\left(x^5 - 3x^4\right)}{\left(x^2 + 7x\right)}$

$$f'(x) = \frac{\left(x^2 + 7x\right)\left(5x^4 - 12x^3\right) - \left(x^5 - 3x^4\right)\left(2x + 7\right)}{\left(x^2 + 7x\right)^2}$$

Example 5: $y = \dfrac{\left(x^{-3} - x^{-8}\right)}{\left(x^{-2} + x^{-6}\right)}$

$$\frac{dy}{dx} = \frac{\left(x^{-2} + x^{-6}\right)\left(-3x^{-4} + 8x^{-9}\right) - \left(x^{-3} - x^{-8}\right)\left(-2x^{-3} - 6x^{-7}\right)}{\left(x^{-2} + x^{-6}\right)^2}$$

We're not going to simplify these, although the Quotient Rule often produces expressions that simplify more readily than those involving the Product Rule. Sometimes it's helpful to simplify, but avoid it otherwise. When you have to find a second derivative, however, you do have to simplify the quotient. If this is the case, the AP Exam usually will give you a simple expression to deal with, such as in the example below.

Example 6: $y = \dfrac{3x + 5}{5x - 3}$

$$\frac{dy}{dx} = \frac{(5x - 3)(3) - (3x + 5)(5)}{(5x - 3)^2} = \frac{(15x - 9) - (15x + 25)}{(5x - 3)^2} = \frac{-34}{(5x - 3)^2}$$

In order to take the derivative of $\dfrac{dy}{dx}$, you have to use the Chain Rule.

THE CHAIN RULE

The most important rule in this chapter (and sometimes the most difficult one) is called the **Chain Rule**. It's used when you're given composite functions—that is, a function inside of another function. You'll always see one of these on the AP Exam, so it's important to know the Chain Rule cold.

A composite function is usually written as: $f(g(x))$.

For example: If $f(x) = \dfrac{1}{x}$ and $g(x) = \sqrt{3x}$, then $f\big(g(x)\big) = \dfrac{1}{\sqrt{3x}}$.

We could also find $g\big(f(x)\big) = \sqrt{\dfrac{3}{x}}$.

When finding the derivative of a composite function, we take the derivative of the "outside" function, with the inside function g considered as the variable, leaving the "inside" function alone. Then, we multiply this by the derivative of the "inside" function, with respect to its variable x.

Here is another way to write the Chain Rule.

> The Chain Rule: If $y = f\big(g(x)\big)$, then $y' = \left(\dfrac{df\big(g(x)\big)}{dg}\right)\left(\dfrac{dg}{dx}\right)$

Chain Rule—always work from the outside in, kind of like Russian dolls or peeling an onion.

This rule is tricky, so here are several examples. The last couple incorporate the Product Rule and the Quotient Rule.

Example 7: If $y = (5x^3 + 3x)^5$, then $\dfrac{dy}{dx} = 5(5x^3 + 3x)^4(15x^2 + 3)$.

We just dealt with the derivative of something to the fifth power, like this:

$$y = (g)^5, \text{ so } \frac{dy}{dg} = 5(g)^4, \text{ where } g = 5x^3 + 3x$$

Then we multiplied by the derivative of g: $(15x^2 + 3)$.

Always do it this way. The process has several successive steps, like peeling away the layers of an onion until you reach the center.

Example 8: If $y = \sqrt{x^3 - 4x}$, then $\dfrac{dy}{dx} = \dfrac{1}{2}\left(x^3 - 4x\right)^{-\frac{1}{2}}\left(3x^2 - 4\right)$.

Again, we took the derivative of the outside function, leaving the inside alone. Then we multiplied by the derivative of the inside.

Example 9: If $y = \sqrt{\left(x^5 - 8x^3\right)\left(x^2 + 6x\right)}$, then

$$\frac{dy}{dx} = \frac{1}{2}\left[\left(x^5 - 8x^3\right)\left(x^2 + 6x\right)\right]^{-\frac{1}{2}}\left[\left(x^5 - 8x^3\right)\left(2x + 6\right) + \left(x^2 + 6x\right)\left(5x^4 - 24x^2\right)\right]$$

Messy, isn't it? That's because we used the Chain Rule and the Product Rule. Now for one with the Chain Rule and the Quotient Rule.

Example 10: If $y = \left(\dfrac{2x + 8}{x^2 - 10x}\right)^5$, then

$$\frac{dy}{dx} = 5\left[\frac{2x + 8}{x^2 - 10x}\right]^4\left[\frac{\left(x^2 - 10x\right)(2) - \left(2x + 8\right)\left(2x - 10\right)}{\left(x^2 - 10x\right)^2}\right]$$

Example 11: If $y = \sqrt{5x^3 + x}$, then $\dfrac{dy}{dx} = \dfrac{1}{2}\left(5x^3 + x\right)^{-\frac{1}{2}}\left(15x^2 + 1\right)$.

Now we use the Product Rule and the Chain Rule to find the second derivative.

$$\frac{d^2 y}{dx^2} = \frac{1}{2}\left(5x^3 + x\right)^{-\frac{1}{2}}(30x) + \left(15x^2 + 1\right)\left[-\frac{1}{4}\left(5x^3 + x\right)^{-\frac{3}{2}}\left(15x^2 + 1\right)\right]$$

You can also simplify this further, if necessary.

There's another representation of the Chain Rule that you need to learn.

$$\text{If } y = y(v) \text{ and } v = v(x), \text{ then } \frac{dy}{dx} = \frac{dy}{dv}\frac{dv}{dx}.$$

Example 12: If $y = 8v^2 - 6v$ and $v = 5x^3 - 11x$, then

$$\frac{dy}{dx} = (16v - 6)(15x^2 - 11)$$

Then substitute for v.

$$\frac{dy}{dx} = \left(16\left(5x^3 - 11x\right) - 6\right)\left(15x^2 - 11\right) = \left(80x^3 - 176x - 6\right)\left(15x^2 - 11\right)$$

As you can see, these grow quite complex, so we simplify these only as a last resort. If you must simplify, the AP Exam will have only a very simple Chain Rule problem.

Here are some solved problems. Cover the answers first, then check your work.

PROBLEM 1. Find $\frac{dy}{dx}$ if $y = \left(5x^4 + 3x^7\right)\left(x^{10} - 8x\right)$.

Answer: $\frac{dy}{dx} = \left(5x^4 + 3x^7\right)\left(10x^9 - 8\right) + \left(x^{10} - 8x\right)\left(20x^3 + 21x^6\right)$

PROBLEM 2. Find $\frac{dy}{dx}$ if $y = \left(x^3 + 3x^2 + 3x + 1\right)\left(x^2 + 2x + 1\right)$.

Answer: $\frac{dy}{dx} = \left(x^3 + 3x^2 + 3x + 1\right)(2x + 2) + \left(x^2 + 2x + 1\right)\left(3x^2 + 6x + 3\right)$

PROBLEM 3. Find $\frac{dy}{dx}$ if $y = \left(\sqrt{x} + \frac{1}{x}\right)\left(\sqrt[3]{x^2} - \frac{1}{x^3}\right)$.

Answer: $\frac{dy}{dx} = \left(\sqrt{x} + \frac{1}{x}\right)\left(\frac{2}{3}x^{-\frac{1}{3}} + \frac{3}{x^4}\right) + \left(\sqrt[3]{x^2} - \frac{1}{x^3}\right)\left(\frac{1}{2\sqrt{x}} - \frac{1}{x^2}\right)$

PROBLEM 4. Find $\frac{dy}{dx}$ if $y = \left(x^3 + 1\right)\left(x^2 + 5x - \frac{1}{x^5}\right)$.

Answer: $\frac{dy}{dx} = \left(x^3 + 1\right)\left(2x + 5 + \frac{5}{x^6}\right) + \left(x^2 + 5x - \frac{1}{x^5}\right)\left(3x^2\right)$

PROBLEM 5. Find $\frac{dy}{dx}$ if $y = \frac{2x - 4}{x^2 - 6}$.

Answer: $\frac{dy}{dx} = \frac{\left(x^2 - 6\right)(2) - (2x - 4)(2x)}{\left(x^2 - 6\right)^2} = \frac{-2x^2 + 8x - 12}{\left(x^2 - 6\right)^2}$

PROBLEM 6. Find $\dfrac{dy}{dx}$ if $y = \dfrac{x^2 + 1}{x^2 + x + 4}$.

Answer: $\dfrac{dy}{dx} = \dfrac{\left(x^2 + x + 4\right)(2x) - \left(x^2 + 1\right)(2x + 1)}{\left(x^2 + x + 4\right)^2} = \dfrac{x^2 + 6x - 1}{\left(x^2 + x + 4\right)^2}$

PROBLEM 7. Find $\dfrac{dy}{dx}$ if $y = \dfrac{x + 5}{x - 5}$.

Answer: $\dfrac{dy}{dx} = \dfrac{(x - 5)(1) - (x + 5)(1)}{(x - 5)^2} = \dfrac{-10}{(x - 5)^2}$

PROBLEM 8. Find $\dfrac{dy}{dx}$ if $y = (x^4 + x)^2$.

Answer: $\dfrac{dy}{dx} = 2\left(x^4 + x\right)\left(4x^3 + 1\right)$

PROBLEM 9. Find $\dfrac{dy}{dx}$ if $y = \left(\dfrac{x + 3}{x - 3}\right)^3$.

Answer: $\dfrac{dy}{dx} = 3\left(\dfrac{x + 3}{x - 3}\right)^2 \left(\dfrac{(x - 3)(1) - (x + 3)(1)}{(x - 3)^2}\right) = -18\left(\dfrac{(x + 3)^2}{(x - 3)^4}\right)$

PROBLEM 10. Find $\dfrac{dy}{dx}$ at $x = 1$ if $y = \left[\left(x^3 + x\right)\left(x^4 - x^2\right)\right]^2$.

Answer: $\dfrac{dy}{dx} = 2\left[\left(x^3 + x\right)\left(x^4 - x^2\right)\right]\left[\left(x^3 + x\right)\left(4x^3 - 2x\right) + \left(x^4 - x^2\right)\left(3x^2 + 1\right)\right]$

Once again, plug in right away. Never simplify until after you've substituted.

At $x = 1$, $\dfrac{dy}{dx} = 0$.

PRACTICE PROBLEM SET 5

Simplify when possible. The answers are in Chapter 19.

1. Find $f'(x)$ if $f(x) = \left(\dfrac{4x^3 - 3x^2}{5x^7 + 1} \right)$.

2. Find $f'(x)$ if $f(x) = (x^2 - 4x + 3)(x + 1)$.

3. Find $f'(x)$ if $f(x) = 8\sqrt{(x^4 - 4x^2)}$.

4. Find $f'(x)$ if $f(x) = \left(\dfrac{x}{x^2 + 1} \right)^3$.

5. Find $f'(x)$ if $f(x) = \sqrt[4]{\left(\dfrac{2x - 5}{5x + 2} \right)}$.

6. Find $f'(x)$ if $f(x) = \left(x + \dfrac{1}{x} \right)\left(x^2 - \dfrac{1}{x^2} \right)$.

7. Find $f'(x)$ if $f(x) = \left(\dfrac{x}{x + 1} \right)^4$.

8. Find $f'(x)$ if $f(x) = (x^2 + x)^{100}$.

9. Find $f'(x)$ if $f(x) = \sqrt{\dfrac{x^2 + 1}{x^2 - 1}}$.

10. Find $f'(x)$ at $x = 2$ if $f(x) = \dfrac{(x + 4)(x - 8)}{(x + 6)(x - 6)}$.

11. Find $f'(x)$ at $x = 1$ if $f(x) = \dfrac{x^6 + 4x^3 + 6}{(x^4 - 2)^2}$.

12. Find $f'(x)$ if $f(x) = \dfrac{x^2 - 3}{(x - 3)}$.

13. Find $f'(x)$ at $x = 1$ if $f(x) = \left(x^4 - x^2\right)\left(2x^3 + x\right)$.

14. Find $f'(x)$ at $x = 2$ if $f(x) = \dfrac{x^2 + 2x}{x^4 - x^3}$.

15. Find $f'(x)$ if $f(x) = \sqrt{x^4 + x^2}$.

16. Find $\dfrac{dy}{dx}$ if $y = u^2 - 1$ and $u = \dfrac{1}{x - 1}$.

17. Find $\dfrac{dy}{dx}$ at $x = 1$ if $y = \dfrac{t^2 + 2}{t^2 - 2}$ and $t = x^3$.

18. Find $\dfrac{dy}{dt}$ if $y = \left(x^6 - 6x^5\right)\left(5x^2 + x\right)$ and $x = \sqrt{t}$.

19. Find $\dfrac{dy}{dx}$ at $x = 1$ if $y = \dfrac{1 + u}{1 + u^2}$ and $u = x^2 - 1$.

20. Find $\dfrac{du}{dv}$ if $u = y^3$, $y = \dfrac{x}{x + 8}$, and $x = v^2$.

DERIVATIVES OF TRIG FUNCTIONS

There are a lot of trigonometry problems in calculus and on the AP Calculus Exam. If you're not sure of your trig, you should definitely go to the Appendix and review the unit on Prerequisite Math. You'll need to remember your trig formulas, the values of the special angles, and the trig ratios, among other stuff.

These formulas can also be found online in your student tools.

In addition, angles are *always* referred to in radians. You can forget all about using degrees.

You should know the derivatives of all six trig functions. The good news is that the derivatives are pretty easy, and all you have to do is memorize them. Because the AP Exam might ask you about this, though, let's use the definition of the derivative to figure out the derivative of sin x.

If $f(x) = \sin x$, then $f(x + h) = \sin(x + h)$.

Substitute this into the definition of the derivative.

$$\lim_{h\to 0}\frac{f(x+h)-f(x)}{h}=\lim_{h\to 0}\frac{\sin(x+h)-\sin x}{h}$$

Remember that sin(x + h) = sin x cos h + cos x sin h. Now simplify it.

$$\lim_{h\to 0}\frac{\sin x\,\cos h+\cos x\,\sin h-\sin x}{h}$$

Next, rewrite this as

$$\lim_{h\to 0}\frac{\sin x\left(\cos h-1\right)+\cos x\sin h}{h}=\lim_{h\to 0}\frac{\sin x\left(\cos h-1\right)}{h}+\lim_{h\to 0}\frac{\cos x\sin h}{h}$$

Next, use some of the trigonometric limits that you memorized back in Chapter 3. Specifically,

$$\lim_{h\to 0}\frac{\left(\cos h-1\right)}{h}=0 \text{ and } \lim_{h\to 0}\frac{\sin h}{h}=1$$

This gives you

$$\lim_{h\to 0}\frac{\sin x\left(\cos h-1\right)}{h}+\lim_{h\to 0}\frac{\cos x\sin h}{h}=\sin x\left(0\right)+\cos x\left(1\right)=\cos x$$

$$\frac{d}{dx}\sin x=\cos x$$

Example 1: Find the derivative of $\sin\left(\dfrac{\pi}{2}-x\right)$.

$$\frac{d}{dx}\sin\left(\frac{\pi}{2}-x\right)=\cos\left(\frac{\pi}{2}-x\right)\left(-1\right)=-\cos\left(\frac{\pi}{2}-x\right)$$

Use some of the rules of trigonometry you remember from last year. Because

$$\sin\left(\frac{\pi}{2} - x\right) = \cos x \text{ and } \cos\left(\frac{\pi}{2} - x\right) = \sin x,$$

you can substitute into the above expression and get

$$\frac{d}{dx}\cos x = -\sin x$$

Now, let's derive the derivatives of the other four trigonometric functions.

Example 2: Find the derivative of $\dfrac{\sin x}{\cos x}$.

Use the Quotient Rule.

$$\frac{d}{dx}\frac{\sin x}{\cos x} = \frac{(\cos x)(\cos x) - (\sin x)(-\sin x)}{(\cos x)^2} = \frac{\cos^2 x + \sin^2 x}{\cos^2 x} = \frac{1}{\cos^2 x} = \sec^2 x$$

Because $\dfrac{\sin x}{\cos x} = \tan x$, you should get

$$\frac{d}{dx}\tan x = \sec^2 x$$

Example 3: Find the derivative of $\dfrac{\cos x}{\sin x}$.

Use the Quotient Rule.

$$\frac{d}{dx}\frac{\cos x}{\sin x} = \frac{(\sin x)(-\sin x) - (\cos x)(\cos x)}{(\sin x)^2} = \frac{-(\cos^2 x + \sin^2 x)}{\sin^2 x} = -\frac{1}{\sin^2 x} = -\csc^2 x$$

Because $\dfrac{\cos x}{\sin x} = \cot x$, you get: $\dfrac{d}{dx}\cot x = -\csc^2 x$.

Example 4: Find the derivative of $\dfrac{1}{\cos x}$.

Use the Reciprocal Rule.

$$\frac{d}{dx}\frac{1}{\cos x} = \frac{-1}{(\cos x)^2}(-\sin x) = \frac{\sin x}{\cos^2 x} = \frac{1}{\cos x}\frac{\sin x}{\cos x} = \sec x \tan x$$

Because $\dfrac{1}{\cos x} = \sec x$, you get

$$\frac{d}{dx}\sec x = \sec x \tan x$$

Example 5: Find the derivative of $\dfrac{1}{\sin x}$.

You get the idea by now.

$$\frac{d}{dx}\frac{1}{\sin x} = \frac{-1}{(\sin x)^2}(\cos x) = \frac{-\cos x}{\sin^2 x} = \frac{-1}{\sin x}\frac{\cos x}{\sin x} = -\csc x \cot x$$

Because $\dfrac{1}{\sin x} = \csc x$, you get

$$\frac{d}{dx}\csc x = -\csc x \cot x$$

There you go. We have now found the derivatives of all six of the trigonometric functions. (A chart of them appears at the end of the book.) Now memorize them. You'll thank us later.

Let's do some more examples.

Example 6: Find the derivative of sin(5x).

$$\frac{d}{dx}\sin(5x) = \cos(5x)(5) = 5\cos(5x)$$

Example 7: Find the derivative of sec(x^2).

$$\frac{d}{dx}\sec(x^2) = \sec(x^2)\tan(x^2)(2x)$$

Example 8: Find the derivative of csc($x^3 - 5x$).

$$\frac{d}{dx}\csc(x^3 - 5x) = -\csc(x^3 - 5x)\cot(x^3 - 5x)(3x^2 - 5) \ .$$

These derivatives are almost like formulas. You just follow the pattern and use the Chain Rule when appropriate.

Here are some solved problems. Do each problem, covering the answer first, then checking your answer.

PROBLEM 1. Find $f'(x)$ if $f(x) = \sin(2x^3)$.

Answer: Follow the rule: $f'(x) = \cos(2x^3)6x^2$

PROBLEM 2. Find $f'(x)$ if $f(x) = \cos\left(\sqrt{3x}\right)$.

Answer: $f'(x) = -\sin\left(\sqrt{3x}\right)\left[\frac{1}{2}(3x)^{-\frac{1}{2}}(3)\right] = \dfrac{-3\sin\left(\sqrt{3x}\right)}{2\sqrt{3x}}$

PROBLEM 3. Find $f'(x)$ if $f(x) = \tan\left(\dfrac{x}{x+1}\right)$.

Answer: $f'(x) = \sec^2\left(\dfrac{x}{x+1}\right)\left[\dfrac{(x+1)-x}{(x+1)^2}\right] = \left(\dfrac{1}{(x+1)^2}\right)\sec^2\left(\dfrac{x}{x+1}\right)$

PROBLEM 4. Find $f'(x)$ if $f(x) = \csc(x^3 + x + 1)$.

Answer: Follow the rule: $f'(x) = -\csc(x^3 + x + 1)\cot(x^3 + x + 1)(3x^2 + 1)$

GRAPHICAL DERIVATIVES

Sometimes, we are given the graph of a function and we are asked to graph the derivative. We do this by analyzing the sign of derivative at various places on the graph and then sketching a graph of the derivative from that information. Let's start with something simple. Suppose we have the graph of $y = f(x)$ below and we are asked to sketch the graph of its derivative.

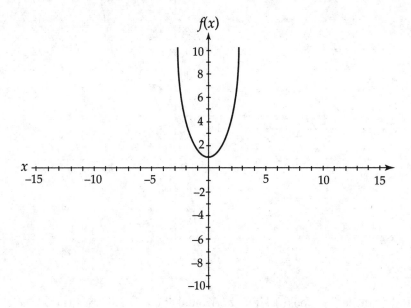

First, note that the derivative is zero at the point (0, 1) because the tangent line is horizontal there. Next, the derivative is negative for all $x < 0$ because the tangent lines to the curve have negative slopes everywhere on the interval $(-\infty, 0)$. Finally, the derivative is positive for all $x > 0$ because the tangent lines to the curve have positive slopes everywhere on the interval $(0, \infty)$. Now we can make a graph of the derivative. It will go through the origin (because the derivative is 0 at $x = 0$), it will be negative on the interval $(-\infty, 0)$ and will be positive on the interval $(0, \infty)$. The graph looks something like the following:

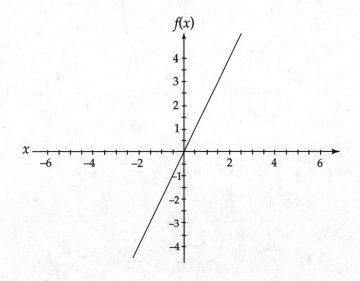

Note that it's not important for your graph to be exact. All we are doing here is sketching the derivative. The important parts of the graph are where the derivative is positive, negative, and zero.

Maxima and minima on the graph of *f* become zeroes on the graph of *f′*.

Now let's try something a little harder. Suppose we have the following graph and we are asked to sketch the graph of the derivative:

Notice that the tangent line looks as if it's horizontal at $x = 1$. This means that the graph of the derivative is zero there. Next, the curve is increasing for all other values of x, so the graph of the derivative will be positive. As we go from left to right on the graph, notice that the slope starts out very steep, so the derivatives are large positive numbers. As we approach $x = 1$, the curve starts to flatten out, so the derivatives will approach zero but will still be positive. Then, the slope is zero at $x = 1$. Then the curve gets steep again. If we sketch the derivative, we get something like the following:

Here is an important one to understand. Suppose we have the graph of $y = \sin x$.

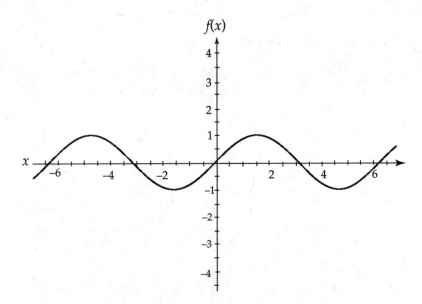

Notice that the slope of the tangent line will be horizontal at all of the maxima and minima of the graph. Because the slope of a horizontal line is zero, this means that the derivative will be zero at those values ($\pm\dfrac{\pi}{2}, \dfrac{3\pi}{2}, \ldots$). Next, notice that the slope of the curve is about 1 as the curve goes through the origin. This should make sense if you recall that $\displaystyle\lim_{x \to 0}\dfrac{\sin x}{x} = 1$. The slope of the curve is about –1 as the curve goes through $x = \pi$, and so on. If we now sketch the derivative, it looks something like the following:

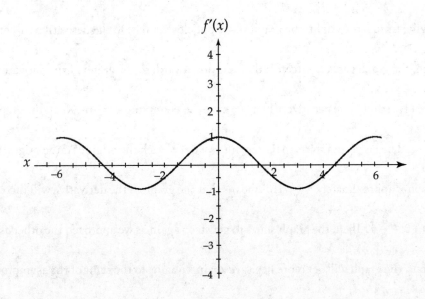

Notice that this is the graph of $y = \cos x$. This should be obvious because the derivative of $\sin x$ is $\cos x$.

Now let's do a hard one. Suppose we have the following graph:

First, notice that we have two vertical asymptotes at $x = 6$ and $x = -6$. This means that the

graph of the derivative will also have vertical asymptotes at $x = 6$ and $x = -6$. Next, notice that

the curve is always decreasing. This means that the graph of the derivative will always be nega-

tive. Moving from left to right, the graph starts out close to flat, so the derivative will be close to

zero. Then, the graph gets very steep and points downward, so the graph of the derivative will be

negative and getting more negative. Then, we have the asymptote $x = -6$. Next, the graph begins

very steep and negative and starts to flatten out as we approach the origin. At the origin, the slope

of the graph is approximately $-\dfrac{1}{2}$. This means that the graph of the derivative will increase until

it reaches $\left(0, -\dfrac{1}{2}\right)$. Then, the graph starts to get steep again as we approach the other asymptote

$x = 6$. Thus, the graph will get more negative again. Finally, to the right of the asymptote $x = 6$, the graph starts out steep and negative and flattens out, approaching zero. This means that the

graph of the derivative will start out very negative and will approach zero. If we now sketch the

derivative, it looks something like the following:

PRACTICE PROBLEM SET 6

Now try these problems. The answers are in Chapter 19.

1. Find $\dfrac{dy}{dx}$ if $y = \sin^2 x$.

2. Find $\dfrac{dy}{dx}$ if $y = \cos x^2$.

3. Find $\dfrac{dy}{dx}$ if $y = (\tan x)(\sec x)$.

4. Find $\dfrac{dy}{dx}$ if $y = \sqrt{\sin 3x}$.

5. Find $\dfrac{dy}{dx}$ if $y = \dfrac{1 + \sin x}{1 - \sin x}$.

6. Find $\dfrac{dy}{dx}$ if $y = \csc^2 x^2$.

7. Find $\dfrac{d^4 y}{dx^4}$ if $y = \sin 2x$.

8. Find $\dfrac{dy}{dx}$ if $y = \sin t - \cos t$ and $t = 1 + \cos^2 x$.

9. Find $\dfrac{dy}{dx}$ if $y = \left(\dfrac{\tan x}{1 - \tan x}\right)^2$.

10. Find $\dfrac{dr}{d\theta}$ if $r = \cos(1 + \sin\theta)$.

11. Find $\dfrac{dr}{d\theta}$ if $r = \dfrac{\sec\theta}{1 + \tan\theta}$.

12. Find $\dfrac{dy}{dx}$ if $y = \left(1 + \cot\left(\dfrac{2}{x}\right)\right)^{-2}$.

End of Chapter 6 Drill

The answers are in Chapter 20.

1. Find $\dfrac{dy}{dx}$ if $y = 5x^3 - 7x^2 + 11x + 2$.

    (A) $\dfrac{dy}{dx} = 15x^2 - 14x$

    (B) $\dfrac{dy}{dx} = 35x^2 - 27x + 11$

    (C) $\dfrac{dy}{dx} = 5x^2 - 7x + 11$

    (D) $\dfrac{dy}{dx} = 15x^2 - 14x + 11$

2. Find $\dfrac{dy}{dx}$ if $y = \tan(4x)$.

    (A) $\dfrac{dy}{dx} = \sec^2(4x)$

    (B) $\dfrac{dy}{dx} = 4\sec^2(4x)$

    (C) $\dfrac{dy}{dx} = 4\tan(4x)$

    (D) $\dfrac{dy}{dx} = 4\cot(4x)$

3. Find $\dfrac{dy}{dx}$ if $y = \cos^2(x^2)$.

    (A) $\dfrac{dy}{dx} = 4x\cos(x^2)$

    (B) $\dfrac{dy}{dx} = \sin^2(x^2)$

    (C) $\dfrac{dy}{dx} = -4x\cos(x^2)\sin(x^2)$

    (D) $\dfrac{dy}{dx} = -4x\sin^2(x^2)$

4. Find $\dfrac{dy}{dx}$ if $y = \dfrac{4}{\sqrt{x-1}}$.

    (A) $\dfrac{dy}{dx} = -\dfrac{2}{(x-1)^{\frac{3}{2}}}$

    (B) $\dfrac{dy}{dx} = \dfrac{2}{(x-1)^{\frac{3}{2}}}$

    (C) $\dfrac{dy}{dx} = -\dfrac{8}{(x-1)^{\frac{3}{2}}}$

    (D) $\dfrac{dy}{dx} = \dfrac{8}{(x-1)^{\frac{3}{2}}}$

5. Find $\dfrac{dy}{dx}$ if $y = \sqrt{\dfrac{x^2}{\sec x}}$.

    (A) $\dfrac{dy}{dx} = \dfrac{1}{2}\left(\dfrac{2x}{\sec x \tan x}\right)^{-\frac{1}{2}}$

    (B) $\dfrac{dy}{dx} = \left(\dfrac{2x}{\sec x \tan x}\right)^{\frac{1}{2}}$

    (C) $\dfrac{dy}{dx} = \dfrac{1}{2}\left(\dfrac{\sec x(2x) - (x^2)\sec x \tan x}{\sec^2 x}\right)^{-\frac{1}{2}}$

    (D) $\dfrac{dy}{dx} = \dfrac{1}{2}\left(\dfrac{x^2}{\sec x}\right)^{-\frac{1}{2}}\left(\dfrac{\sec x(2x) - (x^2)\sec x \tan x}{\sec^2 x}\right)$

REFLECT

Respond to the following questions:

- For which topics discussed in this chapter do you feel you have achieved sufficient mastery to answer multiple-choice questions correctly?

- For which topics discussed in this chapter do you feel you have achieved sufficient mastery to answer open-ended questions correctly?

- For which topics discussed in this chapter do you feel you need more work before you can answer multiple-choice questions correctly?

- For which topics discussed in this chapter do you feel you need more work before you can answer open-ended questions correctly?

- What parts of this chapter are you going to re-review?

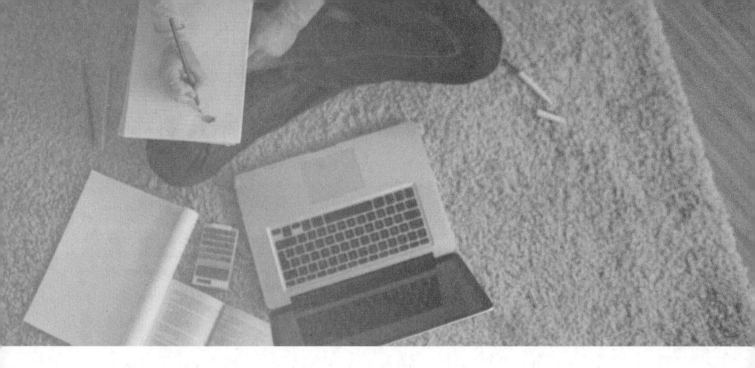

Chapter 7
Implicit
Differentiation

HOW TO DO IT

By now, it should be easy for you to take the derivative of an equation, such as $y = 3x^5 - 7x$. If you're given an equation such as $y^2 = 3x^5 - 7x$, you can still figure out the derivative by taking the square root of both sides, which gives you y in terms of x. This is known as finding the derivative **explicitly**. It's messy, but possible.

If you have to find the derivative of $y^2 + y = 3x^5 - 7x$, you don't have an easy way to get y in terms of x, so you can't differentiate this equation using any of the techniques you've learned so far. That's because each of those previous techniques needs to be used on an equation in which y is in terms of x. When you can't isolate y in terms of x (or if isolating y makes taking the derivative a nightmare), it's time to take the derivative **implicitly**.

Implicit differentiation is one of the simpler techniques you need to learn to do in calculus, but for some reason it gives many students trouble. Suppose you have the equation $y^2 = 3x^5 - 7x$. This means that the value of y is a function of the value of x. When we take the derivative, $\dfrac{dy}{dx}$, we're looking at the rate at which y changes as x changes. Thus, given $y = x^2 + x$, when we write

$$\frac{dy}{dx} = 2x + 1$$

we're saying that "the rate" at which y changes, with respect to how x changes, is $2x + 1$.

Now, suppose you want to find $\dfrac{dx}{dy}$. As you might imagine,

$$\frac{dx}{dy} = \frac{1}{\dfrac{dy}{dx}}$$

So here, $\dfrac{dx}{dy} = \dfrac{1}{2x+1}$. But notice that this derivative is in terms of x, not y, and you need to find the derivative with respect to y. This derivative is an **implicit** one. When you can't isolate the variables of an equation, you often end up with a derivative that is in terms of both variables.

Another way to think of this is that there is a hidden term in the derivative, $\dfrac{dx}{dx}$, and when we take the derivative, what we really get is

$$\frac{dy}{dx} = 2x\left(\frac{dx}{dx}\right) + 1\left(\frac{dx}{dx}\right)$$

A fraction that has the same term in its numerator and denominator is equal to 1, so we write

$$\frac{dy}{dx} = 2x(1) + 1(1) = 2x + 1$$

Every time we take a derivative of a term with x in it, we multiply by the term $\frac{dx}{dx}$, but because this is 1, we ignore it. Suppose however, that we wanted to find out how y changes with respect to t (for time). Then we would have

$$\frac{dy}{dt} = 2x\left(\frac{dx}{dt}\right) + 1\left(\frac{dx}{dt}\right)$$

If we wanted to find out how y changes with respect to r, we would have

$$\frac{dy}{dr} = 2x\left(\frac{dx}{dr}\right) + 1\left(\frac{dx}{dr}\right)$$

and if we wanted to find out how y changes with respect to y, we would have

$$\frac{dy}{dy} = 2x\left(\frac{dx}{dy}\right) + 1\left(\frac{dx}{dy}\right) \text{ or } 1 = 2x\frac{dx}{dy} + \frac{dx}{dy}$$

This is how we really do differentiation. Remember the following:

$$\frac{dx}{dy} = \frac{1}{\frac{dy}{dx}}$$

When you have an equation of x in terms of y, and you want to find the derivative with respect to y, simply differentiate. But if the equation is of y in terms of x, find $\frac{dy}{dx}$ and take its reciprocal to find $\frac{dx}{dy}$. Go back to our original example.

$$y^2 + y = 3x^5 - 7x$$

To take the derivative according to the information in the last paragraph, you get

$$2y\left(\frac{dy}{dx}\right) + 1\left(\frac{dy}{dx}\right) = 15x^4\left(\frac{dx}{dx}\right) - 7\left(\frac{dx}{dx}\right)$$

You should use implicit differentiation any time you can't write a function explicitly in terms of the variable that you want the derivative for.

Notice how each variable is multiplied by its appropriate $\dfrac{d}{dx}$. Now, remembering that $\dfrac{dx}{dx} = 1$, rewrite the expression this way: $2y\left(\dfrac{dy}{dx}\right) + 1\left(\dfrac{dy}{dx}\right) = 15x^4 - 7$.

Next, factor $\dfrac{dy}{dx}$ out of the left-hand side: $\dfrac{dy}{dx}(2y + 1) = 15x^4 - 7$.

Isolating $\dfrac{dy}{dx}$ gives you $\dfrac{dy}{dx} = \dfrac{15x^4 - 7}{(2y + 1)}$.

This is the derivative you're looking for. Notice how the derivative is defined in terms of y **and** x. Up until now, $\dfrac{dy}{dx}$ has been strictly in terms of x. This is why the differentiation is "implicit."

Confused? Let's do a few examples and you will get the hang of it.

Example 1: Find $\dfrac{dy}{dx}$ if $y^3 - 4y^2 = x^5 + 3x^4$.

Using implicit differentiation, you get

$$3y^2\left(\dfrac{dy}{dx}\right) - 8y\left(\dfrac{dy}{dx}\right) = 5x^4\left(\dfrac{dx}{dx}\right) + 12x^3\left(\dfrac{dx}{dx}\right)$$

Remember that $\dfrac{dx}{dx} = 1$: $\dfrac{dy}{dx}\left(3y^2 - 8y\right) = 5x^4 + 12x^3$.

After you factor out $\dfrac{dy}{dx}$, divide both sides by $3y^2 - 8y$.

$$\dfrac{dy}{dx} = \dfrac{5x^4 + 12x^3}{\left(3y^2 - 8y\right)}$$

Note: Now that you understand that the derivative of an x term with respect to x will always be multiplied by $\dfrac{dx}{dx}$, and that $\dfrac{dx}{dx} = 1$, we won't write $\dfrac{dx}{dx}$ anymore. You should understand that the term is implied.

Example 2: Find $\dfrac{dy}{dx}$ if $\sin y^2 - \cos x^2 = \cos y^2 + \sin x^2$.

Use implicit differentiation.

$$\cos y^2 \left(2y \dfrac{dy}{dx} \right) + \sin x^2 \left(2x \right) = -\sin y^2 \left(2y \dfrac{dy}{dx} \right) + \cos x^2 \left(2x \right)$$

Then simplify.

$$2y \cos y^2 \left(\dfrac{dy}{dx} \right) + 2x \sin x^2 = -2y \sin y^2 \left(\dfrac{dy}{dx} \right) + 2x \cos x^2$$

Next, put all of the terms containing $\dfrac{dy}{dx}$ on the left and all of the other terms on the right.

$$2y \cos y^2 \left(\dfrac{dy}{dx} \right) + 2y \sin y^2 \left(\dfrac{dy}{dx} \right) = -2x \sin x^2 + 2x \cos x^2$$

Next, factor out $\dfrac{dy}{dx}$.

$$\dfrac{dy}{dx} \left(2y \cos y^2 + 2y \sin y^2 \right) = -2x \sin x^2 + 2x \cos x^2$$

And isolate $\dfrac{dy}{dx}$.

$$\dfrac{dy}{dx} = \dfrac{-2x \sin x^2 + 2x \cos x^2}{\left(2y \cos y^2 + 2y \sin y^2 \right)}$$

This can be simplified further to the following:

$$\dfrac{dy}{dx} = \dfrac{-x \left(\sin x^2 - \cos x^2 \right)}{y \left(\cos y^2 + \sin y^2 \right)}$$

Example 3: Find $\dfrac{dy}{dx}$ if $3x^2 + 5xy^2 - 4y^3 = 8$.

Implicit differentiation should result in the following:

$$6x + \left[5x \left(2y \dfrac{dy}{dx} \right) + (5) y^2 \right] - 12y^2 \left(\dfrac{dy}{dx} \right) = 0$$

> Did you notice the use of the Product Rule to find the derivative of $5xy^2$? The AP Exam loves to make you do this. All of the same differentiation rules that you've learned up until now still apply. We're just adding another technique.

You can simplify this to

$$6x + 10xy\frac{dy}{dx} + 5y^2 - 12y^2\frac{dy}{dx} = 0$$

Next, put all of the terms containing $\frac{dy}{dx}$ on the left and all of the other terms on the right.

$$10xy\frac{dy}{dx} - 12y^2\frac{dy}{dx} = -6x - 5y^2$$

Next, factor out $\frac{dy}{dx}$.

$$\left(10xy - 12y^2\right)\frac{dy}{dx} = -6x - 5y^2$$

Then, isolate $\frac{dy}{dx}$.

$$\frac{dy}{dx} = \frac{-6x - 5y^2}{\left(10xy - 12y^2\right)}$$

Example 4: Find the derivative of $3x^2 - 4y^2 + y = 9$ at (2, 1).

You need to use implicit differentiation to find $\frac{dy}{dx}$.

$$6x - 8y\left(\frac{dy}{dx}\right) + \left(\frac{dy}{dx}\right) = 0$$

Be smart about your problem solving. Just because you can simplify something doesn't mean that you should. In a case like this, plugging into this form of the derivative is more effective.

Now, instead of rearranging to isolate $\frac{dy}{dx}$, plug in (2, 1) immediately and solve for the derivative.

$$6(2) - 8(1)\left(\frac{dy}{dx}\right) + \left(\frac{dy}{dx}\right) = 0$$

Simplify: $12 - 7\left(\frac{dy}{dx}\right) = 0$, so $\frac{dy}{dx} = \frac{12}{7}$.

Getting the hang of implicit differentiation yet? We hope so, because these next examples are slightly harder.

Example 5: Find the derivative of $\dfrac{2x-5y^2}{4y^3-x^2}=-x$ at (1, 1).

First, cross-multiply.

$$2x-5y^2=-x\left(4y^3-x^2\right)$$

Distribute.

$$2x-5y^2=-4xy^3+x^3$$

Take the derivative.

$$2-10y\frac{dy}{dx}=-4x\left(3y^2\frac{dy}{dx}\right)-4y^3+3x^2$$

Do not simplify now. Rather, plug in (1, 1) right away. This will save you from the algebra.

$$2-10(1)\frac{dy}{dx}=-4(1)\left(3(1)^2\frac{dy}{dx}\right)-4(1)^3+3(1)^2$$

Now solve for $\dfrac{dy}{dx}$.

$$2-10\frac{dy}{dx}=-12\frac{dy}{dx}-1$$

$$2\frac{dy}{dx}=-3$$

$$\frac{dy}{dx}=\frac{-3}{2}$$

SECOND DERIVATIVES

Sometimes, you'll be asked to find a second derivative implicitly.

Example 6: Find $\dfrac{d^2 y}{dx^2}$ if $y^2 + 2y = 4x^2 + 2x$.

Differentiating implicitly, you get

Remember: When it is required to take a second derivative, the first derivative should be simplified first.

$$2y\frac{dy}{dx} + 2\frac{dy}{dx} = 8x + 2$$

Next, simplify and solve for $\dfrac{dy}{dx}$.

$$\frac{dy}{dx} = \frac{4x+1}{y+1}$$

Now, it's time to take the derivative again.

$$\frac{d^2 y}{dx^2} = \frac{4(y+1) - (4x+1)\left(\dfrac{dy}{dx}\right)}{(y+1)^2}$$

Finally, substitute for $\dfrac{dy}{dx}$.

$$\frac{4(y+1) - (4x+1)\left(\dfrac{4x+1}{y+1}\right)}{(y+1)^2} = \frac{4(y+1)^2 - (4x+1)^2}{(y+1)^3}$$

Try these solved problems without looking at the answers. Then check your work.

PROBLEM 1. Find $\dfrac{dy}{dx}$ if $x^2 + y^2 = 6xy$.

Answer: Differentiate with respect to x.

$$2x + 2y\frac{dy}{dx} = 6x\frac{dy}{dx} + 6y$$

Group all of the $\dfrac{dy}{dx}$ terms on the left and the other terms on the right.

$$2y\frac{dy}{dx} - 6x\frac{dy}{dx} = 6y - 2x$$

Now factor out $\dfrac{dy}{dx}$.

$$\frac{dy}{dx}(2y - 6x) = 6y - 2x$$

Therefore, the first derivative is the following:

$$\frac{dy}{dx} = \frac{6y - 2x}{2y - 6x} = \frac{3y - x}{y - 3x}$$

PROBLEM 2. Find $\dfrac{dy}{dx}$ if $x - \cos y = xy$

Answer: Differentiate with respect to x.

$$1 + \sin y\frac{dy}{dx} = x\frac{dy}{dx} + y$$

Grouping the terms, you get

$$\sin y\frac{dy}{dx} - x\frac{dy}{dx} = y - 1$$

Now factor out $\dfrac{dy}{dx}$.

$$\frac{dy}{dx}(\sin y - x) = y - 1$$

The derivative is

$$\frac{dy}{dx} = \frac{y - 1}{\sin y - x}$$

PROBLEM 3. Find the derivative of each variable with respect to t of $x^2 + y^2 = z^2$.

Answer: $2x\dfrac{dx}{dt} + 2y\dfrac{dy}{dt} = 2z\dfrac{dz}{dt}$

PROBLEM 4. Find the derivative of each variable with respect to t of $V = \dfrac{1}{3}\pi r^2 h$.

Answer: $\dfrac{dV}{dt} = \dfrac{1}{3}\pi\left(r^2\dfrac{dh}{dt} + 2r\dfrac{dr}{dt}h \right)$

PROBLEM 5. Find $\dfrac{d^2 y}{dx^2}$ if $y^2 = x^2 - 2x$.

Answer: First, take the derivative with respect to x.

$$2y\dfrac{dy}{dx} = 2x - 2$$

Then, solve for $\dfrac{dy}{dx}$.

$$\dfrac{dy}{dx} = \dfrac{2x-2}{2y} = \dfrac{x-1}{y}$$

The second derivative with respect to x becomes

$$\dfrac{d^2 y}{dx^2} = \dfrac{y(1) - (x-1)\dfrac{dy}{dx}}{y^2}$$

Now substitute for $\dfrac{dy}{dx}$ and simplify.

$$\dfrac{d^2 y}{dx^2} = \dfrac{y - (x-1)\left(\dfrac{x-1}{y}\right)}{y^2} = \dfrac{y^2 - (x-1)^2}{y^3}$$

PRACTICE PROBLEM SET 7

Use implicit differentiation to find the following derivatives. The answers are in Chapter 19.

1. Find $\dfrac{dy}{dx}$ if $x^3 - y^3 = y$.

2. Find $\dfrac{dy}{dx}$ if $x^2 - 16xy + y^2 = 1$.

3. Find $\dfrac{dy}{dx}$ at $(2, 1)$ if $\dfrac{x+y}{x-y} = 3$.

4. Find $\dfrac{dy}{dx}$ if $16x^2 - 16xy + y^2 = 1$ at $(1, 1)$.

5. Find $\dfrac{dy}{dx}$ if $x\sin y + y\sin x = \dfrac{\pi}{2\sqrt{2}}$ at $\left(\dfrac{\pi}{4}, \dfrac{\pi}{4}\right)$.

6. Find $\dfrac{d^2y}{dx^2}$ if $x^2 + 4y^2 = 1$.

7. Find $\dfrac{d^2y}{dx^2}$ if $\sin x + 1 = \cos y$.

8. Find $\dfrac{d^2y}{dx^2}$ if $x^2 - 4x = 2y - 2$.

End of Chapter 7 Drill

The answers are in Chapter 20.

1. Find $\dfrac{dy}{dx}$ if $x^2 + 2y^3 = x^3 - 4y^2$.

   (A) $\dfrac{dy}{dx} = \dfrac{3x^2 + 2x}{6y^2 - 8y}$

   (B) $\dfrac{dy}{dx} = \dfrac{3x^2 - 2x}{6y^2 + 8y}$

   (C) $\dfrac{dy}{dx} = \dfrac{6y^2 + 8y}{3x^2 - 2x}$

   (D) $\dfrac{dy}{dx} = \dfrac{6y^2 - 8y}{3x^2 + 2x}$

2. Find $\dfrac{dy}{dx}$ if $x^2 + 3xy^2 - y^3 = 1$.

   (A) $\dfrac{dy}{dx} = \dfrac{-2x - 3y^2}{6xy - 3y^2}$

   (B) $\dfrac{dy}{dx} = \dfrac{2x + 3y^2}{6xy - 3y^2}$

   (C) $\dfrac{dy}{dx} = 2x + 6xy - 3y^2$

   (D) $\dfrac{dy}{dx} = 2x - 6xy + 3y^2$

3. Find $\dfrac{dy}{dx}$ if $4\sin(x^2) = 3y - y^2$ at $\left(\sqrt{\dfrac{\pi}{6}}, 2\right)$.

   (A) $-4\sqrt{\dfrac{\pi}{6}}$

   (B) $4\sqrt{\dfrac{\pi}{6}}$

   (C) $-4\sqrt{\dfrac{\pi}{2}}$

   (D) $4\sqrt{\dfrac{\pi}{2}}$

4. Find $\dfrac{dy}{dx}$ if $xy^2 - 4x^3y^2 = -3$ at $(1, 1)$.

   (A) $-\dfrac{11}{6}$

   (B) $-\dfrac{6}{11}$

   (C) $\dfrac{10}{13}$

   (D) $\dfrac{13}{10}$

5. Find $\dfrac{d^2y}{dx^2}$ if $y = y^3 - x^2$.

   (A) $\dfrac{\left(3y^2 - 1\right)^2 (2) - \left(24x^2y\right)}{\left(3y^2 - 1\right)^3}$

   (B) $\dfrac{\left(3y^2 - 1\right)(2) - \left(24x^2y\right)}{\left(3y^2 - 1\right)^2}$

   (D) $\dfrac{\left(3y^2 - 1\right)(2) - \left(12x^2\right)}{\left(3y^2 - 1\right)^3}$

   (D) $\dfrac{\left(3y^2 - 1\right)^2 (2) - \left(12x^2\right)}{\left(3y^2 - 1\right)^3}$

REFLECT

Respond to the following questions:

- For which topics discussed in this chapter do you feel you have achieved sufficient mastery to answer multiple-choice questions correctly?

- For which topics discussed in this chapter do you feel you have achieved sufficient mastery to answer open-ended questions correctly?

- For which topics discussed in this chapter do you feel you need more work before you can answer multiple-choice questions correctly?

- For which topics discussed in this chapter do you feel you need more work before you can answer open-ended questions correctly?

- What parts of this chapter are you going to re-review?

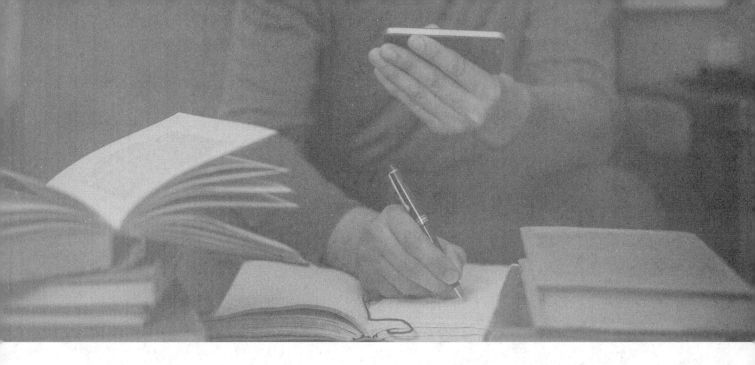

Unit 2

Differential
Calculus
Applications

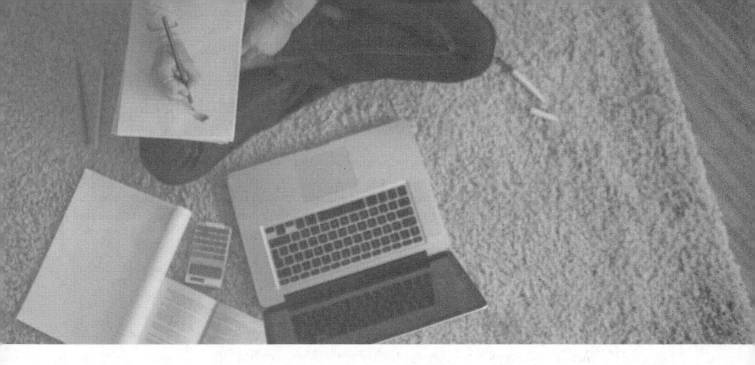

Chapter 8
Basic Applications of the Derivative

EQUATIONS OF TANGENT LINES AND NORMAL LINES

Finding the equation of a line tangent to a certain curve at a certain point is a standard calculus problem. This is because, among other things, the derivative is the slope of a tangent line to a curve at a particular point. Thus, we can find the equation of the tangent line to a curve if we have the equation of the curve and the point at which we want to find the tangent line. Then all we have to do is take the derivative of the equation, plug in the x-coordinate of the point to find the slope, then use the point and the slope to find the equation of the line. Let's take this one step at a time.

Suppose we have a point (x_1, y_1) and a slope m. Then the equation of the line through that point with that slope is

$$(y - y_1) = m(x - x_1)$$

You should remember this formula from algebra. If not, memorize it!

Next, suppose that we have an equation $y = f(x)$, where (x_1, y_1) satisfies that equation. Then $f'(x_1) = m$, and we can plug all of our values into the equation for a line and get the equation of the tangent line. This is much easier to explain with a simple example.

Example 1: Find the equation of the tangent line to the curve $y = 5x^2$ at the point (3, 45).

First of all, notice that the point satisfies the equation: when $x = 3$, $y = 45$. Now, take the derivative of the equation.

$$\frac{dy}{dx} = 10x$$

Now, if you plug in $x = 3$, you'll get the slope of the curve at that point.

$$\frac{dy}{dx}\bigg|_{x=3} = 10(3) = 30$$

By the way, the notation for plugging in a point is $\big|_{x=}$. Learn to recognize it!

Thus, we have the slope and the point, and the equation is

$$(y - 45) = 30(x - 3)$$

It's customary to simplify the equation if it's not too onerous.

$$y = 30x - 45$$

Example 2: Find the equation of the tangent line to $y = x^3 + x^2$ at (3, 36).

The derivative looks like the following:

$$\frac{dy}{dx} = 3x^2 + 2x$$

So, the slope is

$$\left.\frac{dy}{dx}\right|_{x=3} = 3(3)^2 + 2(3) = 33$$

The equation looks like the following:

$$(y - 36) = 33(x - 3), \text{ or } y = 33x - 63$$

Naturally, there are a couple of things that can be done to make the problems harder. First of all, you can be given only the x-coordinate. Second, the equation can be more difficult to differentiate.

In order to find the y-coordinate, all you have to do is plug the x-value into the equation for the curve and solve for y. Remember this: You'll see it again!

Example 3: Find the equation of the tangent line to $y = \dfrac{2x+5}{x^2-3}$ at $x = 1$.

First, find the y-coordinate.

$$y\,(1) = \frac{2(1)+5}{1^2-3} = -\frac{7}{2}$$

Second, take the derivative.

$$\frac{dy}{dx} = \frac{(x^2-3)(2)-(2x+5)(2x)}{(x^2-3)^2}$$

You're probably dreading having to simplify this derivative. Don't waste your time! Plug in $x = 1$ right away.

$$\left.\frac{dy}{dx}\right|_{x=1} = \frac{\left(1^2-3\right)(2)-\left(2(1)+5\right)\left(2(1)\right)}{\left(1^2-3\right)^2} = \frac{-4-14}{4} = -\frac{9}{2}$$

Now, we have a slope and a point, so the equation is

$$y + \frac{7}{2} = -\frac{9}{2}(x-1), \text{ or } 2y = -9x + 2$$

Remember: The slope of a perpendicular line is just the negative reciprocal of the slope of the tangent line.

Sometimes, instead of finding the equation of a tangent line, you will be asked to find the equation of a normal line. A **normal** line is simply the line perpendicular to the tangent line at the same point. You follow the same steps as with the tangent line, but you use the slope that will give you a perpendicular line.

Example 4: Find the equation of the line normal to $y = x^5 - x^4 + 1$ at $x = 2$.

First, find the y-coordinate.

$$y(2) = 2^5 - 2^4 + 1 = 17$$

Second, take the derivative.

$$\frac{dy}{dx} = 5x^4 - 4x^3$$

Third, find the slope at $x = 2$.

$$\frac{dy}{dx}\bigg|_{x=2} = 5(2)^4 - 4(2)^3 = 48$$

Fourth, take the negative reciprocal of 48, which is $-\frac{1}{48}$.

Finally, the equation becomes

$$y - 17 = -\frac{1}{48}(x-2)$$

Try these solved problems. Do each problem, covering the answer first, then checking your answer.

PROBLEM 1. Find the equation of the tangent line to the graph of $y = 4 - 3x - x^2$ at the point $(2, -6)$.

Answer: First, take the derivative of the equation.

$$\frac{dy}{dx} = -3 - 2x$$

Now, plug in $x = 2$ to get the slope of the tangent line.

$$\frac{dy}{dx} = -3 - 2(2) = -7$$

Third, plug the slope and the point into the equation for the line.

$$y - (-6) = -7\ (x - 2)$$

This simplifies to $y = -7x + 8$.

PROBLEM 2. Find the equation of the normal line to the graph of $y = 6 - x - x^2$ at $x = -1$.

Answer: Plug $x = -1$ into the original equation to get the y-coordinate.

$$y = 6 + 1 - 1 = 6$$

Once again, take that derivative.

$$\frac{dy}{dx} = -1 - 2x$$

Now, plug in $x = -1$ to get the slope of the tangent.

$$\frac{dy}{dx} = -1 - 2(-1) = 1$$

Use the negative reciprocal of the slope in the second step to get the slope of the normal line.

$$m = -1$$

Finally, plug the slope and the point into the equation for the line.

$$y - 6 = -1(x + 1)$$

This simplifies to $y = -x + 5$.

PROBLEM 3. Find the equations of the tangent and normal lines to the graph of $y = \dfrac{10x}{x^2+1}$ at the point (2, 4).

Answer: This problem will put your algebra to the test. You have to use the Quotient Rule to take the derivative of this mess.

$$\frac{dy}{dx} = \frac{\left(x^2+1\right)(10) - (10x)(2x)}{\left(x^2+1\right)^2}$$

Second, plug in $x = 2$ to get the slope of the tangent.

$$\frac{dy}{dx} = \frac{(5)(10) - (20)(4)}{5^2} = -\frac{30}{25} = -\frac{6}{5}$$

Now, plug the slope and the point into the equation for the tangent line.

$$y - 4 = -\frac{6}{5}(x - 2)$$

That simplifies to $6x + 5y = 32$. The equation of the normal line must then be

$$y - 4 = \frac{5}{6}(x - 2)$$

That, in turn, simplifies to $-5x + 6y = 14$.

PROBLEM 4. The curve $y = ax^2 + bx + c$ passes through the point (2, 4) and is tangent to the line $y = x + 1$ at (0, 1). Find a, b, and c.

Answer: The curve passes through (2, 4), so if you plug in $x = 2$, you'll get $y = 4$. Therefore,

$$4 = 4a + 2b + c$$

Second, the curve also passes through the point (0, 1), so $c = 1$.

Because the curve is tangent to the line $y = x + 1$ at $(0, 1)$, they must both have the same slope at that point. The slope of the line is 1. The slope of the curve is the first derivative.

$$\frac{dy}{dx} = 2ax + b$$

$$\frac{dy}{dx}\bigg|_{x=0} = 2a(0) + b = b$$

At $(0, 1)$, $\frac{dy}{dx} = b$. Therefore, $b = 1$.

Now that you know b and c, plug them back into the equation from the first step and solve for a.

$$4 = 4a + 2 + 1, \text{ and } a = \frac{1}{4}$$

PROBLEM 5. Find the points on the curve $y = 2x^3 - 3x^2 - 12x + 20$ where the tangent is parallel to the x-axis.

Answer: The x-axis is a horizontal line, so it has a slope of zero. Therefore, you want to know where the derivative of this curve is zero. Take the derivative.

$$\frac{dy}{dx} = 6x^2 - 6x - 12$$

Set it equal to zero and solve for x. Get accustomed to doing this: It's one of the most common questions in differential calculus.

$$\frac{dy}{dx} = 6x^2 - 6x - 12 = 0$$

$$6(x^2 - x - 2) = 0$$

$$6(x - 2)(x + 1) = 0$$

$$x = 2 \text{ or } x = -1$$

Third, find the y-coordinates of these two points.

$$y = 2(8) - 3(4) - 12(2) + 20 = 0$$

$$y = 2(-1) - 3(1) - 12(-1) + 20 = 27$$

Therefore, the points are $(2, 0)$ and $(-1, 27)$.

PRACTICE PROBLEM SET 8

Now try these problems. The answers are in Chapter 19.

1. Find the equation of the tangent to the graph of $y = 3x^2 - x$ at $x = 1$.

2. Find the equation of the tangent to the graph of $y = x^3 - 3x$ at $x = 3$.

3. Find the equation of the tangent to the graph of $y = \dfrac{1}{\sqrt{x^2 + 7}}$ at $x = 3$.

4. Find the equation of the normal to the graph of $y = \dfrac{x+3}{x-3}$ at $x = 4$.

5. Find the equation of the tangent to the graph of $y = 2x^3 - 3x^2 - 12x + 20$ at $x = 2$.

6. Find the equation of the tangent to the graph of $y = \dfrac{x^2 + 4}{x - 6}$ at $x = 5$.

7. Find the equation of the tangent to the graph of $y = \sqrt{x^3 - 15}$ at $(4, 7)$.

8. Find the values of x where the tangent to the graph of $y = 2x^3 - 8x$ has a slope equal to the slope of $y = x$.

9. Find the equation of the normal to the graph of $y = \dfrac{3x+5}{x-1}$ at $x = 3$.

10. Find the values of x where the normal to the graph of $(x-9)^2$ is parallel to the y-axis.

11. Find the coordinates where the tangent to the graph of $y = 8 - 3x - x^2$ is parallel to the x-axis.

12. Find the values of a, b, and c where the curves $y = x^2 + ax + b$ and $y = cx + x^2$ have a common tangent line at $(-1, 0)$.

THE MEAN VALUE THEOREM FOR DERIVATIVES

If $y = f(x)$ is continuous on the interval $[a, b]$, and is differentiable everywhere on the interval (a, b), then there is at least one number c between a and b such that

$$f'(c) = \frac{f(b) - f(a)}{b - a}$$

In other words, there's some point in the interval where the slope of the tangent line equals the slope of the secant line that connects the endpoints of the interval. (The function has to be continuous at the endpoints of the interval, but it doesn't have to be differentiable at the endpoints. Is this important? Maybe to mathematicians, but probably not to you!) You can see this graphically in the following figure:

Remember that in order for The Mean Value Theorem for Derivatives to work, the curve must be continuous on the interval <u>and</u> at the endpoints.

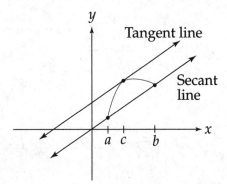

Example 1: Suppose you have the function $f(x) = x^2$, and you're looking at the interval $[1, 3]$. The Mean Value Theorem for Derivatives (this is often abbreviated MVTD) states that there is some number c such that

$$f'(c) = \frac{3^2 - 1^2}{3 - 1} = 4$$

Because $f'(x) = 2x$, plug in c for x and solve: $2c = 4$ so $c = 2$. Notice that 2 is in the interval. This is what the MVTD predicted! If you don't get a value for c within the interval, something went wrong; either the function is not continuous and differentiable in the required interval, or you made a mistake.

Example 2: Consider the function $f(x) = x^3 - 12x$ on the interval $[-2, 2]$. The MVTD states that there is a c such that

$$f'(c) = \frac{(2^3 - 24) - ((-2)^3 + 24)}{2 - (-2)} = -8$$

Then $f'(c) = 3c^2 - 12 = -8$ and $c = \pm\dfrac{2}{\sqrt{3}}$ (which is approximately ± 1.155).

Notice that here there are two values of c that satisfy the MVTD. That's allowed. In fact, there can be infinitely many values, depending on the function.

Example 3: Consider the function $f(x) = \dfrac{1}{x}$ on the interval $[-2, 2]$.

Follow the MVTD.

$$f'(c) = \frac{\dfrac{1}{2} - \left(-\dfrac{1}{2}\right)}{2 - (-2)} = \frac{1}{4}$$

Then

$$f'(c) = \frac{-1}{c^2} = \frac{1}{4}$$

There is no value of c that will satisfy this equation! We expected this. Why? Because $f(x)$ is not continuous at $x = 0$, which is in the interval. Suppose the interval had been $[1, 3]$, eliminating the discontinuity. The result would have been

$$f'(c) = \frac{\dfrac{1}{3} - (1)}{3 - 1} = -\frac{1}{3} \text{ and } f'(c) = \frac{-1}{c^2} = -\frac{1}{3}; \ c = \pm\sqrt{3}$$

Note: This is a great way to self-check your work. Always look at whether your answer makes sense.

$c = -\sqrt{3}$ is not in the interval, but $c = \sqrt{3}$ is. The answer is $c = \sqrt{3}$.

Example 4: Consider the function $f(x) = x^2 - x - 12$ on the interval $[-3, 4]$.

Follow the MVTD.

$$f'(c) = \frac{0 - 0}{7} = 0 \text{ and } f'(c) = 2c - 1 = 0 \text{, so } c = \frac{1}{2}$$

In this last example, you discovered where the derivative of the equation equaled zero. This is going to be the single most common problem you'll encounter in differential calculus. So now, we've got an important tip for you.

> When you don't know what to do, take the derivative of the equation and set it equal to zero!!!

Remember this advice for the rest of AP Calculus.

ROLLE'S THEOREM

Now let's learn Rolle's Theorem, which is a special case of the MVTD.

> If $y = f(x)$ is continuous on the interval $[a, b]$, and is differentiable everywhere on the interval (a, b), and if $f(a) = f(b) = 0$, then there is at least one number c between a and b such that $f'(c) = 0$.

Graphically, this means that a continuous, differentiable curve has a horizontal tangent between any two points where it crosses the x-axis.

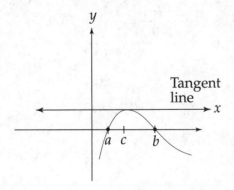

Example 4 was an example of Rolle's Theorem, but let's do another.

Example 5: Consider the function $f(x) = \dfrac{x^2}{2} - 6x$ on the interval $[0, 12]$.

First, show that

$$f(0) = \frac{0}{2} - 6(0) = 0 \text{ and } f(12) = \frac{144}{2} - 6(12) = 0$$

Then find

$$f'(x) = x - 6, \text{ so } f'(c) = c - 6$$

If you set this equal to zero (remember what we told you!), you get $c = 6$. This value of c falls in the interval, so the theorem holds for this example.

As with the MVTD, you'll run into problems with the theorem when the function is not continuous and differentiable over the interval. This is where you need to look out for a trap set by ETS. Otherwise, just follow what we did here and you won't have any trouble with either Rolle's Theorem or the MVTD. Try these example problems, and cover the responses until you check your work.

PROBLEM 1. Find the values of c that satisfy the MVTD for $f(x) = x^2 + 2x - 1$ on the interval $[0, 1]$.

Answer: First, find $f(0)$ and $f(1)$.

$$f(0) = 0^2 + 2(0) - 1 = -1 \text{ and } f(1) = 1^2 + 2(1) - 1 = 2$$

Then,

$$\frac{2 - (-1)}{1 - 0} = \frac{3}{1} = 3 = f'(c)$$

Next, find $f'(x)$.

$$f'(x) = 2x + 2$$

Thus, $f'(c) = 2c + 2 = 3$, and $c = \dfrac{1}{2}$.

PROBLEM 2. Find the values of c that satisfy the MVTD for $f(x) = x^3 + 1$ on the interval $[1, 2]$.

Answer: Find $f(1) = 1^3 + 1 = 2$ and $f(2) = 2^3 + 1 = 9$. Then,

$$\frac{9 - 2}{2 - 1} = 7 = f'(c)$$

Next, $f'(x) = 3x^2$, so $f'(c) = 3c^2 = 7$ and $c = \pm\sqrt{\dfrac{7}{3}}$.

Notice that there are two answers for c, but only one of them is in the interval. The answer is $c = \sqrt{\dfrac{7}{3}}$.

PROBLEM 3. Find the values of c that satisfy the MVTD for $f(x) = x + \dfrac{1}{x}$ on the interval $[-4, 4]$.

Answer: First, because the function is not continuous on the interval, there may not be a solution for c. Let's show that this is true. Find $f(-4) = -4 - \dfrac{1}{4} = -\dfrac{17}{4}$ and $f(4) = 4 + \dfrac{1}{4} = \dfrac{17}{4}$. Then,

$$\frac{\dfrac{17}{4} - \left(-\dfrac{17}{4}\right)}{4 - (-4)} = \frac{17}{16} = f'(c)$$

Next, $f'(x) = 1 - \dfrac{1}{x^2}$. Therefore, $f'(c) = 1 - \dfrac{1}{c^2} = \dfrac{17}{16}$.

There's no solution to this equation.

PROBLEM 4. Find the values of c that satisfy Rolle's Theorem for $f(x) = x^4 - x$ on the interval $[0, 1]$.

Answer: Show that $f(0) = 0^4 - 0 = 0$ and that $f(1) = 1^4 - 1 = 0$.

Next, find $f'(x) = 4x^3 - 1$. By setting $f'(c) = 4c^3 - 1 = 0$ and solving, you'll see that $c = \sqrt[3]{\dfrac{1}{4}}$, which is in the interval.

PRACTICE PROBLEM SET 9

Now try these problems. The answers are in Chapter 19.

1. Find the values of c that satisfy the MVTD for $f(x) = 3x^2 + 5x - 2$ on the interval $[-1, 1]$.

2. Find the values of c that satisfy the MVTD for $f(x) = x^3 + 24x - 16$ on the interval $[0, 4]$.

3. Find the values of c that satisfy the MVTD for $f(x) = \dfrac{6}{x} - 3$ on the interval $[1, 2]$.

4. Find the values of c that satisfy the MVTD for $f(x) = \dfrac{6}{x} - 3$ on the interval $[-1, 2]$.

5. Find the values of c that satisfy Rolle's Theorem for $f(x) = x^2 - 8x + 12$ on the interval $[2, 6]$.

6. Find the values of c that satisfy Rolle's Theorem for $f(x) = x(1 - x)$ on the interval $[0, 1]$.

7. Find the values of c that satisfy Rolle's Theorem for $f(x) = 1 - \dfrac{1}{x^2}$ on the interval $[-1, 1]$.

8. Find the values of c that satisfy Rolle's Theorem for $f(x) = x^{\frac{2}{3}} - x^{\frac{1}{3}}$ on the interval $[0, 1]$.

End of Chapter 8 Drill

The answers are in Chapter 20.

1. Find the equation of the line tangent to $y = 5x^3 - 20x + 10$ at $x = 2$.

   (A) $y - 10 = 10(x-2)$
   (B) $y + 10 = 40(x+2)$
   (C) $y - 10 = 40(x-2)$
   (D) $y - 10 = 90(x+2)$

2. Find the equation of the line normal to $y = 4\sec(2x)$ at $x = \dfrac{\pi}{8}$.

   (A) $y + 4\sqrt{2} = 8\sqrt{2}\left(x - \dfrac{\pi}{8}\right)$

   (B) $y - 4\sqrt{2} = -8\sqrt{2}\left(x - \dfrac{\pi}{8}\right)$

   (C) $y + 4\sqrt{2} = -\dfrac{1}{8\sqrt{2}}\left(x + \dfrac{\pi}{8}\right)$

   (D) $y - 4\sqrt{2} = -\dfrac{1}{8\sqrt{2}}\left(x - \dfrac{\pi}{8}\right)$

3. Find the value of c that satisfies the MVTD for $f(x) = x^2 - 6x + 5$ on the interval [3, 8].

   (A) 5
   (B) $\dfrac{11}{2}$
   (C) 11
   (D) There is no value of c.

4. Find the value of c that satisfies the MVTD for $f(x) = \dfrac{2}{x-4}$ on the interval [2, 6].

   (A) –2
   (B) 6
   (C) 8
   (D) There is no value of c.

5. Find all of the values of c that satisfy Rolle's Theorem for $f(x) = x^4 - 5x^2 + 4$ on the interval [–3, 3].

   (A) $c = 0$
   (B) $c = \pm\sqrt{\dfrac{5}{2}}$
   (C) $c = 0, \pm 2$
   (D) $c = 0, \pm\sqrt{\dfrac{5}{2}}$

REFLECT

Respond to the following questions:

- For which topics discussed in this chapter do you feel you have achieved sufficient mastery to answer multiple-choice questions correctly?

- For which topics discussed in this chapter do you feel you have achieved sufficient mastery to answer open-ended questions correctly?

- For which topics discussed in this chapter do you feel you need more work before you can answer multiple-choice questions correctly?

- For which topics discussed in this chapter do you feel you need more work before you can answer open-ended questions correctly?

- What parts of this chapter are you going to re-review?

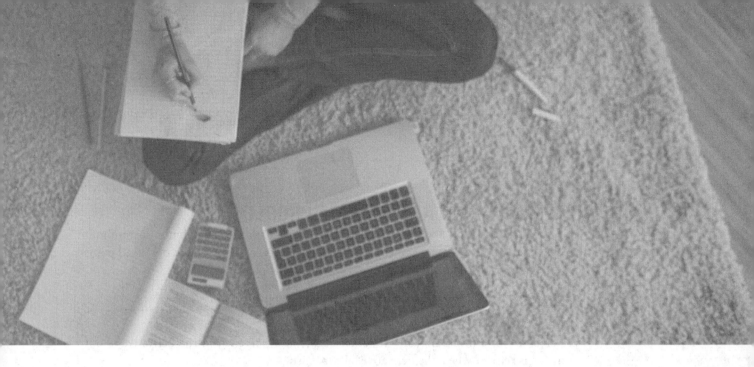

Chapter 9
Maxima, Minima, and Curve Stretching

Here's another chapter of material involving more ways to apply the derivative to several other types of problems. This stuff focuses mainly on using the derivative to aid in graphing a function, etc.

APPLIED MAXIMA AND MINIMA PROBLEMS

One of the most common applications of the derivative is to find a maximum or minimum value of a function. These values can be called extreme values, optimal values, or critical points. Each of these problems involves the same, very simple principle.

> A maximum or a minimum of a function occurs at a point where the derivative of a function is zero, or where the derivative fails to exist.

At a point where the first derivative equals zero, the curve has a horizontal tangent line, at which point it could be reaching either a "peak" (maximum) or a "valley" (minimum).

There are a few exceptions to every rule. This rule is no different.

If the derivative of a function is zero at a certain point, it is usually a maximum or minimum—but not always.

There are two different kinds of maxima and minima: relative and absolute. A **relative** or **local** maximum or minimum means that the curve has a horizontal tangent line at that point, but it is not the highest or lowest value that the function attains. In the figure to the right, the two indicated points are relative maxima/minima.

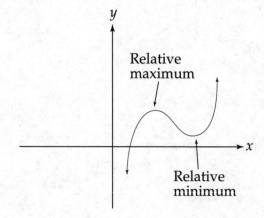

> An **absolute** maximum or minimum occurs either at an artificial point or an end point. In the figure below, the two indicated points are absolute maxima/minima. A relative maximum can also be an absolute maximum.

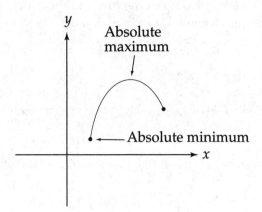

A typical word problem will ask you to find a maximum or a minimum value of a function, as it pertains to a certain situation. Sometimes you're given the equation; other times, you have to figure it out for yourself. Once you have the equation, you find its derivative and set it equal to zero. The values you get are called critical values. That is, if $f'(c) = 0$ or $f'(c)$ does not exist, then c is a critical value. Then, test these values to determine whether each value is a maximum or a minimum. The simplest way to do this is with the second derivative test.

> If a function has a critical value at $x = c$, then that value is a relative maximum if $f''(c) < 0$ and it is a relative minimum if $f''(c) > 0$.

If the second derivative is also zero at $x = c$, then the point is neither a maximum nor a minimum but a point of inflection. More about that later.

It's time to do some examples.

Example 1: Find the minimum value on the curve $y = ax^2$, if $a > 0$.

Take the derivative and set it equal to zero.

$$\frac{dy}{dx} = 2ax = 0$$

The first derivative is equal to zero at $x = 0$. By plugging 0 back into the original equation, we can solve for the y-coordinate of the minimum (the y-coordinate is also 0, so the point is at the origin).

In order to determine if this is a maximum or a minimum, take the second derivative.

$$\frac{d^2y}{dx^2} = 2a$$

Extreme Learning
The Extreme Value Theorem says that if a function is continuous on the interval [a, b], then somewhere in the interval (a, b), there exists a maximum or minimum value of $f(c)$ between $f(a)$ and $f(b)$ inclusive. In other words, if the function is continuous in some interval, then you always have a maximum or minimum somewhere in the interval, or the curve is flat.

Because *a* is positive, the second derivative is positive and the critical point we obtained from the first derivative is a minimum point. Had *a* been negative, the second derivative would have been negative and a maximum would have occurred at the critical point.

Example 2: A manufacturing company has determined that the total cost of producing an item can be determined from the equation $C = 8x^2 - 176x + 1800$, where *x* is the number of units that the company makes. How many units should the company manufacture in order to minimize the cost?

Once again, take the derivative of the cost equation and set it equal to zero.

$$\frac{dC}{dx} = 16x - 176 = 0$$

$$x = 11$$

This tells us that 11 is a critical point of the equation. Now we need to figure out if this is a maximum or a minimum using the second derivative.

$$\frac{d^2C}{dx^2} = 16$$

Because 16 is always positive, any critical value is going to be a minimum. Therefore, the company should manufacture 11 units in order to minimize its cost.

Example 3: A rocket is fired into the air, and its height in meters at any given time *t* can be calculated using the formula $h(t) = 1600 + 196t - 4.9t^2$. Find the maximum height of the rocket and the time at which it occurs.

Take the derivative and set it equal to zero.

$$\frac{dh}{dt} = 196 - 9.8t$$

$$t = 20$$

Now that we know 20 is a critical point of the equation, use the second derivative test.

$$\frac{d^2h}{dt^2} = -9.8$$

This is always negative, so any critical value is a maximum. To determine the maximum height of the rocket, plug *t* = 20 into the equation.

$$h(20) = 1,600 + 196(20) - 4.9(20^2) = 3,560 \text{ meters}$$

The technique is always the same: (a) take the derivative of the equation; (b) set it equal to zero; and (c) use the second derivative test.

The hardest part of these word problems is when you have to set up the equation yourself. The following is a classic AP problem:

Example 4: Max wants to make a box with no lid from a rectangular sheet of cardboard that is 18 inches by 24 inches. The box is to be made by cutting a square of side x from each corner of the sheet and folding up the sides (see figure below). Find the value of x that maximizes the volume of the box.

After we cut out the squares of side x and fold up the sides, the dimensions of the box will be

width: $18 - 2x$
length: $24 - 2x$
depth: x

Using the formula for the volume of a rectangular prism, we can get an equation for the volume in terms of x.

$$V = x(18 - 2x)(24 - 2x)$$

Multiply the terms together (and be careful with your algebra).

$$V = x(18 - 2x)(24 - 2x) = 4x^3 - 84x^2 + 432x$$

Now take the derivative.

$$\frac{dV}{dx} = 12x^2 - 168x + 432$$

Set the derivative equal to zero, and solve for x.

$$12x^2 - 168x + 432 = 0$$

$$x^2 - 14x + 36 = 0$$

$$x = \frac{14 \pm \sqrt{196 - 144}}{2} = 7 \pm \sqrt{13} \approx 3.4, 10.6$$

At the end of the day, no matter how complex the math might get, if a problem is based on a real world example, like this cardboard box, then the answer will make sense in reality.

Common sense tells us that you can't cut out two square pieces that measure 10.6 inches to a side (the sheet's only 18 inches wide!), so the maximizing value has to be 3.4 inches. Here's the second derivative test, just to be sure.

$$\frac{d^2V}{dx^2} = 24x - 168$$

At $x = 3.4$,

$$\frac{d^2V}{dx^2} = -86.4$$

So, the volume of the box will be maximized when $x = 3.4$.

Therefore, the dimensions of the box that maximize the volume are approximately: 11.2 in. × 17.2 in. × 3.4 in.

Sometimes, particularly when the domain of a function is restricted, you have to test the endpoints of the interval as well. This is because the highest or lowest value of a function may be at an endpoint of that interval; the critical value you obtained from the derivative might be just a local maximum or minimum. For the purposes of the AP Exam, however, endpoints are considered separate from critical values.

Example 5: Find the absolute maximum and minimum values of $y = x^3 - x$ on the interval $[-3, 3]$.

Take the derivative and set it equal to zero.

$$\frac{dy}{dx} = 3x^2 - 1 = 0$$

Solve for *x*.

$$x = \pm \frac{1}{\sqrt{3}}$$

Test the critical points.

$$\frac{d^2 y}{dx^2} = 6x$$

At $x = \frac{1}{\sqrt{3}}$, we have a minimum. At $x = -\frac{1}{\sqrt{3}}$, we have a maximum.

$$\text{At } x = -\frac{1}{\sqrt{3}}, \ y = -\frac{1}{3\sqrt{3}} + \frac{1}{\sqrt{3}} = \frac{2}{3\sqrt{3}} \approx 0.385$$

$$\text{At } x = \frac{1}{\sqrt{3}}, \ y = \frac{1}{3\sqrt{3}} - \frac{1}{\sqrt{3}} = -\frac{2}{3\sqrt{3}} \approx -0.385$$

Now it's time to check the endpoints of the interval.

At *x* = –3, *y* = –24

At *x* = 3, *y* = 24

We can see that the function actually has a *lower* value at *x* = –3 than at its "minimum" when $x = \frac{1}{\sqrt{3}}$. Similarly, the function has a *higher* value at *x* = 3 than at its "maximum" of $x = -\frac{1}{\sqrt{3}}$. This means that the function has a "local minimum" at $x = \frac{1}{\sqrt{3}}$ and an "absolute minimum" when *x* = –3. And, the function has a "local maximum" at $x = -\frac{1}{\sqrt{3}}$ and an "absolute maximum" at *x* = 3.

Example 6: A rectangle is to be inscribed in a semicircle with radius 4, with one side on the semicircle's diameter. What is the largest area this rectangle can have?

Let's look at this on the coordinate axes. The equation for a circle of radius 4, centered at the origin, is $x^2 + y^2 = 16$; a semicircle has the equation $y = \sqrt{16 - x^2}$. Our rectangle can then be expressed as a function of x, where the height is $\sqrt{16 - x^2}$ and the base is $2x$. See the following figure:

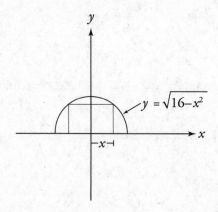

The area of the rectangle is: $A = 2x\sqrt{16 - x^2}$. Let's take the derivative of the area.

$$\frac{dA}{dx} = 2\sqrt{16 - x^2} - \frac{2x^2}{\sqrt{16 - x^2}}$$

The derivative is not defined at $x = \pm 4$. Setting the derivative equal to zero, we get

$$2\sqrt{16 - x^2} - \frac{2x^2}{\sqrt{16 - x^2}} = 0$$

$$2\sqrt{16 - x^2} = \frac{2x^2}{\sqrt{16 - x^2}}$$

$$2\left(16 - x^2\right) = 2x^2$$

$$32 - 2x^2 = 2x^2$$

$$32 = 4x^2$$

$$x = \pm\sqrt{8}$$

> If you're wondering why we don't use the negative root, it's because there is no such thing as a negative area.

Note that the domain of this function is $-4 \leq x \leq 4$, so these numbers serve as endpoints of the interval. Let's compare the critical values and the endpoints.

When $x = -4$, $y = 0$ and the area is 0.

When $x = 4$, $y = 0$ and the area is 0.

When $x = \sqrt{8}$, $y = \sqrt{8}$ and the area is 16.

Thus, the maximum area occurs when $x = \sqrt{8}$ and the area equals 16.

Try some of these solved problems on your own. As always, cover the answers as you work.

PROBLEM 1. A rectangular field, bounded on one side by a building, is to be fenced in on the other three sides. If 3,000 feet of fence is to be used, find the dimensions of the largest field that can be fenced in.

Answer: First, let's make a rough sketch of the situation.

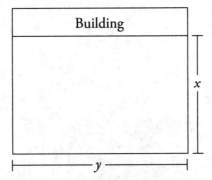

If we call the length of the field y and the width of the field x, the formula for the area of the field becomes

$$A = xy$$

The perimeter of the fencing is equal to the sum of two widths and the length.

$$2x + y = 3,000$$

Now solve this second equation for y.

$$y = 3,000 - 2x$$

When you plug this expression into the formula for the area, you get a formula for A in terms of x.

$$A = x(3,000 - 2x) = 3,000x - 2x^2$$

Next, take the derivative, set it equal to zero, and solve for x.

$$\frac{dA}{dx} = 3,000 - 4x = 0$$

$$x = 750$$

Let's check to make sure it's a maximum. Find the second derivative.

$$\frac{d^2 A}{dx^2} = -4$$

Because we have a negative result, $x = 750$ is a maximum. Finally, if we plug in $x = 750$ and solve for y, we find that $y = 1,500$. The largest field will measure 750 feet by 1,500 feet.

PROBLEM 2. A poster is to contain 100 square inches of picture surrounded by a 4-inch margin at the top and bottom and a 2-inch margin on each side. Find the overall dimensions that will minimize the total area of the poster.

Answer: First, make a sketch.

Let the area of the picture be $xy = 100$. The total area of the poster is $A = (x + 4)(y + 8)$. Then, expand the equation.

$$A = xy + 4y + 8x + 32$$

Substitute $xy = 100$ and $y = \dfrac{100}{x}$ into the area equation, and we get

$$A = 132 + \frac{400}{x} + 8x$$

Now take the derivative and set it equal to zero.

$$\frac{dA}{dx} = 8 - \frac{400}{x^2} = 0$$

Solving for x, we find that $x = \sqrt{50}$. Now solve for y by plugging $x = \sqrt{50}$ into the area equation: $y = 2\sqrt{50}$. Then check that these dimensions give us a minimum.

$$\frac{d^2 A}{dx^2} = \frac{800}{x^3}$$

This is positive when x is positive, so the minimum area occurs when $x = \sqrt{50}$. Thus, the overall dimensions of the poster are $4 + \sqrt{50}$ inches by $8 + 2\sqrt{50}$ inches.

PROBLEM 3. An open-top box with a square bottom and rectangular sides is to have a volume of 256 cubic inches. Find the dimensions that require the minimum amount of material.

Answer: First, make a sketch of the situation.

The amount of material necessary to make the box is equal to the surface area.

$$S = x^2 + 4xy$$

The formula for the volume of the box is $x^2 y = 256$.

If we solve the latter equation for y, $y = \dfrac{256}{x^2}$, and plug it into the former equation, we get

$$S = x^2 + 4x\frac{256}{x^2} = x^2 + \frac{1,024}{x}$$

Now take the derivative and set it equal to zero.

$$\frac{dS}{dx} = 2x - \frac{1,024}{x^2} = 0$$

If we solve this for x, we get $x^3 = 512$ and $x = 8$. Solving for y, we get $y = 4$.

Check that these dimensions give us a minimum.

$$\frac{d^2A}{dx^2} = 2 + \frac{2,048}{x^3}$$

This is positive when x is positive, so the minimum surface area occurs when $x = 8$. The dimensions of the box should be 8 inches by 8 inches by 4 inches.

PROBLEM 4. Find the point on the curve $y = \sqrt{x}$ that is a minimum distance from the point $(4, 0)$.

Answer: First, make that sketch.

Using the distance formula, we get

$$D^2 = (x - 4)^2 + (y - 0)^2 = x^2 - 8x + 16 + y^2$$

Because $y = \sqrt{x}$,

$$D^2 = x^2 - 8x + 16 + x = x^2 - 7x + 16$$

Next, let $L = D^2$. We can do this because the minimum value of D^2 will occur at the same value of x as the minimum value of D. Therefore, it's simpler to minimize D^2 rather than D (because we won't have to take a square root!).

$$L = x^2 - 7x + 16$$

Now, take the derivative and set it equal to zero.

$$\frac{dL}{dx} = 2x - 7 = 0$$

$$x = \frac{7}{2}$$

Solving for y, we get $y = \sqrt{\dfrac{7}{2}}$.

Finally, because $\dfrac{d^2L}{dx^2} = 2$, the point $\left(\dfrac{7}{2}, \sqrt{\dfrac{7}{2}}\right)$ is the minimum distance from the point (4, 0).

PRACTICE PROBLEM SET 10

Now try these problems on your own. The answers are in Chapter 19.

1. A rectangle has its base on the x-axis and its two upper corners on the parabola $y = 12 - x^2$. What is the largest possible area of the rectangle?

2. An open rectangular box is to be made from a 9 × 12 inch piece of tin by cutting squares of side x inches from the corners and folding up the sides. What should x be to maximize the volume of the box?

3. A 384-square-meter plot of land is to be enclosed by a fence and divided into two equal parts by another fence parallel to one pair of sides. What dimensions of the outer rectangle will minimize the amount of fence used?

4. What is the radius of a cylindrical soda can with volume of 512 cubic inches that will use the minimum material?

5. A swimmer is at a point 500 m from the closest point on a straight shoreline. She needs to reach a cottage located 1,800 m down shore from the closest point. If she swims at 4 m/s and she walks at 6 m/s, how far from the cottage should she come ashore so as to arrive at the cottage in the shortest time?

6. Find the closest point on the curve $x^2 + y^2 = 1$ to the point (2, 1).

7. A window consists of an open rectangle topped by a semicircle and is to have a perimeter of 288 inches. Find the radius of the semicircle that will maximize the area of the window.

8. The range of a projectile is $R = \dfrac{v_0^2 \sin 2\theta}{g}$, where v_0 is its initial velocity, g is the acceleration due to gravity and is a constant, and θ is its firing angle. Find the angle that maximizes the projectile's range.

9. A box with a square base, vertical sides, and an open top contains 500 ft^3 of material. Find the dimensions of the base that minimize the surface area.

10. Where on the curve $y = \dfrac{1}{1 + x^2}$ does the tangent line have the greatest slope?

CURVE SKETCHING

Another topic on which students spend a lot of time in calculus is curve sketching. In the old days, whole courses (called "Analytic Geometry") were devoted to the subject, and students had to master a wide variety of techniques to learn how to sketch a curve accurately.

Fortunately (or unfortunately, depending on your point of view), students no longer need to be as good at analytic geometry. There are two reasons for this: (1) The AP Exam tests only a few types of curves; and (2) you can use a graphing calculator. Because of the calculator, you can get an idea of the shape of the curve, and all you need to do is find important points to label the graph. We use calculus to find some of these points.

When it's time to sketch a curve, we'll show you a four-part analysis that'll give you all the information you need.

Step 1: Test the Function

Find where $f(x) = 0$. This tells you the function's x-intercepts (or roots). By setting $x = 0$, we can determine the y-intercepts. Then find any horizontal and/or vertical asymptotes.

Step 2: Test the First Derivative

Find where $f'(x) = 0$. This tells you the critical points. We can determine whether the curve is rising or falling, as well as where the maxima and minima are. It's also possible to determine if the curve has any points where it's nondifferentiable.

Step 3: Test the Second Derivative

Find where $f''(x) = 0$. This shows you where any points of inflection are. (These are points where the graph of a function changes concavity.) Then we can determine where the graph curves upward and where it curves downward.

Step 4: Test End Behavior

Look at what the general shape of the graph will be, based on the values of y for very large values of $\pm x$. Using this analysis, we can always come up with a sketch of a curve.

And now, the rules.

> (1)   When $f'(x) > 0$, the curve is rising; when $f'(x) < 0$, the curve is falling; when $f'(x) = 0$, the curve is at a critical point.

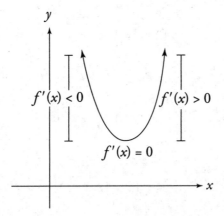

> (2)   When $f''(x) > 0$, the curve is "concave up"; when $f''(x) < 0$, the curve is "concave down"; and when $f''(x) = 0$, the curve is at a point of inflection.

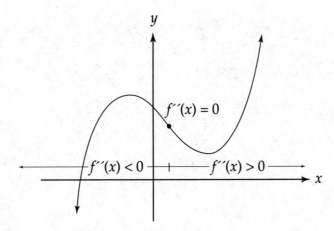

> (3)   The y-coordinates of each critical point are found by plugging the x-value into the original equation.

As always, this stuff will sink in better if we try a few examples.

Example 1: Sketch the equation $y = x^3 - 12x$.

Step 1: Find the x-intercepts.

$$x^3 - 12x = 0$$

$$x(x^2 - 12) = 0$$

$$x\left(x - \sqrt{12}\right)\left(x + \sqrt{12}\right) = 0$$

$$x = 0, \pm\sqrt{12}$$

The curve has x-intercepts at $\left(\sqrt{12},\ 0\right), \left(-\sqrt{12},\ 0\right)$, and $(0, 0)$.

Next, find the y-intercepts.

$$y = (0)^3 - 12(0) = 0$$

The curve has a y-intercept at $(0, 0)$.

There are no asymptotes, because there's no place where the curve is undefined (you won't have asymptotes for curves that are polynomials).

Step 2: Take the derivative of the function to find the critical points.

$$\frac{dy}{dx} = 3x^2 - 12$$

Set the derivative equal to zero, and solve for x.

$$3x^2 - 12 = 0$$

$$3(x^2 - 4) = 0$$

$$3(x - 2)(x + 2) = 0$$

so $x = 2, -2$.

Next, plug $x = 2, -2$ into the original equation to find the y-coordinates of the critical points.

$$y = (2)^3 - 12(2) = -16$$

$$y = (-2)^3 - 12(-2) = 16$$

Thus, we have critical points at $(2, -16)$ and $(-2, 16)$.

Step 3: Now, take the second derivative to find any points of inflection.

$$\frac{d^2 y}{dx^2} = 6x$$

This equals zero at $x = 0$. We already know that when $x = 0$, $y = 0$, so the curve has a point of inflection at $(0, 0)$.

Now, plug the critical values into the second derivative to determine whether each is a maximum or a minimum. $f''(2) = 6(2) = 12$. This is positive, so the curve has a minimum at $(2, -16)$, and the curve is concave up at that point. $f''(-2) = 6(-2) = -12$. This value is negative, so the curve has a maximum at $(-2, 16)$ and the curve is concave down there.

Armed with this information, we can now plot the graph.

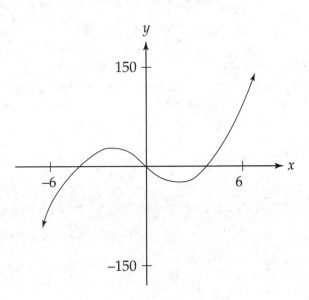

Example 2: Sketch the graph of $y = x^4 + 2x^3 - 2x^2 + 1$.

Step 1: First, let's find the x-intercepts.

$$x^4 + 2x^3 - 2x^2 + 1 = 0$$

If the equation doesn't factor easily, it's best not to bother to find the function's roots. Convenient, huh?

Next, let's find the y-intercepts.

$$y = (0)^4 + 2(0)^3 - 2(0)^2 + 1$$

> The good news is that if the roots aren't easy to find, ETS won't ask you to find them, or you can find them with your calculator.

The curve has a y-intercept at $(0, 1)$.

There are no vertical asymptotes because there is no place where the curve is undefined.

Step 2: Now we take the derivative to find the critical points.

$$\frac{dy}{dx} = 4x^3 + 6x^2 - 4x$$

Set the derivative equal to zero.

$$4x^3 + 6x^2 - 4x = 0$$

$$2x(2x^2 + 3x - 2) = 0$$

$$2x(2x - 1)(x + 2) = 0$$

$$x = 0, \frac{1}{2}, -2$$

Next, plug these three values into the original equation to find the y-coordinates of the critical points. We already know that when $x = 0$, $y = 1$.

$$\text{When} \quad x = \frac{1}{2}, y = \left(\frac{1}{2}\right)^4 + 2\left(\frac{1}{2}\right)^3 - 2\left(\frac{1}{2}\right)^2 + 1 = \frac{13}{16}$$

$$\text{When } x = -2, y = (-2)^4 + 2(-2)^3 - 2(-2)^2 + 1 = -7$$

Thus, we have critical points at $(0, 1)$, $\left(\frac{1}{2}, \frac{13}{16}\right)$, and $\left(-2, -7\right)$.

Step 3: Take the second derivative to find any points of inflection.

$$\frac{d^2y}{dx^2} = 12x^2 + 12x - 4$$

Set this equal to zero.

$$12x^2 + 12x - 4 = 0$$

$$3x^2 + 3x - 1 = 0$$

$$x = \frac{-3 \pm \sqrt{21}}{6} \approx 0.26, -1.26$$

Therefore, the curve has points of inflection at $x = \dfrac{-3 \pm \sqrt{21}}{6}$.

Now solve for the y-coordinates.

$$(0.26, 0.90) \text{ and } (-1.26, -3.66)$$

We can now plug the critical values into the second derivative to determine whether each is a maximum or a minimum.

$$12(0)^2 + 12(0) - 4 = -4$$

This is negative, so the curve has a maximum at $(0, 1)$; the curve is concave down there.

$$12\left(\frac{1}{2}\right)^2 + 12\left(\frac{1}{2}\right) - 4 = 5$$

This is positive, so the curve has a minimum at $\left(\dfrac{1}{2}, \dfrac{13}{16}\right)$; the curve is concave up there.

$$12(-2)^2 + 12(-2) - 4 = 20$$

This is positive, so the curve has a minimum at $(-2, -7)$ and the curve is also concave up there.

We can now plot the graph.

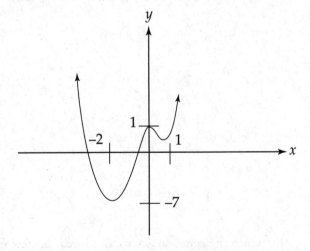

Finding a Cusp

If the derivative of a function approaches ∞ from one side of a point and $-\infty$ from the other, and if the function is continuous at that point, then the curve has a "cusp" at that point. In order to find a cusp, you need to look at points where the first derivative is undefined, as well as where it's zero.

Example 3: Sketch the graph of $y = 2 - x^{\frac{2}{3}}$.

Step 1: Find the x-intercepts.

$$2 - x^{\frac{2}{3}} = 0$$

$$x^{\frac{2}{3}} = 2 \qquad x = \pm 2^{\frac{3}{2}} = \pm 2\sqrt{2}$$

The x-intercepts are at $\left(\pm 2\sqrt{2},\, 0\right)$.

Next, find the y-intercepts.

$$y = 2 - (0)^{\frac{2}{3}} = 2$$

The curve has a y-intercept at $(0, 2)$.

There are no asymptotes because there is no place where the curve is undefined.

Step 2: Now, take the derivative to find the critical points.

$$\frac{dy}{dx} = -\frac{2}{3}x^{-\frac{1}{3}}$$

What's next? You guessed it! Set the derivative equal to zero.

$$-\frac{2}{3}x^{-\frac{1}{3}} = 0$$

There are no values of x for which the equation is zero. But here's the new stuff to deal with: at $x = 0$, the derivative is undefined. If we look at the limit as x approaches 0 from both sides, we can determine whether the graph has a cusp.

$$\lim_{x \to 0^+} -\frac{2}{3}x^{-\frac{1}{3}} = -\infty \quad \text{and} \quad \lim_{x \to 0^-} -\frac{2}{3}x^{-\frac{1}{3}} = \infty$$

Therefore, the curve has a cusp at (0, 2).

There aren't any other critical points. But we can see that when $x < 0$, the derivative is positive (which means that the curve is rising to the left of zero), and when $x > 0$, the derivative is negative (which means that the curve is falling to the right of zero).

Step 3: Now, we take the second derivative to find any points of inflection.

$$\frac{d^2 y}{dx^2} = \frac{2}{9} x^{-\frac{4}{3}}$$

Again, there's no x-value where this is zero. In fact, the second derivative is positive at all values of x except 0. Therefore, the graph is concave up everywhere.

Now it's time to graph this.

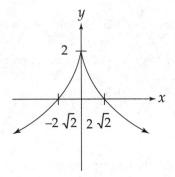

There's one other type of graph you should know about: a rational function. In order to graph a rational function, you need to know how to find that function's asymptotes.

How to Find Asymptotes

A line $y = c$ is a horizontal asymptote of the graph of $y = f(x)$ if

$$\lim_{x \to \infty} f(x) = c \text{ or if } \lim_{x \to -\infty} f(x) = c$$

A line $x = k$ is a vertical asymptote of the graph of $y = f(x)$ if

$$\lim_{x \to k^+} f(x) = \pm\infty \text{ or if } \lim_{x \to k^-} f(x) = \pm\infty$$

Example 4: Sketch the graph of $y = \dfrac{3x}{x+2}$.

Step 1: Find the x-intercepts. A fraction can be equal to zero only when its numerator is equal to zero (provided that the denominator is not also zero there). All we have to do is set $3x = 0$, and you get $x = 0$. Thus, the graph has an x-intercept at $(0, 0)$. Note: This is also the y-intercept.

Next, look for asymptotes. The denominator is undefined at $x = -2$, and if we take the left- and right-hand limits of the function, we see the following:

$$\lim_{x \to -2^+} \frac{3x}{x+2} = -\infty \quad \text{and} \quad \lim_{x \to -2^-} \frac{3x}{x+2} = \infty$$

The curve has a vertical asymptote at $x = -2$.

If we take $\displaystyle\lim_{x \to \infty} \frac{3x}{x+2} = 3$ and $\displaystyle\lim_{x \to -\infty} \frac{3x}{x+2} = 3$, the curve has a horizontal asymptote at $y = 3$.

Step 2: Now, take the derivative to figure out the critical points.

$$\frac{dy}{dx} = \frac{(x+2)(3) - (3x)(1)}{(x+2)^2} = \frac{6}{(x+2)^2}$$

There are no values of x that make the derivative equal to zero. Because the numerator is 6 and the denominator is squared, the derivative will always be positive (the curve is always rising). You should note that the derivative is undefined at $x = -2$, but you already know that there's an asymptote at $x = -2$, so you don't need to examine this point further.

Step 3: Now, it's time for the second derivative.

$$\frac{d^2 y}{dx^2} = \frac{-12}{(x+2)^3}$$

This is never equal to zero. The expression is positive when $x < -2$, so the graph is concave up when $x < -2$. The second derivative is negative when $x > -2$, so it's concave down when $x > -2$.

Now plot the graph.

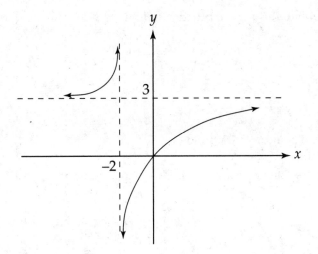

Now it's time to practice some problems. Do each problem, covering the answer first, then check your answer.

PROBLEM 1. Sketch the graph of $y = x^3 - 9x^2 + 24x - 10$. Plot all extrema, points of inflection, and asymptotes.

Answer: Follow the three steps.

First, see if the x-intercepts are easy to find. This is a cubic equation that isn't easily factored. So skip this step.

Next, find the y-intercepts by setting $x = 0$.

$$y = (0)^3 - 9(0)^2 + 24(0) - 10 = -10$$

The curve has a y-intercept at $(0, -10)$.

There are no asymptotes, because the curve is a simple polynomial.

Next, find the critical points using the first derivative.

$$\frac{dy}{dx} = 3x^2 - 18x + 24$$

Set the derivative equal to zero and solve for x.

$$3x^2 - 18x + 24 = 0$$

$$3(x^2 - 6x + 8) = 0$$

$$3(x - 4)(x - 2) = 0$$

$$x = 2, 4$$

Plug $x = 2$ and $x = 4$ into the original equation to find the y-coordinates of the critical points.

When $x = 2$, $y = 10$

When $x = 4$, $y = 6$

Thus, we have critical points at $(2, 10)$ and $(4, 6)$.

In our third step, the second derivative indicates any points of inflection.

$$\frac{d^2 y}{dx^2} = 6x - 18$$

This equals zero at $x = 3$.

Next, plug $x = 3$ into the original equation to find the y-coordinates of the point of inflection, which is at $(3, 8)$. Plug the critical values into the second derivative to determine whether each is a maximum or a minimum.

$$6(2) - 18 = -6$$

This is negative, so the curve has a maximum at $(2, 10)$, and the curve is concave down there.

$$6(4) - 18 = 6$$

This is positive, so the curve has a minimum at $(4, 6)$, and the curve is concave up there.

It's graph-plotting time.

PROBLEM 2. Sketch the graph of $y = 8x^2 - 16x^4$. Plot all extrema, points of inflection, and asymptotes.

Answer: Factor the polynomial.

$$8x^2\left(1 - 2x^2\right) = 0$$

Solving for x, we get $x = 0$ (a double root), $x = \dfrac{1}{\sqrt{2}}$, and $x = -\dfrac{1}{\sqrt{2}}$.

Find the y-intercepts: when $x = 0$, $y = 0$.

There are no asymptotes, because the curve is a simple polynomial.

Find the critical points using the first derivative.

$$\frac{dy}{dx} = 16x - 64x^3$$

Set the derivative equal to zero and solve for x. You get $x = 0$, $x = \dfrac{1}{2}$, and $x = -\dfrac{1}{2}$.

Next, plug $x = 0$, $x = \dfrac{1}{2}$, and $x = -\dfrac{1}{2}$ into the original equation to find the y-coordinates of the critical points.

When $x = 0$, $y = 0$

When $x = \dfrac{1}{2}$, $y = 1$

When $x = -\dfrac{1}{2}$, $y = 1$

Thus, there are critical points at $(0, 0)$, $\left(\dfrac{1}{2}, 1\right)$, and $\left(-\dfrac{1}{2}, 1\right)$.

Take the second derivative to find any points of inflection.

$$\frac{d^2 y}{dx^2} = 16 - 192x^2$$

This equals zero at $x = \dfrac{1}{\sqrt{12}}$ and $x = -\dfrac{1}{\sqrt{12}}$.

Next, plug $x = \dfrac{1}{\sqrt{12}}$ and $x = -\dfrac{1}{\sqrt{12}}$ into the original equation to find the y-coordinates of the points of inflection, which are at $\left(\dfrac{1}{\sqrt{12}}, \dfrac{5}{9} \right)$ and $\left(-\dfrac{1}{\sqrt{12}}, \dfrac{5}{9} \right)$. Now determine whether the points are maxima or minima.

At $x = 0$, we have a minimum; the curve is concave up there.

At $x = \dfrac{1}{2}$, it's a maximum, and the curve is concave down.

At $x = -\dfrac{1}{2}$, it's also a maximum (still concave down).

Now plot.

PROBLEM 3. Sketch the graph of $y = \left(\dfrac{x-4}{x+3} \right)^{2}$. Plot all extrema, points of inflection, and asymptotes.

Answer: This should seem rather routine by now.

Find the x-intercepts by setting the numerator equal to zero; $x = 4$. The graph has an x-intercept at $(4, 0)$. (It's a double root.)

Next, find the y-intercept by plugging in $x = 0$.

$$y = \frac{16}{9}$$

The denominator is undefined at $x = -3$, so there's a vertical asymptote at that point.

Look at the limits.

$$\lim_{x \to \infty} \left(\frac{x-4}{x+3} \right)^2 = 1 \text{ and } \lim_{x \to -\infty} \left(\frac{x-4}{x+3} \right)^2 = 1$$

The curve has a horizontal asymptote at $y = 1$.

It's time for the first derivative.

$$\frac{dy}{dx} = 2 \left(\frac{x-4}{x+3} \right) \frac{(x+3)(1) - (x-4)(1)}{(x+3)^2} = \frac{14x - 56}{(x+3)^3}$$

The derivative is zero when $x = 4$, and the derivative is undefined at $x = -3$. (There's an asymptote there, so we can ignore the point. If the curve were *defined* at $x = -3$, then it would be a critical point, as you'll see in the next example.)

Now for the second derivative.

$$\frac{d^2y}{dx^2} = \frac{(x+3)^3(14) - (14x - 56)3(x+3)^2}{(x+3)^6} = \frac{-28x + 210}{(x+3)^4}$$

This is zero when $x = \frac{15}{2}$. The second derivative is positive (and the graph is concave up) when $x < \frac{15}{2}$, and it's negative (and the graph is concave down) when $x > \frac{15}{2}$.

We can now plug $x = 4$ into the second derivative. It's positive there, so $(4, 0)$ is a minimum.

Your graph should look like the following:

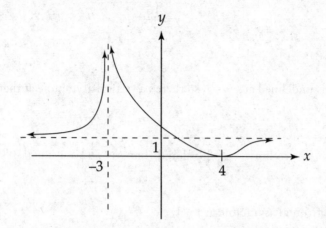

PROBLEM 4. Sketch the graph of $y = (x-4)^{\frac{2}{3}}$. Plot all extrema, points of inflection, and asymptotes.

Answer: By inspection, the x-intercept is at $x = 4$.

Next, find the y-intercepts. When $x = 0$, $y = \sqrt[3]{16} \approx 2.52$.

No asymptotes exist because there's no place where the curve is undefined.

The first derivative is

$$\frac{dy}{dx} = \frac{2}{3}(x-4)^{-\frac{1}{3}}$$

Set it equal to zero.

$$\frac{2}{3}(x-4)^{-\frac{1}{3}} = 0$$

This can never equal zero. But, at $x = 4$ the derivative is undefined, so this is a critical point. If you look at the limit as x approaches 4 from both sides, you can see if there's a cusp.

$$\lim_{x \to 4^+} \frac{2}{3}(x-4)^{-\frac{1}{3}} = \infty \text{ and } \lim_{x \to 4^-} \frac{2}{3}(x-4)^{-\frac{1}{3}} = -\infty$$

The curve has a cusp at (4, 0).

There were no other critical points. But, we can see that when $x > 4$, the derivative is positive and the curve is rising; when $x < 4$ the derivative is negative, and the curve is falling.

The second derivative is

$$\frac{d^2y}{dx^2} = -\frac{2}{9}(x-4)^{-\frac{4}{3}}$$

No value of x can set this equal to zero. In fact, the second derivative is negative at all values of x except 4. Therefore, the graph is concave down everywhere.

Your graph should look like the following:

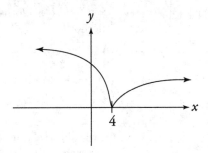

PRACTICE PROBLEM SET 11

It's time for you to try some of these on your own. Sketch each of the graphs below and check the answers in Chapter 19.

1. $y = x^3 - 9x - 6$

2. $y = -x^3 - 6x^2 - 9x - 4$

3. $y = \left(x^2 - 4\right)\left(9 - x^2\right)$

4. $y = \dfrac{x - 3}{x + 8}$

5. $y = \dfrac{x^2 - 4}{x - 3}$

6. $y = 3 + x^{\frac{2}{3}}$

7. $y = x^{\frac{2}{3}}\left(3 - 2x^{\frac{1}{3}}\right)$

End of Chapter 9 Drill

The answers are in Chapter 20.

1. A rectangle has its base on the x-axis and its two upper corners on the semicircle $y = \sqrt{200 - x^2}$. Find the value of x that maximizes the area of the rectangle.

    (A) 0
    (B) 5
    (C) 10
    (D) $\sqrt{200}$

2. An open rectangular box with a square base is to have a volume of 108 in³. Find the dimensions of the box that minimize its surface area (base by height).

    (A) 3 in. by 6 in.
    (B) 6 in. by 6 in.
    (C) 3 in. by 3 in.
    (D) 6 in. by 3 in.

3. Find the coordinates of the maximum and minimum (if they exist) of $y = x^3 - 3x^2 - 24x + 6$.

	Minimum	Maximum
(A)	(−4, −10)	(2, 46)
(B)	(4, −74)	(−2, 34)
(C)	(1, −20)	None
(D)	(1, −20)	(2, 46)

4. At what value(s) of x does the curve $y = x^3 - 12x^2 + 120$ have a point (or points) of inflection?

    (A) (4, −8)
    (B) (0, 120)
    (C) (8, −136)
    (D) (−4, −136)

5. On what interval is the curve $y = \sin x + \sqrt{3} \cos x$ decreasing on $[0, \pi]$?

    (A) $\left[0, \dfrac{\pi}{6}\right]$

    (B) $\left(\dfrac{\pi}{6}, \pi\right)$

    (C) $\left(\dfrac{\pi}{2}, 1\right)$

    (D) The function is always increasing.

REFLECT

Respond to the following questions:

- For which topics discussed in this chapter do you feel you have achieved sufficient mastery to answer multiple-choice questions correctly?

- For which topics discussed in this chapter do you feel you have achieved sufficient mastery to answer open-ended questions correctly?

- For which topics discussed in this chapter do you feel you need more work before you can answer multiple-choice questions correctly?

- For which topics discussed in this chapter do you feel you need more work before you can answer open-ended questions correctly?

- What parts of this chapter are you going to re-review?

Chapter 10
Motion

This chapter deals with two different types of word problems that involve motion: related rates and the relationship between velocity and acceleration of a particle. The subject matter might seem arcane, but once you get the hang of them, you'll see that these aren't so hard, either. Besides, the AP Exam tests only a few basic problem types.

RELATED RATES

The idea behind these problems is very simple. In a typical problem, you'll be given an equation relating two or more variables. These variables will change with respect to time, and you'll use derivatives to determine how the rates of change are related. (Hence the name: related rates.) Sounds easy, doesn't it?

Example 1: A circular pool of water is expanding at the rate of 16π in.^{2}/sec. At what rate is the radius expanding when the radius is 4 inches?

Note: The pool is expanding in square inches per second. We've been given the rate that the area is changing, and we need to find the rate of change of the radius. What equation relates the area of a circle to its radius? $A = \pi r^2$.

Step 1: Set up the equation and take the derivative of this equation with respect to t (time).

$$\frac{dA}{dt} = 2\pi r \frac{dr}{dt}$$

In this equation, $\frac{dA}{dt}$ represents the rate at which the area is changing, and $\frac{dr}{dt}$ is the rate at which the radius is changing. The simplest way to explain this is that whenever you have a variable in an equation (r, for example), the derivative with respect to time $\left(\frac{dr}{dt}\right)$ represents the rate at which that variable is increasing or decreasing.

Step 2: Now we can plug in the values for the rate of change of the area and for the radius. (Never plug in the values until after you have taken the derivative or you will get nonsense!)

$$16\pi = 2\pi (4) \frac{dr}{dt}$$

Solving for $\frac{dr}{dt}$, we get

$$16\pi = 8\pi \frac{dr}{dt} \text{ and } \frac{dr}{dt} = 2$$

The radius is changing at a rate of 2 in./sec. It's important to note that this is the rate only when the radius is 4 inches. As the circle gets bigger and bigger, the radius will expand at a slower and slower rate.

Example 2: A 25-foot long ladder is leaning against a wall and sliding toward the floor. If the foot of the ladder is sliding away from the base of the wall at a rate of 15 ft/sec, how fast is the top of the ladder sliding down the wall when the top of the ladder is 7 feet from the ground?

Here's another classic related rates problem. As always, a picture is worth 1,000 words.

You can see that the ladder forms a right triangle with the wall. Let x stand for the distance from the foot of the ladder to the base of the wall, and let y represent the distance from the top of the ladder to the ground. What's our favorite theorem that deals with right triangles? The Pythagorean Theorem tells us here that $x^2 + y^2 = 25^2$. Now we have an equation that relates the variables to each other.

Now take the derivative of the equation with respect to t.

$$2x \frac{dx}{dt} + 2y \frac{dy}{dt} = 0$$

Just plug in what you know and solve. Because we're looking for the rate at which the vertical distance is changing, we're going to solve for $\frac{dy}{dt}$.

Let's see what we know. We're given the rate at which the ladder is sliding away from the wall: $\frac{dx}{dt} = 15$. The distance from the ladder to the top of the wall is 7 feet ($y = 7$). To find x, use the Pythagorean Theorem. If we plug in $y = 7$ to the equation $x^2 + y^2 = 25^2$, $x = 24$.

Now plug all this information into the derivative equation.

$$2(24)(15) + 2(7)\frac{dy}{dt} = 0$$

$$\frac{dy}{dt} = \frac{-360}{7} \frac{\text{feet}}{\text{sec}}$$

Example 3: A spherical balloon is expanding at a rate of 60π in.^{3}/sec. How fast is the surface area of the balloon expanding when the radius of the balloon is 4 inches?

Step 1: You're given the rate at which the volume's expanding, and you know the equation that relates volume to radius. But you have to relate radius to surface area as well, because you have to find the surface area's rate of change. This means that you'll need the equations for volume and surface area of a sphere.

$$V = \frac{4}{3}\pi r^3$$

$$A = 4\pi r^2$$

You're trying to find $\dfrac{dA}{dt}$, but A is given in terms of r, so you have to get $\dfrac{dr}{dt}$ first. Because we know the volume, if we work with the equation that gives us volume in terms of radius, we can find $\dfrac{dr}{dt}$. From there, work with the other equation to find $\dfrac{dA}{dt}$. If we take the derivative of the equation with respect to t, we get $\dfrac{dV}{dt} = 4\pi r^2 \dfrac{dr}{dt}$. Plugging in for $\dfrac{dV}{dt}$ and for r, we get $60\pi = 4\pi(4)^2 \dfrac{dr}{dt}$.

Solving for $\dfrac{dr}{dt}$, we get

$$\frac{dr}{dt} = \frac{15}{16} \text{ in./sec}$$

Step 2: Now we take the derivative of the other equation with respect to t.

$$\frac{dA}{dt} = 8\pi r \frac{dr}{dt}$$

We can plug in for r and $\dfrac{dr}{dt}$ from the previous step to get

$$\frac{dA}{dt} = 8\pi (4)\frac{15}{16} = \frac{480\pi}{16}\frac{\text{in.}^2}{\text{sec}} = 30\pi \text{ in.}^2/\text{sec}$$

One final example.

Example 4: An underground conical tank, standing on its vertex, is being filled with water at the rate of 18π ft^3/min. If the tank has a height of 30 feet and a radius of 15 feet, how fast is the water level rising when the water is 12 feet deep?

This "cone" problem is also typical. The key point to getting these right is knowing that the ratio of the height of a right circular cone to its radius is constant. By telling us that the height of the cone is 30 and the radius is 15, we know that at any level, the height of the water will be twice its radius, or $h = 2r$.

> The volume of a cone is $V = \frac{1}{3}\pi r^2 h$. (You'll learn to derive this formula through integration in Chapter 17.)

You must find the rate at which the water is rising (the height is changing), or $\dfrac{dh}{dt}$.

Therefore, you want to eliminate the radius from the volume. By substituting $\dfrac{h}{2} = r$ into the equation for volume, we get

$$V = \frac{1}{3}\pi \left(\frac{h}{2}\right)^2 h = \frac{\pi h^3}{12}$$

Differentiate both sides with respect to t.

$$\frac{dV}{dt} = \frac{\pi}{12} 3h^2 \frac{dh}{dt}$$

Now we can plug in and solve for $\dfrac{dh}{dt}$.

$$18\pi = \frac{\pi}{12} 3(12)^2 \frac{dh}{dt}$$

$$\frac{dh}{dt} = \frac{1}{2} \text{ feet/min}$$

In order to solve related rates problems, you have to be good at determining relationships between variables. Once you figure that out, the rest is a piece of cake. Many of these problems involve geometric relationships, so review the formulas for the volumes and areas of cones, spheres, boxes, and other solids. Once you get the hang of setting up the problems, you'll see that these problems follow the same predictable patterns. Look through these sample problems.

PROBLEM 1. A circle is increasing in area at the rate of 16π in.^{2}/s. How fast is the radius increasing when the radius is 2 in.?

Answer: Use the expression that relates the area of a circle to its radius: $A = \pi r^2$.

Next, take the derivative of the expression with respect to t.

$$\frac{dA}{dt} = 2\pi r \frac{dr}{dt}$$

Now, plug in $\frac{dA}{dt} = 16\pi$ and $r = 2$.

$$16\pi = 2\pi(2)\frac{dr}{dt}$$

When you solve for $\frac{dr}{dt}$, you'll get $\frac{dr}{dt} = 4$ in./sec.

Notice that velocity is given in miles per hour and the answer asks for radians per second. In situations like this one, you have to be sure to convert the units properly, or you'll get nailed.

PROBLEM 2. A rocket is rising vertically at a rate of 5,400 miles per hour. An observer on the ground is standing 20 miles from the rocket's launch point. How fast (in radians per second) is the angle of elevation between the ground and the observer's line of sight of the rocket increasing when the rocket is at an elevation of 40 miles?

Answer: First, draw a picture.

Now, find the equation that relates the angle of elevation to the rocket's altitude.

$$\tan\theta = \frac{h}{20}$$

If we take the derivative of both sides of this expression with respect to t, we get

$$\sec^2\theta\,\frac{d\theta}{dt} = \frac{1}{20}\frac{dh}{dt}$$

We know that $\frac{dh}{dt} = 5{,}400$ miles per hour, but the problem asks for time in seconds, so we need to convert this number. There are 3,600 seconds in an hour, so $\frac{dh}{dt} = \frac{3}{2}$ miles per second. Next, we know that $\tan\theta = \frac{h}{20}$, so when $h = 40$, $\tan\theta = 2$. Because $1 + \tan^2\theta = \sec^2\theta$, we get $\sec^2\theta = 5$.

Plug in the following information:

$$5\frac{d\theta}{dt} = \frac{1}{20}\left(\frac{3}{2}\right) \quad \text{and} \quad \frac{d\theta}{dt} = \frac{3}{200} \text{ radians per second}$$

PRACTICE PROBLEM SET 12

Now try these problems on your own. The answers are in Chapter 19.

1. Oil spilled from a tanker spreads in a circle whose circumference increases at a rate of 40 ft/sec. How fast is the area of the spill increasing when the circumference of the circle is 100π feet?

2. A spherical balloon is inflating at a rate of 27π in.³/sec. How fast is the radius of the balloon increasing when the radius is 3 inches?

3. Cars A and B leave a town at the same time. Car A heads due south at a rate of 80 km/hr and car B heads due west at a rate of 60 km/hr. How fast is the distance between the cars increasing after three hours?

4. The sides of an equilateral triangle are increasing at the rate of 27 in./sec. How fast is the triangle's area increasing when the sides of the triangle are each 18 inches long?

5. An inverted conical container has a diameter of 42 inches and a depth of 15 inches. If water is flowing out of the vertex of the container at a rate of 35π in.³/sec, how fast is the depth of the water dropping when the height is 5 inches?

6. A boat is being pulled toward a dock by a rope attached to its bow through a pulley on the dock 7 feet above the bow. If the rope is hauled in at a rate of 4 ft/sec, how fast is the boat approaching the dock when 25 feet of rope is out?

7. A 6-foot-tall woman is walking at the rate of 4 ft/sec away from a street lamp that is 24 feet tall. How fast is the length of her shadow changing?

8. The minute hand of a clock is 6 inches long. Starting from noon, how fast is the area of the sector swept out by the minute hand increasing in in.²/min at any instant?

POSITION, VELOCITY, AND ACCELERATION

Almost every AP Exam has a question on position, velocity, or acceleration. It's one of the traditional areas of physics where calculus comes in handy. Some of these problems require the use of integral calculus, which we won't talk about until the second half of this book. So this unit is divided in half; you'll see the other half later.

If you have a function that gives you the position of an object (usually called a "particle") at a specified time, then the derivative of that function with respect to time is the velocity of the object, and the second derivative is the acceleration. These are usually represented by the following:

Position: $x(t)$ or sometimes $s(t)$

Velocity: $v(t)$, which is $x'(t)$

Acceleration: $a(t)$, which is $x''(t)$ or $v'(t)$

Please note that these equations are usually functions of time (t). Typically, t is greater than zero, but it doesn't have to be.

By the way, speed is the absolute value of velocity.

Example 1: If the position of a particle at a time t is given by the equation $x(t) = t^3 - 11t^2 + 24t$, find the velocity and the acceleration of the particle at time $t = 5$.

First, take the derivative of $x(t)$.

$$x'(t) = 3t^2 - 22t + 24 = v(t)$$

Second, plug in $t = 5$ to find the velocity at that time.

$$v(5) = 3(5^2) - 22(5) + 24 = -11$$

Third, take the derivative of $v(t)$ to find $a(t)$.

$$v'(t) = 6t - 22 = a(t)$$

Finally, plug in $t = 5$ to find the acceleration at that time.

$$a(5) = 6(5) - 22 = 8$$

See the negative velocity? The sign of the velocity is important because it indicates the direction of the particle. Make sure that you know the following:

> When the velocity is negative, the particle is moving to the left.
>
> When the velocity is positive, the particle is moving to the right.
>
> When the velocity and acceleration of the particle have the same signs, the particle's speed is increasing.
>
> When the velocity and acceleration of the particle have opposite signs, the particle's speed is decreasing (or slowing down).
>
> When the velocity is zero and the acceleration is not zero, the particle is momentarily stopped and changing direction.

Example 2: If the position of a particle is given by $x(t) = t^3 - 12t^2 + 36t + 18$, where $t > 0$, find the point at which the particle changes direction.

The derivative is

$$x'(t) = v(t) = 3t^2 - 24t + 36$$

Set it equal to zero and solve for t.

$$x'(t) = 3t^2 - 24t + 36 = 0$$

$$t^2 - 8t + 12 = 0$$

$$(t - 2)(t - 6) = 0$$

So we know that $t = 2$ or $t = 6$.

You need to check that the acceleration is not 0: $x''(t) = 6t - 24$. This equals 0 at $t = 4$. Therefore, the particle is changing direction at $t = 2$ and $t = 6$.

Example 3: Given the same position function as in Example 2, find the interval of time during which the particle is slowing down.

When $0 < t < 2$ and $t > 6$, the particle's velocity is positive; when $2 < t < 6$, the particle's velocity is negative. You can verify this by graphing the function and seeing when it's above or below the x-axis. Or, try some points in the regions between the roots and outside the roots. Now, we need to determine the same information about the acceleration.

$$a(t) = v'(t) = 6t - 24$$

So the acceleration will be negative when $t < 4$, and positive when $t > 4$.

So we have

Time	Velocity	Acceleration
$0 < t < 2$	Positive	Negative
$2 < t < 4$	Negative	Negative
$4 < t < 6$	Negative	Positive
$t > 6$	Positive	Positive

Whenever the velocity and acceleration have opposite signs, the particle is slowing down. Here the particle is slowing down during the first two seconds ($0 < t < 2$) and between the fourth and sixth seconds ($4 < t < 6$).

Another typical question you'll be asked is to find the distance a particle has traveled from one time to another. This is the distance that the particle has covered without regard to the sign, not just the displacement. In other words, if the particle had an odometer on it, what would it read? Usually, all you have to do is plug the two times into the position function and find the difference.

Example 4: How far does a particle travel between the eighth and tenth seconds if its position function is $x(t) = t^2 - 6t$?

Find $x(10) - x(8) = (100 - 60) - (64 - 48) = 24$.

Be careful about one very important thing: **if the velocity changes sign during the problem's time interval**, you'll get the wrong answer if you simply follow the method in the paragraph above. For example, suppose we had the same position function as above but we wanted to find the distance that the particle travels from $t = 2$ to $t = 4$.

$$x(4) - x(2) = (-8) - (-8) = 0$$

This is wrong. The particle travels from −8 back to −8, but it hasn't stood still. To fix this problem, divide the time interval into the time when the velocity is negative and the time when the velocity is positive, and add the absolute values of each distance. Here the velocity is $v(t) = 2t - 6$. The velocity is negative when $t < 3$ and positive when $t > 3$. So we find the absolute value of the distance traveled from $t = 2$ to $t = 3$, and add to that the absolute value of the distance traveled from $t = 3$ to $t = 4$.

Because $x(t) = t^2 - 6t$,

$$\left|x(3) - x(2)\right| + \left|x(4) - x(3)\right| = \left|-9 + 8\right| + \left|-8 + 9\right| = 2$$

This is the distance that the particle traveled.

Example 5: Given the position function $x(t) = t^4 - 8t^2$, find the distance that the particle travels from $t = 0$ to $t = 4$.

First, find the first derivative ($v(t) = 4t^3 - 16t$) and set it equal to zero.

$$4t^3 - 16t = 0 \quad 4t(t^2 - 4) = 0 \quad t = 0, 2, -2$$

So we need to divide the time interval into $t = 0$ to $t = 2$ and $t = 2$ to $t = 4$.

$$\left|x(2) - x(0)\right| + \left|x(4) - x(2)\right| = 16 + 144 = 160$$

Here are some solved problems. Do each problem, covering the answer first, then check your answer.

PROBLEM 1. Find the velocity and acceleration of a particle whose position function is $x(t) = 2t^3 - 21t^2 + 60t + 3$, for $t > 0$.

Answer: Find the first two derivatives.

$$v(t) = 6t^2 - 42t + 60$$

$$a(t) = 12t - 42$$

PROBLEM 2. Given the position function in problem 1, find when the particle's speed is increasing.

Answer: First, set $v(t) = 0$.

$$6t^2 - 42t + 60 = 0$$

$$t^2 - 7t + 10 = 0$$

$$(t - 2)(t - 5) = 0$$

$$t = 2, t = 5$$

You should be able to determine that the velocity is positive from $0 < t < 2$, negative from $2 < t < 5$, and positive again from $t > 5$.

Now, set $a(t) = 0$.

$$12t - 42 = 0$$

$$t = \frac{7}{2}$$

You should be able to determine that the acceleration is negative from $0 < t < \frac{7}{2}$ and positive from $t > \frac{7}{2}$.

The intervals where the velocity and the acceleration have the same sign are $2 < t < \frac{7}{2}$ and $t > 5$.

PROBLEM 3. Given that the position of a particle is found by $x(t) = t^3 - 6t^2 + 1$, $t > 0$, find the distance that the particle travels from $t = 2$ to $t = 5$.

Answer: First, find $v(t)$.

$$v(t) = 3t^2 - 12t$$

Second, set $v(t) = 0$ and find the critical values.

$$3t^2 - 12t = 0 \qquad\qquad 3t(t - 4) = 0 \qquad\qquad t = \{0, 4\}$$

Because the particle changes direction after four seconds, you have to figure out two time intervals separately (from $t = 2$ to $t = 4$ and from $t = 4$ to $t = 5$) and add the absolute values of the distances.

$$\left|x(4) - x(2)\right| + \left|x(5) - x(4)\right| = \left|(-31) - (-15)\right| + \left|(-24) - (-31)\right| = 23$$

PRACTICE PROBLEM SET 13

Now try these problems. The answers are in Chapter 19.

1. Find the velocity and acceleration of a particle whose position function is $x(t) = t^3 - 9t^2 + 24t$, $t > 0$.

2. Find the velocity and acceleration of a particle whose position function is $x(t) = \sin(2t) + \cos(t)$.

3. If the position function of a particle is $x(t) = \sin\left(\dfrac{t}{2}\right)$, $0 < t < 4\pi$, find when the particle is changing direction.

4. If the position function of a particle is $x(t) = 3t^2 + 2t + 4$, $t > 0$, find the distance that the particle travels from $t = 2$ to $t = 5$.

5. If the position function of a particle is $x(t) = t^2 + 8t$, $t > 0$, find the distance that the particle travels from $t = 0$ to $t = 4$.

6. If the position function of a particle is $x(t) = 2\sin^2 t + 2\cos^2 t$, $t > 0$, find the velocity and acceleration of the particle.

7. If the position function of a particle is $x(t) = t^3 + 8t^2 - 2t + 4$, $t > 0$, find when the particle is changing direction.

8. If the position function of a particle is $x(t) = 2t^3 - 6t^2 + 12t - 18$, $t > 0$, find when the particle is changing direction.

End of Chapter 10 Drill

The answers are in Chapter 20.

1. A 25 ft ladder is sliding down a wall at –6 ft/s. How fast is the bottom of the ladder sliding out when the top of the ladder is 20 ft from the ground?

    (A)   2 ft/s
    (B)   6 ft/s
    (C)   8 ft/s
    (D)   15 ft/s

2. A baseball diamond has right angles at each of the bases, which are 90 ft apart. A baseball player is running from second base to third base at a rate of 30 ft/s. How fast is the distance between the player and home plate changing when the player is halfway from second base to third base?

    (A) $-\dfrac{30}{\sqrt{5}}$ ft/s

    (B) -30 ft/s

    (C) $-\dfrac{135}{\sqrt{5}}$ ft/s

    (D) -135 ft/s

3. A spherical balloon is deflating at -24π in.^{3}/s. How fast is the surface area of the balloon decreasing when the radius of the balloon is 6 in. ?

    (A) -6π in.^{2}/s
    (B) -8π in.^{2}/s
    (C) -16π in.^{2}/s
    (D) -24π in.^{2}/s

4. Find where the velocity of a particle is zero if its position function is $x(t) = t \ln t$.

    (A) $-e$

    (B) 1

    (C) $\dfrac{1}{e}$

    (D) e

5. If the position function of a particle is $x(t) = 4t^3 - 12t^2 + 18$, $t > 0$, at what time is the particle changing direction?

    (A) $t = 0$

    (B) $t = \dfrac{1}{2}$

    (C) $t = 1$

    (D) $t = 2$

REFLECT

Respond to the following questions:

*   For which topics discussed in this chapter do you feel you have achieved sufficient mastery to answer multiple-choice questions correctly?

*   For which topics discussed in this chapter do you feel you have achieved sufficient mastery to answer open-ended questions correctly?

*   For which topics discussed in this chapter do you feel you need more work before you can answer multiple-choice questions correctly?

*   For which topics discussed in this chapter do you feel you need more work before you can answer open-ended questions correctly?

*   What parts of this chapter are you going to re-review?

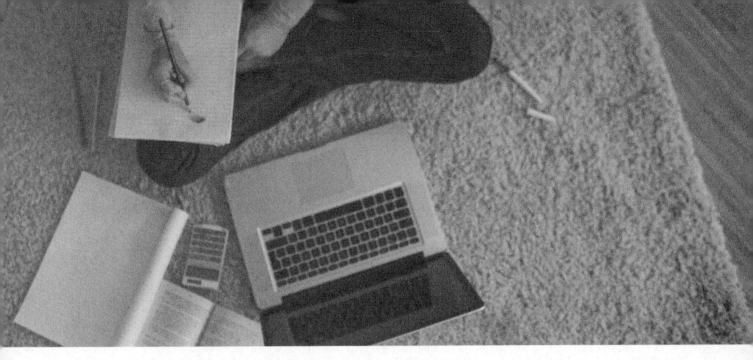

Chapter 11

Derivatives of Exponential and Logarithmic Functions

As with trigonometric functions, you'll be expected to remember all of the logarithmic and exponential functions you've studied in the past. If you're not sure about any of this stuff, review the unit on Prerequisite Mathematics. Also, this is only part one of our treatment of exponents and logs. Much of what you need to know about these functions requires knowledge of integrals (the second half of the book), so we'll discuss them again later.

THE DERIVATIVE OF ln *x*

When you studied logs in the past, you probably concentrated on common logs (that is, those with a base of 10), and avoided natural logarithms (base *e*) as much as possible. Well, we have bad news for you: most of what you'll see from now on involves natural logs. In fact, common logs almost never show up in calculus. But that's okay. All you have to do is memorize a bunch of rules, and you'll be fine.

Rule No. 1: If $y = \ln x$, then $\dfrac{dy}{dx} = \dfrac{1}{x}$

This rule has a corollary that incorporates the Chain Rule and is actually a more useful rule to memorize.

Rule No. 2: If $y = \ln u$, then $\dfrac{dy}{dx} = \dfrac{1}{u}\dfrac{du}{dx}$

Remember: *u* is a function of *x*, and $\dfrac{du}{dx}$ is its derivative.

You'll see how simple this rule is after we try a few examples.

Example 1: Find the derivative of $f(x) = \ln(x^3)$.

$$f'(x) = \frac{3x^2}{x^3} = \frac{3}{x}$$

If you recall your rules of logarithms, you could have done this another way.

$$\ln(x^3) = 3 \ln x$$

Therefore, $f'(x) = 3\left(\dfrac{1}{x}\right) = \dfrac{3}{x}$.

Example 2: Find the derivative of $f(x) = \ln(5x - 3x^6)$.

$$f'(x) = \frac{\left(5 - 18x^5\right)}{\left(5x - 3x^6\right)}$$

Example 3: Find the derivative of $f(x) = \ln(\cos x)$.

$$f'(x) = \frac{-\sin x}{\cos x} = -\tan x$$

Finding the derivative of a natural logarithm is just a matter of following a simple formula.

THE DERIVATIVE OF e^x

As you'll see in Rule No. 3, the derivative of e^x is probably the easiest thing that you'll ever have to do in calculus.

> Rule No. 3: If $y = e^x$, then $\dfrac{dy}{dx} = e^x$

That's not a typo. The derivative is the same as the original function! Incorporating the Chain Rule, we get a good formula for finding the derivative.

> Rule No. 4: If $y = e^u$, then $\dfrac{dy}{dx} = e^u \dfrac{du}{dx}$

And you were worried that all of this logarithm and exponential stuff was going to be hard!

Example 4: Find the derivative of $f(x) = e^{3x}$.

$$f'(x) = e^{3x}(3) = 3e^{3x}$$

Example 5: Find the derivative of $f(x) = e^{x^3}$.

$$f'(x) = e^{x^3}\left(3x^2\right) = 3x^2 e^{x^3}$$

Example 6: Find the derivative of $f(x) = e^{\tan x}$.

$$f'(x) = \left(\sec^2 x\right)e^{\tan x}$$

Example 7: Find the second derivative of $f(x) = e^{x^2}$.

$$f'(x) = 2xe^{x^2}$$
$$f''(x) = 2e^{x^2} + 4x^2 e^{x^2}$$

Once again, it's just a matter of following a formula.

THE DERIVATIVE OF $\log_a x$

This derivative is actually a little trickier than the derivative of a natural log. First, if you remember your logarithm rules about change of base, we can rewrite $\log_a x$.

$$\log_a x = \frac{\ln x}{\ln a}$$

Review the unit on Prerequisite Mathematics if this leaves you scratching your head. Anyway, because $\ln a$ is a constant, we can take the derivative.

$$\frac{1}{\ln a}\frac{1}{x}$$

This leads us to our next rule.

> **Rule No. 5:** If $y = \log_a x$, then $\dfrac{dy}{dx} = \dfrac{1}{x \ln a}$

Once again, incorporating the Chain Rule gives us a more useful formula.

> **Rule No. 6:** If $y = \log_a u$, then $\dfrac{dy}{dx} = \dfrac{1}{u \ln a}\dfrac{du}{dx}$

Example 8: Find the derivative of $f(x) = \log_{10} x$.

$$f'(x) = \frac{1}{x \ln 10}$$

Note: We refer to the $\log_{10} x$ as $\log x$.

Example 9: Find the derivative of $f(x) = \log_8 (x^2 + x)$.

$$f'(x) = \frac{2x + 1}{(x^2 + x) \ln 8}$$

Example 10: Find the derivative of $f(x) = \log_e x$.

$$f'(x) = \frac{1}{x \ln e} = \frac{1}{x}$$

You can expect this result from Rules 1 and 2 involving natural logs.

THE DERIVATIVE OF a^x

You should recall from your precalculus days that we can rewrite a^x as $e^{x \ln a}$. Keep in mind that $\ln a$ is just a constant, which gives us the next rule.

> **Rule No. 7:** If $y = a^x$, then $\dfrac{dy}{dx} = \left(e^{x \ln a}\right) \ln a = a^x (\ln a)$

Given the pattern of this chapter, you can guess what's coming: another rule that incorporates the Chain Rule.

> **Rule No. 8:** If $y = a^u$, then $\dfrac{dy}{dx} = a^u (\ln a) \dfrac{du}{dx}$

And now, some examples.

Example 11: Find the derivative of $f(x) = 3^x$.

$$f'(x) = 3^x \ln 3$$

Example 12: Find the derivative of $f(x) = 8^{4x^5}$.

$$f'(x) = 8^{4x^5}\left(20x^4\right)\ln 8$$

Example 13: Find the derivative of $f(x) = \pi^{\sin x}$.

$$f'(x) = \pi^{\sin x}\left(\cos x\right)\ln \pi$$

Finally, here's every nasty teacher's favorite exponential derivative.

Example 14: Find the derivative of $f(x) = x^x$.

First, rewrite this as $f(x) = e^{x\ln x}$. Then, take the derivative.

$$f'(x) = e^{x\ln x}\left(\ln x + \frac{x}{x}\right) = e^{x\ln x}\left(\ln x + 1\right) = x^x\left(\ln x + 1\right)$$

Would you have thought of that? Remember this trick. It might come in handy! Okay. Ready for some practice? Here are some more solved problems. Cover the solutions and get cracking.

PROBLEM 1. Find the derivative of $y = 3\ln(5x^2 + 4x)$.

Answer: Use Rule No. 2.

$$\frac{dy}{dx} = 3\frac{10x + 4}{5x^2 + 4x} = \frac{30x + 12}{5x^2 + 4x}$$

PROBLEM 2. Find the derivative of $f(x) = \ln\left(\sin\left(x^5\right)\right)$.

Answer: Use Rule No. 2 in addition to the Chain Rule.

$$f'(x) = \frac{5x^4 \cos\left(x^5\right)}{\sin\left(x^5\right)} = 5x^4 \cot\left(x^5\right)$$

PROBLEM 3. Find the derivative of $f(x) = e^{3x^7 - 4x^2}$.

Answer: Use Rule No. 4.

$$f'(x) = \left(21x^6 - 8x\right)e^{3x^7 - 4x^2}$$

PROBLEM 4. Find the derivative of $f(x) = \log_4(\tan x)$.

Answer: Use Rule No. 6.

$$f'(x) = \frac{1}{\ln 4} \frac{\sec^2 x}{\tan x}$$

PROBLEM 5. Find the derivative of $y = \log_8 \sqrt{\dfrac{x^3}{1+x^2}}$.

Answer: First, use the rules of logarithms to rewrite the equation.

$$y = \frac{1}{2}\left[3\log_8 x - \log_8\left(1+x^2\right) \right]$$

Now it's much easier to find the derivative.

$$\frac{dy}{dx} = \frac{1}{2}\left[3\frac{1}{x\ln 8} \right] - \frac{1}{2}\left[\frac{2x}{\ln 8(1+x^2)} \right] = \frac{1}{2\ln 8}\left[\frac{3}{x} - \frac{2x}{(1+x^2)} \right]$$

PROBLEM 6. Find the derivative of $y = 5^{\sqrt{x}}$.

Answer: Use Rule No. 8.

$$\frac{dy}{dx} = 5^{\sqrt{x}} \frac{1}{2\sqrt{x}} \ln 5 = \frac{5^{\sqrt{x}} \ln 5}{2\sqrt{x}}$$

PROBLEM 7. Find the derivative of $y = \dfrac{e^{x^3}}{5^{\cos x}}$.

Answer: Here, you need to use the Quotient Rule and Rules Nos. 4 and 8.

$$\frac{dy}{dx} = \frac{5^{\cos x}\left(3x^2 e^{x^3}\right) - e^{x^3}\left(5^{\cos x}\ln 5(-\sin x)\right)}{\left(5^{\cos x}\right)^2} = 5^{\cos x} e^{x^3} \frac{(3x^2)+(\sin x \ln 5)}{5^{2\cos x}} = e^{x^3} \frac{3x^2 + \sin x \ln 5}{5^{\cos x}}$$

PRACTICE PROBLEM SET 14

Now find the derivative of each of the following functions. The answers are in Chapter 19.

1. $f(x) = \ln(x^4 + 8)$

2. $f(x) = \ln(3x\sqrt{3+x})$

3. $f(x) = \ln(\cot x - \csc x)$

4. $f(x) = \ln\left(\dfrac{5x^2}{\sqrt{5+x^2}}\right)$

5. $f(x) = e^{x\cos x}$

6. $f(x) = e^{-3x}\sin 5x$

7. $f(x) = e^{\pi x} - \ln e^{\pi x}$

8. $f(x) = \log_{12}(x^3)$

9. $f(x) = \dfrac{\log_4 x}{e^{4x}}$

10. $f(x) = \log\sqrt{10^{3x}}$

11. $f(x) = e^{3x} - 3^{ex}$

12. $f(x) = 10^{\sin x}$

13. $f(x) = \ln(10^x)$

14. $f(x) = x^5 5^x$

End of Chapter 11 Drill

The answers are in Chapter 20.

1. Find $\dfrac{dy}{dx}$ if $y = \ln\left(x^3 - 7x + 5\right)$.

   (A) $\dfrac{1}{x^3 - 7x + 5}$

   (B) $\dfrac{x^3 - 7x + 5}{3x^2 - 7}$

   (C) $\dfrac{3x^2 - 7}{x^3 - 7x + 5}$

   (D) $\dfrac{1}{3x^2 - 7}$

2. Find $\dfrac{dy}{dx}$ if $y = \ln\left(\dfrac{\sqrt{x^2 - 11}}{x^2 + 9}\right)$.

   (A) $\dfrac{x}{x^2 - 11} - \dfrac{2x}{x^2 + 9}$

   (B) $\dfrac{2x}{\sqrt{x^2 - 11}} - \dfrac{2x}{x^2 + 9}$

   (C) $\dfrac{1}{2}\dfrac{x^2 + 9}{x^2 - 11}$

   (D) $\dfrac{x^2 + 9}{\sqrt{x^2 - 11}}$

3. Find $\dfrac{dy}{dx}$ if $y = e^{2\sin x}$.

   (A) $2e^{2\sin x}$

   (B) $e^{2\cos x}$

   (C) $2\sin x e^{2\sin x}$

   (D) $2\cos x e^{2\sin x}$

4. Find $\dfrac{dy}{dx}$ if $y = \log_8\left(x^4 + \tan x\right)$.

   (A) $\dfrac{1}{x^4 + \tan x}\dfrac{1}{\ln 8}$

   (B) $\dfrac{4x^3 + \sec^2 x}{x^4 + \tan x}\dfrac{1}{\ln 8}$

   (C) $\dfrac{4x^3 + \sec^2 x}{x^4 + \tan x}\dfrac{1}{\log 8}$

   (D) $\dfrac{1}{x^4 + \tan x}\dfrac{1}{\log 8}$

5. Find $\dfrac{dy}{dx}$ if $y = 7^{x^2 + 1}$.

   (A) $7^{x^2 + 1}\left(2x \log 7\right)$

   (B) $7^{x^2 + 1}\left(2x\right)$

   (C) $7^{x^2 + 1}\left(2x \ln 7\right)$

   (D) $7^{x^2 + 1}\left(\ln 7\right)$

REFLECT

Respond to the following questions:

- For which topics discussed in this chapter do you feel you have achieved sufficient mastery to answer multiple-choice questions correctly?

- For which topics discussed in this chapter do you feel you have achieved sufficient mastery to answer open-ended questions correctly?

- For which topics discussed in this chapter do you feel you need more work before you can answer multiple-choice questions correctly?

- For which topics discussed in this chapter do you feel you need more work before you can answer open-ended questions correctly?

- What parts of this chapter are you going to re-review?

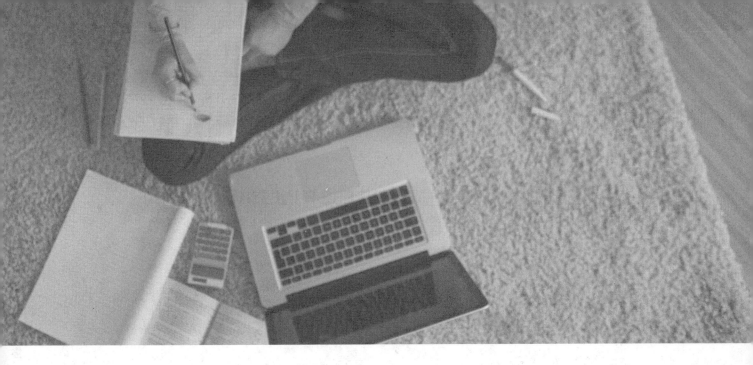

Chapter 12
Other Topics in Differential Calculus

This chapter is devoted to other topics involving differential calculus that don't fit into a specific category.

THE DERIVATIVE OF AN INVERSE FUNCTION

ETS occasionally asks a question about finding the derivative of an inverse function. To do this, you need to learn only this simple formula.

Suppose we have a function $x = f(y)$ that is defined and differentiable at $y = a$ where $x = c$. Suppose we also know that the $f^{-1}(x)$ exists at $x = c$. Thus, $f(a) = c$ and $f^{-1}(c) = a$. Then, because

$$\frac{dy}{dx} = \frac{1}{\frac{dx}{dy}},$$

$$\frac{d}{dx} f^{-1}(x)\Big|_{x=c} = \frac{1}{\left[\frac{d}{dy} f(y)\right]_{y=a}}$$

To translate, we can find the derivative of a function's inverse at a particular point by taking the reciprocal of the derivative at that point's corresponding y-value. These examples should help clear up any confusion.

Example 1: If $f(x) = x^2$, find a derivative of $f^{-1}(x)$ at $x = 9$.

First, notice that $f(3) = 9$. One of the most confusing parts of finding the derivative of an inverse function is that when you're asked to find the derivative at a value of x, they're *really* asking you for the derivative of the inverse of the function at the value that *corresponds* to $f(x) = 9$. This is because x-values of the inverse correspond to $f(x)$-values of the original function.

The rule is very simple: when you're asked to find the derivative of $f^{-1}(x)$ at $x = c$, you take the reciprocal of the derivative of $f(x)$ at $x = a$, where $f(a) = c$.

We know that $\frac{d}{dx} f(x) = 2x$. This means that we're going to plug $x = 3$ into the formula (because $f(3) = 9$). This gives us

$$\frac{1}{2x}\Big|_{x=3} = \frac{1}{6}$$

We can verify this by finding the inverse of the function first and then taking the derivative. The inverse of the function $f(x) = x^2$ is the function $f^{-1}(x) = \sqrt{x}$. Now we find the derivative and evaluate it at $f(3) = 9$.

$$\frac{d}{dx}\sqrt{x} = \frac{1}{2\sqrt{x}}\bigg|_{x=9} = \frac{1}{6}$$

Remember the rule: Find the value, a, of $f(x)$ that gives you the value of x that the problem asks for. Then plug that value, a, into the reciprocal of the derivative of the *inverse* function.

Example 2: Find a derivative of the inverse of $y = x^3 - 1$ when $y = 7$.

First, we need to find the x-value that corresponds to $y = 7$. A little algebra tells us that this is $x = 2$. Then,

$$\frac{dy}{dx} = 3x^2 \text{ and } \frac{1}{\dfrac{dy}{dx}} = \frac{1}{3x^2}$$

Therefore, the derivative of the inverse is

$$\frac{1}{3x^2}\bigg|_{x=2} = \frac{1}{12}$$

Verify it: The inverse of the function $y = x^3 - 1$ is the function $y = \sqrt[3]{x+1}$. The derivative of this latter function is

$$\frac{1}{3\sqrt[3]{(x+1)^2}}\bigg|_{x=7} = \frac{1}{12}$$

Let's do one more.

Example 3: Find a derivative of the inverse of $y = x^2 + 4$ when $y = 29$.

At $y = 29$, $x = 5$, the derivative of the function is

$$\frac{dy}{dx} = 2x$$

So, a derivative of the inverse is

$$\frac{1}{2x}\bigg|_{x=5} = \frac{1}{10}$$

Note that $x = -5$ also gives us $y = 29$, so $-\dfrac{1}{10}$ is also a derivative. It's not that hard, once you get the hang of it.

This is all you'll be required to know involving derivatives of inverses. Naturally, there are ways to create harder problems, but the AP Exam stays away from them and sticks to simpler stuff.

Here are some solved problems. Do each problem, cover the answer first, and then check your answer.

PROBLEM 1. Find a derivative of the inverse of $f(x) = 2x^3 + 5x + 1$ at $y = 8$.

Answer: First, we take the derivative of $f(x)$.

$$f'(x) = 6x^2 + 5$$

A possible value of x is $x = 1$.

Then, we use the formula to find the derivative of the inverse.

$$\frac{1}{f'(1)} = \frac{1}{11}$$

PROBLEM 2. Find a derivative of the inverse of $f(x) = 3x^3 - x + 7$ at $y = 9$.

Answer: First, take the derivative of $f(x)$.

$$f'(x) = 9x^2 - 1$$

A possible value of x is $x = 1$.

Then, use the formula to find the derivative of the inverse.

$$\frac{1}{f'(1)} = \frac{1}{8}$$

PROBLEM 3. Find a derivative of the inverse of $y = \dfrac{8}{x^3}$ at $y = 1$.

Answer: Take the derivative of y.

$$y' = -\frac{24}{x^4}$$

Find the value of x where $y = 1$.

$$1 = \frac{8}{x^3}$$
$$x = 2$$

Use the formula.

$$\frac{1}{\left.\frac{dy}{dx}\right|_{x=2}} = \frac{1}{\left.\left(-\frac{24}{x^4}\right)\right|_{x=2}} = -\frac{2}{3}$$

Here's one more.

PROBLEM 4. Find a derivative of the inverse of $y = 2x - x^3$ at $y = 1$.

Answer: The derivative of the function is

$$\frac{dy}{dx} = 2 - 3x^2$$

Next, find the value of x where $y = 1$. By inspection, $y = 1$ when $x = 1$.

Then, we use the formula to find the derivative of the inverse.

$$\frac{1}{\left.\frac{dy}{dx}\right|_{x=1}} = \frac{1}{\left.\left(2 - 3x^2\right)\right|_{x=1}} = -1$$

PRACTICE PROBLEM SET 15

Find a derivative of the inverse of each of the following functions. The answers are in Chapter 19.

1. $y = x + \dfrac{1}{x}$ at $y = \dfrac{17}{4}$; where $x > 1$

2. $y = 3x - 5x^3$ at $y = 2$

3. $y = e^x$ at $y = e$

4. $y = x + x^3$ at $y = -2$

5. $y = 4x - x^3$ at $y = 3$

6. $y = \ln x$ at $y = 0$

LINEARIZATION

Sometimes this is called "linearization." A differential is a very small quantity that corresponds to a change in a number. We use the symbol Δx to denote a differential. What are differentials used for? The AP Exam mostly wants you to use them to approximate the value of a function or to find the error of an approximation.

Recall the formula for the definition of the derivative.

$$f'(x) = \lim_{h \to 0} \frac{f(x+h) - f(x)}{h}$$

Replace h with Δx, which also stands for a very small increment of x, and get rid of the limit.

$$f'(x) \approx \frac{f(x + \Delta x) - f(x)}{\Delta x}$$

Notice that this is no longer equal to the derivative, but an approximation of it. If Δx is kept small, the approximation remains fairly accurate. Next, rearrange the equation as follows:

$$f(x + \Delta x) \approx f(x) + f'(x)\Delta x$$

This is our formula for differentials. It says that "the value of a function (at x plus a little bit) equals the value of the function (at x) plus the product of the derivative of the function (at x) and the little bit."

Example 1: Use differentials to approximate $\sqrt{9.01}$.

You can start by letting $x = 9$, $\Delta x = +0.01$, $f(x) = \sqrt{x}$. Next, we need to find $f'(x)$.

$$f'(x) = \frac{1}{2\sqrt{x}}$$

Now, plug in to the formula.

$$f(x + \Delta x) \approx f(x) + f'(x)\Delta x$$
$$\sqrt{x + \Delta x} \approx \sqrt{x} + \frac{1}{2\sqrt{x}}\Delta x$$

Now, if we plug in $x = 9$ and $\Delta x = + 0.01$,

$$\sqrt{9.01} \approx \sqrt{9} + \frac{1}{2\sqrt{9}}(0.01) \approx 3.001666666$$

If you enter $\sqrt{9.01}$ into your calculator, you get: 3.001666204. As you can see, our answer is a pretty good approximation. It's not so good, however, when Δx is too big. How big is too big? Good question.

Example 2: Use differentials to approximate $\sqrt{9.5}$.

Let $x = 9$, $\Delta x = +.5$, $f(x) = \sqrt{x}$ and plug in to what you found in Example 1.

$$\sqrt{9.5} \approx \sqrt{9} + \frac{1}{2\sqrt{9}}(.5) \approx 3.083333333$$

However, $\sqrt{9.5}$ equals 3.082207001 on a calculator. This is good to only two decimal places. As the ratio of $\frac{\Delta x}{x}$ grows larger, the approximation gets less accurate, and we start to get away from the actual value.

There's another approximation formula that you'll need to know for the AP Exam. This formula is used to estimate the error in a measurement, or to find the effect on a formula when a small change in measurement is made. The formula is:

$$dy = f'(x)dx$$

> Note that this equation is simply a rearrangement of $\frac{dy}{dx} = f'(x)$.

This notation may look a little confusing. It says that the change in a measurement dy, due to a differential dx, is found by multiplying the derivative of the equation for y by the differential. Let's do an example.

Example 3: The radius of a circle is increased from 3 to 3.04. Estimate the change in area.

Let $A = \pi r^2$. Then our formula says that $dA = A'dr$, where A' is the derivative of the area with respect to r, and $dr = 0.04$ (the change). First, find the derivative of the area: $A' = 2\pi r$. Now, plug in to the formula.

$$dA = 2\pi r \, dr = 2\pi(3)(0.04) = 0.754$$

The actual change in the area is from 9π to 9.2416π, which is approximately 0.759. As you can see, this approximation formula is pretty accurate.

Here are some sample problems involving this differential formula. Try them out, then check your work against the answers directly beneath.

PROBLEM 1. Use differentials to approximate $(3.98)^4$.

Answer: Let $f(x) = x^4$, $x = 4$, and $\Delta x = -0.02$. Next, find $f'(x)$, which is: $f'(x) = 4x^3$.

Now, plug in to the formula.

$$f(x + \Delta x) \approx f(x) + f'(x)\Delta x$$
$$(x + \Delta x)^4 \approx x^4 + 4x^3\Delta x$$

If you plug in $x = 4$ and $\Delta x = -0.02$, you get

$$(3.98)^4 \approx 4^4 + 4(4)^3(-0.02) \approx 250.88$$

Check $(3.98)^4$ by using your calculator; you should get 250.9182722. Not a bad approximation.

You're probably asking yourself, why can't I just use my calculator every time? Because most math teachers are dedicated to teaching you several complicated ways to calculate things without your calculator.

PROBLEM 2. Use differentials to approximate sin 46°.

Answer: This is a tricky question. The formula doesn't work if you use degrees. Here's why: Let $f(x) = \sin x$, $x = 45°$, and $\Delta x = 1°$. The derivative is $f'(x) = \cos x$.

If you plug this information into the formula, you get $\sin 46° \approx \sin 45° + \cos 45°(1°) = \sqrt{2}$. You should recognize that this is nonsense for two reasons: (1) the sine of any angle is between −1 and 1; and (2) the answer should be close to $\sin 45° = \dfrac{1}{\sqrt{2}}$.

What went wrong? You have to use radians! As we mentioned before, angles in calculus problems are measured in radians, not degrees.

Let $f(x) = \sin x$, $x = \dfrac{\pi}{4}$, and $\Delta x = \dfrac{\pi}{180}$. Now plug in to the formula.

$$\sin\left(\frac{46\pi}{180}\right) \approx \sin\frac{\pi}{4} + \left(\cos\frac{\pi}{4}\right)\left(\frac{\pi}{180}\right) = 0.7194$$

PROBLEM 3. The radius of a sphere is measured to be 4 cm with an error of ±0.01 cm. Use differentials to approximate the error in the surface area.

Answer: Now it's time for the other differential formula. The formula for the surface area of a sphere is

$$S = 4\pi r^2$$

The formula says that $dS = S'dr$, so first, we find the derivative of the surface area, $(S' = 8\pi r)$ and plug away.

$$dS = 8\pi r\, dr = 8\pi (4)(\pm 0.01) = \pm 1.0053$$

This looks like a big error, but given that the surface area of a sphere with radius 4 is approximately 201 cm², the error is quite small.

PRACTICE PROBLEM SET 16

Use the differential formulas in this chapter to solve these problems. The answers are in Chapter 19.

1. Approximate $\sqrt{25.02}$.

2. Approximate $\sqrt[3]{63.97}$.

3. Approximate $\tan 61°$.

4. The side of a cube is measured to be 6 in. with an error of ± 0.02 in. Estimate the error in the volume of the cube.

5. When a spherical ball bearing is heated, its radius increases by 0.01 mm. Estimate the change in volume of the ball bearing when the radius is 5 mm.

6. A cylindrical tank is constructed to have a diameter of 5 meters and a height of 20 meters. Find the error in the volume if

(a) the diameter is exact, but the height is 20.1 meters; and

(b) the height is exact, but the diameter is 5.1 meters.

L'HÔPITAL'S RULE

L'Hôpital's Rule is a way to find the limit of certain kinds of expressions that are indeterminate forms. If the limit of an expression results in $\dfrac{0}{0}$ or $\dfrac{\infty}{\infty}$, the limit is called "indeterminate" and you can use L'Hôpital's Rule to evaluate these expressions.

> If $f(c) = g(c) = 0$, and if $f'(c)$ and $g'(c)$ exist, and if $g'(c) \neq 0$, then
>
> $$\lim_{x \to c} \frac{f(x)}{g(x)} = \frac{f'(c)}{g'(c)}.$$

Remember that you can only use L'Hôpital's Rule for indeterminate forms.

Similarly,

> If $f(c) = g(c) = \infty$, and if $f'(c)$ and $g'(c)$ exist, and if $g'(c) \neq 0$, then
>
> $$\lim_{x \to c} \frac{f(x)}{g(x)} = \frac{f'(c)}{g'(c)}.$$

In other words, if the limit of the function gives us an undefined expression, like $\dfrac{0}{0}$ or $\dfrac{\infty}{\infty}$, L'Hôpital's Rule says we can take the derivative of the top and the derivative of the bottom and see if we get a determinate expression. If not, we can repeat the process.

Example 1: Find $\displaystyle\lim_{x \to 0} \frac{\sin x}{x}$.

First, notice that plugging in 0 results in $\dfrac{0}{0}$, which is indeterminate. Take the derivative of the top and of the bottom.

$$\lim_{x \to 0} \frac{\cos x}{1}$$

The limit equals 1.

Example 2: Find $\lim\limits_{x \to 0} \dfrac{2x - \sin x}{x}$.

If you plug in 0, you get $\dfrac{0-0}{0}$, which is indeterminate. Now, take the derivative of the top and of the bottom.

$$\lim\limits_{x \to 0} \dfrac{2 - \cos x}{1}$$

This limit also equals 1.

Example 3: Find $\lim\limits_{x \to 0} \dfrac{\sqrt{4+x} - 2}{2x}$.

Again, plugging in 0 is no help; you get $\dfrac{0}{0}$. Take the derivative of the top and bottom.

$$\lim\limits_{x \to 0} \dfrac{\dfrac{1}{2\sqrt{4+x}}}{2} = \dfrac{1}{8}$$

Example 4: Find $\lim\limits_{x \to 0} \dfrac{\sqrt{4+x} - 2 - \dfrac{x}{4}}{2x^2}$.

Take the derivative of the top and bottom.

$$\lim\limits_{x \to 0} \dfrac{\dfrac{1}{2\sqrt{4+x}} - \dfrac{1}{4}}{4x}$$

Now if you take the limit, we still get $\dfrac{0}{0}$. So do it again.

$$\lim\limits_{x \to 0} \dfrac{-\dfrac{1}{4}(4+x)^{-\frac{3}{2}}}{4} = -\dfrac{1}{128}$$

Now let's try a couple of $\dfrac{\infty}{\infty}$ forms.

Example 5: Find $\lim\limits_{x \to \infty} \dfrac{5x - 8}{3x + 1}$.

Now, the limit is $\dfrac{\infty}{\infty}$. The derivative of the top and bottom is

$$\lim\limits_{x \to \infty} \dfrac{5}{3} = \dfrac{5}{3}$$

Don't you wish that you had learned this back when you first did limits?

Example 6: Find $\lim\limits_{x \to \frac{\pi}{2}} \dfrac{\sec x}{1 + \sec x}$.

The derivative of the top and bottom is

$$\lim\limits_{x \to \frac{\pi}{2}} \dfrac{\sec x \tan x}{\sec x \tan x} = 1$$

Example 7: Find $\lim\limits_{x \to 0^+} x \cot x$.

Taking this limit results in $(0)(\infty)$, which is also indeterminate. (But you can't use the rule yet!)

If you rewrite this expression as $\lim\limits_{x \to 0^+} \dfrac{x}{\tan x}$, it's of the form $\dfrac{0}{0}$, and we can use L'Hôpital's Rule.

$$\lim\limits_{x \to 0^+} \dfrac{1}{\sec^2 x} = 1$$

That's all that you need to know about L'Hôpital's Rule. Just check to see if the limit results in an indeterminate form. If it does, use the rule until you get a determinate form.

Here are some more examples.

PROBLEM 1. Find $\lim\limits_{x \to 0} \dfrac{\sin 8x}{x}$.

Answer: First, notice that plugging in 0 gives us an indeterminate result: $\dfrac{0}{0}$. Now, take the derivative of the top and of the bottom.

$$\lim\limits_{x \to 0} \dfrac{8 \cos 8x}{1} = 8$$

PROBLEM 2. Find $\lim\limits_{x \to \infty} xe^{-2x}$.

Answer: First, rewrite this expression as $\dfrac{x}{e^{2x}}$. Notice that when x nears infinity, the expression becomes $\dfrac{\infty}{\infty}$, which is indeterminate.

Take the derivative of the top and bottom.

$$\lim\limits_{x \to \infty} \dfrac{1}{2e^{2x}} = 0$$

PROBLEM 3. Find $\lim\limits_{x\to\infty}\dfrac{5x^3-4x^2+1}{7x^3+2x-6}$.

Answer: We learned this in Chapter 3, remember? Now we'll use L'Hôpital's Rule. At first glance, the limit is indeterminate: $\dfrac{\infty}{\infty}$. Let's take some derivatives.

$$\frac{15x^2-8x}{21x^2+2}$$

This is still indeterminate, so it's time to take the derivative of the top and bottom again.

$$\frac{30x-8}{42x}$$

It's still indeterminate! If you try it one more time, you'll get a fraction with no variables: $\dfrac{30}{42}$, which can be simplified to $\dfrac{5}{7}$ (as we expected).

PROBLEM 4. Find $\lim\limits_{x\to\frac{\pi}{2}}\dfrac{x-\dfrac{\pi}{2}}{\cos x}$.

Answer: Plugging in $\dfrac{\pi}{2}$ gives you the indeterminate response of $\dfrac{0}{0}$. The derivative is

$$-\frac{1}{\sin x}$$

When we take the limit of this expression, we get -1.

PRACTICE PROBLEM SET 17

Now find these limits using L'Hôpital's Rule. The answers are in Chapter 19.

1. Find $\lim\limits_{x \to 0} \dfrac{\sin 3x}{\sin 4x}$.

2. Find $\lim\limits_{x \to \pi} \dfrac{x - \pi}{\sin x}$.

3. Find $\lim\limits_{x \to 0} \dfrac{x - \sin x}{x^3}$.

4. Find $\lim\limits_{x \to 0} \dfrac{e^{3x} - e^{5x}}{x}$.

5. Find $\lim\limits_{x \to 0} \dfrac{\tan x - x}{\sin x - x}$.

6. Find $\lim\limits_{x \to \infty} \dfrac{x^5}{e^{5x}}$.

7. Find $\lim\limits_{x \to \infty} \dfrac{x^5 + 4x^3 - 8}{7x^5 - 3x^2 - 1}$.

8. Find $\lim\limits_{x \to 0^+} \dfrac{\ln(\sin x)}{\ln(\tan x)}$.

9. Find $\lim\limits_{x \to 0^+} \dfrac{\cot 2x}{\cot x}$.

Study Break!

You're about halfway through your review! Before you dive into Unit 3, be sure to give yourself some downtime to let your brain absorb the information you've been studying.

10. Find $\lim\limits_{x \to 0^+} \dfrac{x}{\ln(x + 1)}$.

End of Chapter 12 Drill

The answers are in Chapter 20.

1. Find the derivative of the inverse of $y = 5x^3 + x - 7$ at $y = -7$.

   (A) $-\dfrac{1}{7}$

   (B) $\dfrac{1}{7}$

   (C) 1

   (D) 5131

2. Find the equation of the tangent line to $x = 4\sin^2 t$ and $y = 2\cos t$ at $t = \dfrac{\pi}{3}$.

   (A) $y - 1 = -\dfrac{1}{4\sqrt{3}}(x-3)$

   (B) $y - \sqrt{3} = -\dfrac{1}{4\sqrt{3}}(x-1)$

   (C) $y - \sqrt{3} = -\dfrac{1}{2}(x-1)$

   (D) $y - 1 = -\dfrac{1}{2}(x-3)$

3. Use L'Hôpital's Rule to find $\lim\limits_{x \to \infty} \dfrac{x^{-\frac{4}{3}}}{\sin\left(\dfrac{1}{x}\right)}$.

   (A) $-\infty$
   (B) 0
   (C) ∞
   (D) The limit does not exist.

4. Approximate $(2.03)^5$.

   (A) 29.6
   (B) 32.03
   (C) 34.4
   (D) 80.03

5. Find the derivative of $y = \sqrt[4]{\dfrac{(8x+1)^3}{(2x-11)^5}}$.

   (A) $y\left(\dfrac{3}{4}\dfrac{8}{8x+1} - \dfrac{5}{4}\dfrac{2}{2x-11}\right)$

   (B) $y\left(\dfrac{3}{4}\dfrac{1}{8x+1} - \dfrac{5}{4}\dfrac{1}{2x-11}\right)$

   (C) $y\left(\dfrac{3}{4}\dfrac{1}{8x+1} + \dfrac{5}{4}\dfrac{1}{2x-11}\right)$

   (D) $y\left(\dfrac{3}{4}\dfrac{8}{8x+1} + \dfrac{5}{4}\dfrac{2}{2x-11}\right)$

REFLECT

Respond to the following questions:

- For which topics discussed in this chapter do you feel you have achieved sufficient mastery to answer multiple-choice questions correctly?

- For which topics discussed in this chapter do you feel you have achieved sufficient mastery to answer open-ended questions correctly?

- For which topics discussed in this chapter do you feel you need more work before you can answer multiple-choice questions correctly?

- For which topics discussed in this chapter do you feel you need more work before you can answer open-ended questions correctly?

- What parts of this chapter are you going to re-review?

Unit 3

Integral Calculus
Essentials

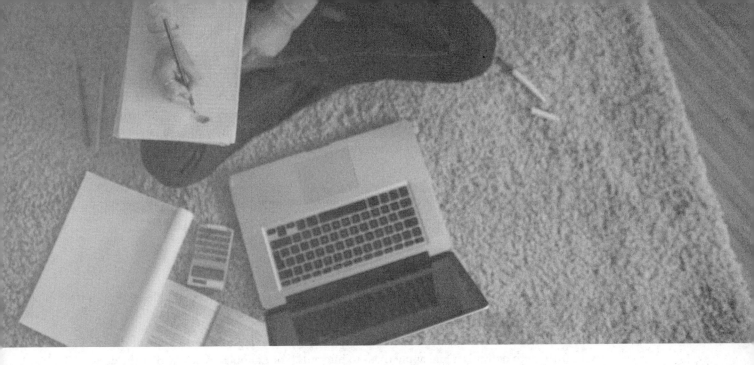

Chapter 13
The Integral

Welcome to the other half of calculus! This, unfortunately, is the more difficult half, but don't worry. We'll get you through it. In differential calculus, you learned all of the fun things that you can do with the derivative. Now you'll learn to do the reverse: how to take an integral. As you might imagine, there's a bunch of new fun things that you can do with integrals too.

It's also time for a new symbol $\int$, which stands for integration. An integral actually serves several different purposes, but the first, and most basic, is that of the antiderivative.

THE ANTIDERIVATIVE

An antiderivative is a derivative in reverse. Therefore, we're going to reverse some of the rules we learned with derivatives and apply them to integrals. For example, we know that the derivative of x^2 is $2x$. If we are given the derivative of a function and have to figure out the original function, we use antidifferentiation. Thus, the antiderivative of $2x$ is x^2. (Actually, the answer is slightly more complicated than that, but we'll get into that in a few moments.)

Now we need to add some info here to make sure that you get this absolutely correct. First, as far as notation goes, it is traditional to write the antiderivative of a function using its uppercase letter, so the antiderivative of $f(x)$ is $F(x)$, the antiderivative of $g(x)$ is $G(x)$, and so on.

The second idea is very important: each function has more than one antiderivative. In fact, there are an infinite number of antiderivatives of a function. Let's go back to our example to help illustrate this.

Remember that the antiderivative of $2x$ is x^2? Well, consider this. If you take the derivative of $x^2 + 1$, you get $2x$. The same is true for $x^2 + 2$, $x^2 - 1$, and so on. In fact, if *any* constant is added to x^2, the derivative is still $2x$ because the derivative of a constant is zero.

Because of this, we write the antiderivative of $2x$ as $x^2 + C$, where C stands for any constant.

Finally, whenever you take the integral (or antiderivative) of a function of x, you always add the term dx (or dy if it's a function of y, etc.) to the integrand (the thing inside the integral). You'll learn why later.

Just remember that you must always use the *dx* symbol, and teachers love to take points off for forgetting the *dx*. Don't ask why, but they do!

Here is the **Power Rule** for antiderivatives.

$$\text{If } f(x) = x^n \text{, then } \int f(x)dx = \frac{x^{n+1}}{n+1} + C \text{ (except when } n = -1\text{).}$$

Example 1: Find $\int x^3 dx$.

Using the Power Rule, we get

$$\int x^3 dx = \frac{x^4}{4} + C$$

Don't forget the constant C, or your teachers will take points off for that too!

Example 2: Find $\int x^{-3} dx$.

The Power Rule works with negative exponents too.

$$\int x^{-3} dx = \frac{x^{-2}}{-2} + C$$

> Think of a constant as something that is always there, and you won't forget it.

Not terribly hard, is it? Now it's time for a few more rules that look remarkably similar to the rules for derivatives that we saw in Chapter 6.

$$\int kf(x)dx = k\int f(x)dx$$

$$\int \left[f(x) + g(x) \right] dx = \int f(x)dx + \int g(x)dx$$

$$\int k\,dx = kx + C$$

Here are a few more examples to make you an expert.

Example 3: $\int 5\,dx = 5x + C$

Example 4: $\int 7x^3 dx = \frac{7x^4}{4} + C$

Example 5: $\int \left(3x^2 + 2x\right) dx = x^3 + x^2 + C$

Example 6: $\int \sqrt{x}\,dx = \frac{x^{\frac{3}{2}}}{\frac{3}{2}} + C = \frac{2x^{\frac{3}{2}}}{3} + C$

$$\int \sin\,ax\,dx = -\frac{\cos\,ax}{a} + C$$

$$\int \cos\,ax\,dx = \frac{\sin\,ax}{a} + C$$

$$\int \sec\,ax\,\tan\,ax\,dx = \frac{\sec\,ax}{a} + C$$

$$\int \sec^2\,ax\,dx = \frac{\tan\,ax}{a} + C$$

$$\int \csc\,ax\,\cot\,ax\,dx = -\frac{\csc\,ax}{a} + C$$

$$\int \csc^2\,ax\,dx = -\frac{\cot\,ax}{a} + C$$

INTEGRALS OF TRIG FUNCTIONS

The integrals of some trigonometric functions follow directly from the derivative formulas in Chapter 6.

We didn't mention the integrals of tangent, cotangent, secant, and cosecant, because you need to know some rules about logarithms to figure them out. We'll get to them in a few chapters. Notice also that each of the answers is divided by a constant. This is to account for the Chain Rule. Let's do some examples.

Example 7: Check the integral $\int \sin 5x\,dx = -\dfrac{\cos 5x}{5} + C$

by differentiating the answer.

$$\frac{d}{dx}\left[-\frac{\cos 5x}{5} + C \right] = -\frac{1}{5}(-\sin 5x)(5) = \sin 5x$$

Notice how the constant is accounted for in the answer?

Example 8: $\int \sec^2 3x\,dx = \dfrac{\tan 3x}{3} + C$

Example 9: $\int \cos \pi x\,dx = \dfrac{\sin \pi x}{\pi} + C$

Example 10: $\int \sec\left(\dfrac{x}{2}\right) \tan\left(\dfrac{x}{2}\right) dx = 2\sec\left(\dfrac{x}{2}\right) + C$

If you're not sure if you have the correct answer when you take an integral, you can always check by differentiating the answer and seeing if you get what you started with. Try to get in the habit of doing that at the beginning, because it'll help you build confidence in your ability to find integrals properly. You'll see that, although you can differentiate just about any expression that you'll normally encounter, you won't be able to integrate many of the functions you see.

ADDITION AND SUBTRACTION

By using the rules for addition and subtraction, we can integrate most polynomials.

Example 11: Find $\int \left(x^3 + x^2 - x \right) dx$.

We can break this into separate integrals, which gives us

$$\int x^3\,dx + \int x^2\,dx - \int x\,dx$$

Now you can integrate each of these individually.

$$\frac{x^4}{4} + C + \frac{x^3}{3} + C - \frac{x^2}{2} + C$$

You can combine the constants into one constant (it doesn't matter how many C's we use, because their sum is one collective constant whose derivative is zero).

$$\frac{x^4}{4} + \frac{x^3}{3} - \frac{x^2}{2} + C$$

Sometimes you'll be given information about the function you're seeking that will enable you to solve for the constant. Often, this is an "initial value," which is the value of the function when the variable is zero. As we've seen, normally there are an infinite number of solutions for an integral, but when we solve for the constant, there's only one.

Example 12: Find the equation of y where $\dfrac{dy}{dx} = 3x + 5$ and $y = 6$ when $x = 0$.

Let's put this in integral form.

$$y = \int (3x + 5)\ dx$$

Integrating, we get

$$y = \frac{3x^2}{2} + 5x + C$$

Now we can solve for the constant because we know that $y = 6$ when $x = 0$.

$$6 = \frac{3(0)^2}{2} + 5(0) + C$$

Therefore, $C = 6$ and the equation is

$$y = \frac{3x^2}{2} + 5x + 6$$

Example 13: Find $f(x)$ if $f'(x) = \sin x - \cos x$ and $f(\pi) = 3$.

Integrate $f'(x)$.

$$f(x) = \int (\sin x - \cos x)\ dx = -\cos x - \sin x + C$$

Now solve for the constant.

$$3 = -\cos(\pi) - \sin(\pi) + C$$

$$C = 2$$

Therefore, the equation becomes

$$f(x) = -\cos x - \sin x + 2$$

Now we've covered the basics of integration. However, integration is a very sophisticated topic and there are many types of integrals that will cause you trouble. We will need several techniques to learn how to evaluate these integrals. The first and most important is called u-substitution, which we will cover in the second half of this chapter.

In the meantime, here are some solved problems. Do each problem, covering the answer first, then check your answer.

PROBLEM 1. Evaluate $\int x^{\frac{3}{5}} dx$.

Answer: Here's the Power Rule again.

$$\int x^n dx = \frac{x^{n+1}}{n+1} + C$$

Using the rule,

$$\int x^{\frac{3}{5}} dx = \frac{x^{\frac{8}{5}}}{\frac{8}{5}} + C$$

You can rewrite it as $\dfrac{5x^{\frac{8}{5}}}{8} + C$.

PROBLEM 2. Evaluate $\int \left(5x^3 + x^2 - 6x + 4\right) dx$.

Answer: We can break this up into several integrals.

$$\int \left(5x^3 + x^2 - 6x + 4\right) dx = \int 5x^3 dx + \int x^2 dx - \int 6x\, dx + 4\int dx$$

Each of these can be integrated according to the Power Rule.

$$\frac{5x^4}{4} + C + \frac{x^3}{3} + C - \frac{6x^2}{2} + C + 4x + C$$

This can be rewritten as

$$\frac{5x^4}{4} + \frac{x^3}{3} - 3x^2 + 4x + C$$

Notice that we combine the constant terms into one constant term C.

PROBLEM 3. Evaluate $\int \left(3 - x^2\right)^2 dx$.

Answer: First, expand the integrand.

$$\int \left(9 - 6x^2 + x^4\right) dx$$

Break this up into several integrals.

$$\int 9\, dx - 6\int x^2 dx + \int x^4 dx$$

And integrate according to the Power Rule.

$$9x - 2x^3 + \frac{x^5}{5} + C$$

PROBLEM 4. Evaluate $\int \left(4 \sin x - 3 \cos x\right) dx$.

Answer: Break this problem into two integrals.

$$4\int \sin x\, dx - 3\int \cos x\, dx$$

Each of these trig integrals can be evaluated according to its rule.

$$-4 \cos x - 3 \sin x + C$$

PROBLEM 5. Evaluate $\int \left(2 \sec^2 x - 5 \csc^2 x\right) dx$.

Answer: Break the integral in two.

$$2\int \sec^2 x\, dx - 5\int \csc^2 x\, dx$$

Each of these trig integrals can be evaluated according to its rule.

$$2 \tan x + 5 \cot x + C$$

PRACTICE PROBLEM SET 18

Now evaluate the following integrals. The answers are in Chapter 19.

1. $\displaystyle\int \frac{1}{x^4}\,dx$

2. $\displaystyle\int \frac{5}{\sqrt{x}}\,dx$

3. $\displaystyle\int \left(5x^4 - 3x^2 + 2x + 6\right)dx$

4. $\displaystyle\int \left(3x^{-3} - 2x^{-2} + x^4 + 16x^7\right)dx$

5. $\displaystyle\int x^{\frac{1}{3}}\left(2 + x\right)dx$

6. $\displaystyle\int \left(x^3 + x\right)^2 dx$

7. $\displaystyle\int \frac{x^6 - 2x^4 + 1}{x^2}\,dx$

8. $\displaystyle\int x(x-1)^3\,dx$

9. $\displaystyle\int \sec x\left(\sec x + \tan x\right)dx$

10. $\displaystyle\int \left(\sec^2 x + x\right)dx$

11. $\displaystyle\int \frac{\cos^3 x + 4}{\cos^2 x}\,dx$

12. $\displaystyle\int \frac{\sin 2x}{\cos x}\,dx$

13. $\displaystyle\int \left(1 + \cos^2 x \sec x\right)dx$

14. $\displaystyle\int \frac{1}{\csc x}\,dx$

15. $\displaystyle\int \left(x - \frac{2}{\cos^2 x}\right)dx$

u-SUBSTITUTION

When we discussed differentiation, one of the most important techniques we mastered was the Chain Rule. Now, you'll learn the integration corollary of the Chain Rule (called *u*-substitution), which we use when the integrand is a composite function. All you do is replace the function with *u*, and then you can integrate the simpler function using the Power Rule (as shown below).

$$\int u^n\,du = \frac{u^{n+1}}{n+1} + C$$

Suppose you have to integrate $\displaystyle\int (x - 4)^{10}\,dx$. You could expand out this function and integrate each term, but that'll take a while. Instead, you can follow these four steps.

Step 1: Let $u = x - 4$. Then $\dfrac{du}{dx} = 1$ (rearrange this to get $du = dx$).

Step 2: Substitute $u = x - 4$ and $du = dx$ into the integrand.

$$\int u^{10}\,du$$

Step 3: Integrate.

$$\int u^{10}\,du = \frac{u^{11}}{11} + C$$

Step 4: Substitute back for u.

$$\frac{(x-4)^{11}}{11} + C$$

That's u-substitution. The main difficulty you'll have will be picking the appropriate function to set equal to u. The best way to get better is to practice. The object is to pick a function and replace it with u, then take the derivative of u to find du. If we can't replace *all* of the terms in the integrand, we *can't* do the substitution.

Let's do some examples.

Example 1: $\int 10x \left(5x^2 - 3\right)^6 dx =$

Once again, you could expand this out and integrate each term, but that would be difficult. Use u-substitution.

Let $u = 5x^2 - 3$. Then $\dfrac{du}{dx} = 10x$ and $du = 10x\, dx$. Now you can substitute.

$$\int u^6 du$$

And integrate.

$$\int u^6 du = \frac{u^7}{7} + C$$

Substituting back gives you

$$\frac{\left(5x^2 - 3\right)^7}{7} + C$$

Confirm that this is the integral by differentiating $\dfrac{\left(5x^2 - 3\right)^7}{7} + C$.

$$\frac{d}{dx}\left[\frac{\left(5x^2 - 3\right)^7}{7} + C\right] = \frac{7\left(5x^2 - 3\right)^6}{7}(10x) = \left(5x^2 - 3\right)^6 (10x)$$

Example 2: $\int 2x\sqrt{x^2 - 5}\, dx =$

If $u = x^2 - 5$, then $\dfrac{du}{dx} = 2x$ and $du = 2x\, dx$. Substitute u into the integrand.

$$\int u^{\frac{1}{2}}\, du$$

Integrate.

$$\int u^{\frac{1}{2}} \, du = \frac{u^{\frac{3}{2}}}{\frac{3}{2}} + C = \frac{2u^{\frac{3}{2}}}{3} + C$$

And substitute back.

$$\frac{2\left(x^2 - 5\right)^{\frac{3}{2}}}{3} + C$$

Note: From now on, we're not going to rearrange $\frac{du}{dx}$; we'll go directly to "$du =$" format. You should be able to do that step on your own.

Example 3: $\int 3\sin\left(3x - 1\right) dx =$

Let $u = 3x - 1$. Then $du = 3dx$. Substitute the u in the integral.

$$\int \sin u \, du$$

Figure out the integral.

$$\int \sin u \, du = -\cos u + C$$

And throw the x's back in.

$$-\cos(3x - 1) + C$$

So far, this is only the simplest kind of u-substitution; naturally, the process can get worse when the substitution isn't as easy. Usually, you'll have to insert a constant term to put the integrand into a workable form.

Example 4: $\int \left(5x + 7\right)^{20} dx =$

Let $u = 5x + 7$. Then $du = 5 \, dx$. Notice that we can't do the substitution immediately because we need to substitute for dx and we have $5 \, dx$. No problem! Because 5 is a constant, just solve for dx.

$$\frac{1}{5} du = dx$$

Now you can substitute.

$$\int \left(5x + 7\right)^{20} dx = \int u^{20} \left(\frac{1}{5}\right) du$$

Rearrange the integral and solve.

$$\int u^{20}\left(\frac{1}{5}\right)du = \frac{1}{5}\int u^{20}\,du = \frac{1}{5}\frac{u^{21}}{21}+C = \frac{u^{21}}{105}+C$$

And now it's time to substitute back.

$$\frac{u^{21}}{105}+C = \frac{(5x+7)^{21}}{105}+C$$

Example 5: $\int x\cos\left(3x^2+1\right)dx =$

Let $u = 3x^2 + 1$. Then $du = 6x\,dx$. We need to substitute for $x\,dx$, so we can rearrange the du term.

$$\frac{1}{6}du = x\,dx$$

Now substitute.

$$\int\frac{1}{6}\cos u\,du$$

Evaluate the integral.

$$\int\frac{1}{6}\cos u\,du = \frac{1}{6}\sin u + C$$

And substitute back.

$$\frac{1}{6}\sin u + C = \frac{1}{6}\sin\left(3x^2+1\right)+C$$

Example 6: $\int x\sec^2\left(x^2\right)dx =$

Let $u = x^2$. Then $du = 2x\,dx$ and $\frac{1}{2}du = x\,dx$.

Substitute.

$$\int\frac{1}{2}\sec^2 u\,du$$

Evaluate the integral.

$$\int\frac{1}{2}\sec^2 u\,du = \frac{1}{2}\tan u + C$$

Now the original function goes back in.

$$\frac{1}{2}\tan\left(x^2\right)+C$$

This is a good technique to master, so practice on the following solved problems. Do each problem, covering the answer first, then check your answer.

PROBLEM 1. Evaluate $\int \sec^2 3x \, dx$.

Answer: Let $u = 3x$ and $du = 3dx$. Then $\frac{1}{3}du = dx$.

Substitute and integrate.

$$\frac{1}{3}\int \sec^2 u \, du = \frac{1}{3}\tan u + C$$

Then substitute back.

$$\frac{1}{3}\tan 3x + C$$

PROBLEM 2. Evaluate $\int \sqrt{5x-4} \, dx$.

Answer: Let $u = 5x - 4$ and $du = 5dx$. Then $\frac{1}{5}du = dx$.

Substitute and integrate.

$$\frac{1}{5}\int u^{\frac{1}{2}} \, du = \frac{2}{15}u^{\frac{3}{2}} + C$$

Then substitute back.

$$\frac{2}{15}(5x-4)^{\frac{3}{2}} + C$$
.

PROBLEM 3. Evaluate $\int x\left(4x^2 - 7\right)^{10} dx$.

Answer: Let $u = 4x^2 - 7$ and $du = 8x \, dx$. Then $\frac{1}{8}du = x \, dx$.

Substitute and integrate.

$$\frac{1}{8}\int u^{10} \, du = \frac{1}{88}u^{11} + C$$

Then substitute back.

$$\frac{1}{88}\left(4x^2 - 7\right)^{11} + C$$

PROBLEM 4. Evaluate $\int \tan\frac{x}{3} \sec^2\frac{x}{3} dx$.

Answer: Let $u = \tan\frac{x}{3}$ and $du = \frac{1}{3}\sec^2\frac{x}{3} dx$. Then $3du = \sec^2\frac{x}{3} dx$.

Substituting, we get

$$3\int u \, du = \frac{3}{2}u^2 + C$$

Then substitute back.

$$\frac{3}{2}\tan^2\frac{x}{3} + C$$

PRACTICE PROBLEM SET 19

Now evaluate the following integrals. The answers are in Chapter 19.

1. $\int \sin 2x \cos 2x \, dx$

2. $\int \dfrac{3x \, dx}{\sqrt[3]{10 - x^2}}$

3. $\int x^3 \sqrt{5x^4 + 20} \, dx$

4. $\int \left(x^2 + 1\right)\left(x^3 + 3x\right)^{-5} dx$

5. $\int \dfrac{1}{\sqrt{x}} \sin\sqrt{x} \, dx$

6. $\int x^2 \sec^2 x^3 \, dx$

7. $\int \dfrac{\cos\left(\frac{3}{x}\right)}{x^2} dx$

8. $\int \sin(\sin x)\cos x \, dx$

End of Chapter 13 Drill

The answers are in Chapter 20.

1. Evaluate $\int \left(3x^4 - 6x^3 + 2x + 7\right) dx$.

(A) $12x^3 - 18x^2 + 2 + C$

(B) $3x^5 - 6x^4 + x^2 + 7x + C$

(C) $\dfrac{3x^5}{5} - \dfrac{6x^4}{4} + x^2 + 7x + C$

(D) $\dfrac{3x^4}{4} - \dfrac{6x^3}{3} + \dfrac{x^2}{2} + 7x + C$

2. Evaluate $\int 3\cos x - \dfrac{4}{\sqrt{x}}\, dx$.

(A) $3\sin x - 2\sqrt{x} + C$

(B) $3\sin x - 8\sqrt{x} + C$

(C) $3\sin x + 2\sqrt{x} + C$

(D) $-3\sin x + 8\sqrt{x} + C$

3. Evaluate $\int \dfrac{9x^2}{\left(1-x^3\right)^2}\, dx$.

(A) $\dfrac{3}{1-x^3} + C$

(B) $\dfrac{-3}{1-x^3} + C$

(C) $\dfrac{3}{\left(1-x^3\right)^3} + C$

(D) $\dfrac{-3}{\left(1-x^3\right)^3} + C$

4. Evaluate $\int \dfrac{\cos\left(\dfrac{1}{x}\right)}{x^2}\, dx$.

(A) $\dfrac{\sin\left(\dfrac{1}{x}\right)}{x} + C$

(B) $\dfrac{-\sin\left(\dfrac{1}{x}\right)}{x} + C$

(C) $\sin\left(\dfrac{1}{x}\right) + C$

(D) $-\sin\left(\dfrac{1}{x}\right) + C$

5. Evaluate $\int \cos^3(3x)\sin(3x)\, dx$.

(A) $\dfrac{\cos^4(3x)}{12} + C$

(B) $-\dfrac{\cos^4(3x)}{12} + C$

(C) $-\dfrac{\sin^4(3x)}{12} + C$

(D) $\dfrac{\sin^4(3x)}{12} + C$

REFLECT

Respond to the following questions:

- For which topics discussed in this chapter do you feel you have achieved sufficient mastery to answer multiple-choice questions correctly?

- For which topics discussed in this chapter do you feel you have achieved sufficient mastery to answer open-ended questions correctly?

- For which topics discussed in this chapter do you feel you need more work before you can answer multiple-choice questions correctly?

- For which topics discussed in this chapter do you feel you need more work before you can answer open-ended questions correctly?

- What parts of this chapter are you going to re-review?

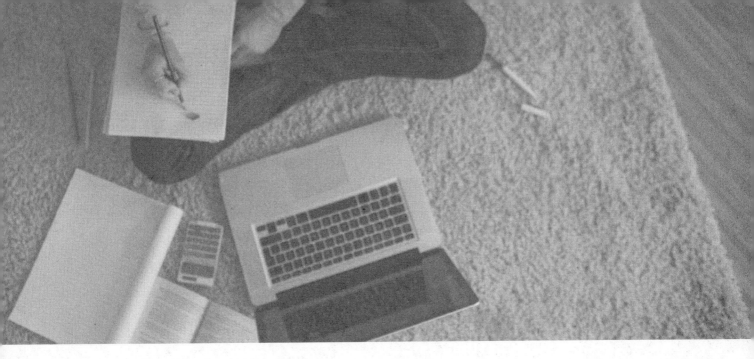

Chapter 14
Definite Integrals

AREA UNDER A CURVE

It's time to learn one of the most important uses of the integral. We've already discussed how integration can be used to "antidifferentiate" a function; now you'll see that you can also use it to find the area under a curve. First, here's a little background about how to find the area without using integration.

Suppose you have to find the area under the curve $y = x^2 + 2$ from $x = 1$ to $x = 3$. The graph of the curve looks like the following:

Don't panic yet. Nothing you've learned in geometry thus far has taught you how to find the area of something like this. You have learned how to find the area of a rectangle, though, and we're going to use rectangles to approximate the area between the curve and the x-axis.

Let's divide the region into two rectangles, one from $x = 1$ to $x = 2$ and one from $x = 2$ to $x = 3$, where the top of each rectangle comes just under the curve. It looks like the following:

Notice that the width of each rectangle is 1. The height of the left rectangle is found by plugging 1 into the equation $y = x^2 + 2$ (yielding 3); the height of the right rectangle is found by plugging 2 into the same equation (yielding 6). The combined area of the two rectangles is $(1)(3) + (1)(6) = 9$. So we could say that the area under the curve is approximately 9 square units.

Naturally, this is a pretty rough approximation that significantly underestimates the area. Look at how much of the area we missed by using two rectangles! How do you suppose we could make the approximation better? Divide the region into more, thinner rectangles.

This time, cut up the region into four rectangles, each with a width of $\frac{1}{2}$. It looks like the following:

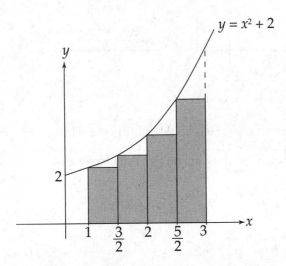

Now find the height of each rectangle the same way as before. Notice that the values we use are the left endpoints of each rectangle. The heights of the rectangles are, respectively,

$$(1)^2 + 2 = 3; \quad \left(\frac{3}{2}\right)^2 + 2 = \frac{17}{4}; \quad (2)^2 + 2 = 6 \text{ and } \left(\frac{5}{2}\right)^2 + 2 = \frac{33}{4}$$

Now, multiply each height by the width of $\frac{1}{2}$ and add up the areas.

$$\left(\frac{1}{2}\right)(3) + \left(\frac{1}{2}\right)\left(\frac{17}{4}\right) + \left(\frac{1}{2}\right)(6) + \left(\frac{1}{2}\right)\left(\frac{33}{4}\right) = \frac{43}{4}$$

This is a much better approximation of the area, but there's still a lot of space that isn't accounted for. We're still underestimating the area. The rectangles need to be thinner. But before we do that, let's do something else.

Notice how each of the rectangles is inscribed in the region. Suppose we used circumscribed rectangles instead—that is, we could determine the height of each rectangle by the higher of the two y-values, not the lower. The following graph shows what the region would look like:

To find the area of the rectangles, we would still use the width of $\frac{1}{2}$, but the heights would change. The heights of each rectangle are now found by plugging in the right endpoint of each rectangle.

$$\left(\frac{3}{2}\right)^2 + 2 = \frac{17}{4};\ (2)^2 + 2 = 6;\ \left(\frac{5}{2}\right)^2 + 2 = \frac{33}{4};\ \text{and } (3)^2 + 2 = 11$$

Once again, multiply each height by the width of $\frac{1}{2}$ and add up the areas.

$$\left(\frac{1}{2}\right)\left(\frac{17}{4}\right) + \left(\frac{1}{2}\right)(6) + \left(\frac{1}{2}\right)\left(\frac{33}{4}\right) + \left(\frac{1}{2}\right)(11) = \frac{59}{4}$$

The area under the curve using four left endpoint rectangles is $\frac{43}{4}$, and the area using four right endpoint rectangles is $\frac{59}{4}$, so why not average the two? This gives us $\frac{51}{4}$, which is a better approximation of the area.

Now that we've found the area using the rectangles a few times, let's turn the method into a formula. Call the left endpoint of the interval a and the right endpoint of the interval b, and set the number of rectangles we use equal to n. So the width of each rectangle is $\dfrac{b-a}{n}$. The height of the first inscribed rectangle is y_0, the height of the second rectangle is y_1, the height of the third rectangle is y_2, and so on, up to the last rectangle, which is y_{n-1}. If we use the left endpoint of each rectangle, the area under the curve is

$$\left(\frac{b-a}{n}\right)\left[y_0 + y_1 + y_2 + y_3 \ldots + y_{n-1}\right]$$

If we use the right endpoint of each rectangle, then the formula is

$$\left(\frac{b-a}{n}\right)\left[y_1 + y_2 + y_3 \ldots + y_n\right]$$

Now for the fun part. Remember how we said that we could make the approximation better by making more, thinner rectangles? By letting n approach infinity, we create an infinite number of rectangles that are infinitesimally thin. The formula for "left-endpoint" rectangles becomes

$$\lim_{n\to\infty}\left(\frac{b-a}{n}\right)\left[y_0 + y_1 + y_2 + y_3 \ldots + y_{n-1}\right]$$

For "right-endpoint" rectangles, the formula becomes

$$\lim_{n\to\infty}\left(\frac{b-a}{n}\right)\left[y_1 + y_2 + y_3 \ldots + y_n\right]$$

Note: Sometimes these are called "inscribed" and "circumscribed" rectangles, but that restricts the use of the formula. It's more exact to evaluate these rectangles using right and left endpoints.

We could also find the area using the midpoint of each interval. Let's once again use the above example, dividing the region into four rectangles. The region would look like the following:

To find the area of the rectangles, we would still use the width of $\frac{1}{2}$, but the heights would now be found by plugging the midpoint of each interval into the equation.

$$\left(\frac{5}{4}\right)^2 + 2 = \frac{57}{16}; \left(\frac{7}{4}\right)^2 + 2 = \frac{81}{16}; \left(\frac{9}{4}\right)^2 + 2 = \frac{113}{16}; \left(\frac{11}{4}\right)^2 + 2 = \frac{153}{16}$$

Once again, multiply each height by the width of $\frac{1}{2}$ and add up the areas.

$$\left(\frac{1}{2}\right)\left(\frac{57}{16}\right) + \left(\frac{1}{2}\right)\left(\frac{81}{16}\right) + \left(\frac{1}{2}\right)\left(\frac{113}{16}\right) + \left(\frac{1}{2}\right)\left(\frac{153}{16}\right) = \frac{404}{32} = \frac{101}{8} = 12.625$$

The general formula for approximating the area under a curve using midpoints is

$$\left(\frac{b-a}{n}\right)\left[y_{\frac{1}{2}} + y_{\frac{3}{2}} + y_{\frac{5}{2}} + \ldots + y_{\frac{2n-1}{2}}\right]$$

(Note: The fractional subscript means to evaluate the function at the number halfway between each integral pair of values of n.)

When you take the limit of this infinite sum, you get the integral. (You knew the integral had to show up sometime, didn't you?) Actually, we write the integral as

$$\int_1^3 \left(x^2 + 2\right)dx$$

It is called a **definite integral**, and it means that we're finding the area under the curve $x^2 + 2$ from $x = 1$ to $x = 3$. (We'll discuss how to evaluate this in a moment.) On the AP Exam, you'll only be asked to divide the region into a small number of rectangles, so it won't be very hard. Let's do an example.

Example 1: Approximate the area under the curve $y = x^3$ from $x = 2$ to $x = 3$ using four left-endpoint rectangles.

Draw four rectangles that look like the following:

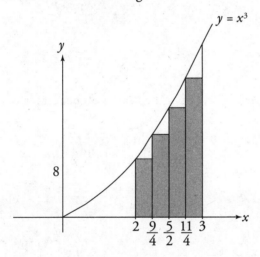

The width of each rectangle is $\frac{1}{4}$. The heights of the rectangles are

$$2^3, \left(\frac{9}{4}\right)^3, \left(\frac{5}{2}\right)^3, \text{ and } \left(\frac{11}{4}\right)^3$$

Therefore, the area is

$$\left(\frac{1}{4}\right)\left(2^3\right)+\left(\frac{1}{4}\right)\left(\frac{9}{4}\right)^3+\left(\frac{1}{4}\right)\left(\frac{5}{2}\right)^3+\left(\frac{1}{4}\right)\left(\frac{11}{4}\right)^3 = \frac{893}{64} \approx 13.953$$

Example 2: Repeat Example 1 using four right-endpoint rectangles.

Now draw four rectangles that look like the following:

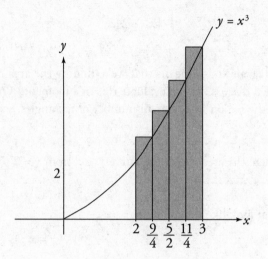

The width of each rectangle is still $\dfrac{1}{4}$, but the heights of the rectangles are now

$$\left(\frac{9}{4}\right)^3, \left(\frac{5}{2}\right)^3, \left(\frac{11}{4}\right)^3, \text{ and } \left(3^3\right)$$

The area is now

$$\left(\frac{1}{4}\right)\left(\frac{9}{4}\right)^3 + \left(\frac{1}{4}\right)\left(\frac{5}{2}\right)^3 + \left(\frac{1}{4}\right)\left(\frac{11}{4}\right)^3 + \left(\frac{1}{4}\right)\left(3^3\right) = \frac{1197}{64} \approx 18.703$$

Example 3: Repeat Example 1 using four midpoint rectangles.

Now draw four rectangles that look like the following:

The width of each rectangle is still $\dfrac{1}{4}$, but the heights of the rectangles are now

$$\left(\frac{17}{8}\right)^3, \left(\frac{19}{8}\right)^3, \left(\frac{21}{8}\right)^3, \text{ and } \left(\frac{23}{8}\right)^3$$

The area is now

$$\left(\frac{1}{4}\right)\left(\frac{17}{8}\right)^3 + \left(\frac{1}{4}\right)\left(\frac{19}{8}\right)^3 + \left(\frac{1}{4}\right)\left(\frac{21}{8}\right)^3 + \left(\frac{1}{4}\right)\left(\frac{23}{8}\right)^3 = \frac{2075}{128} \approx 16.211$$

TABULAR RIEMANN SUMS

Sometimes the AP Exam will ask you to find a Riemann sum, or to approximate an integral (same thing, right?), but won't give you a function to work with. Instead, they will give you a table of values for x and $f(x)$. These are quite simple to evaluate. All you do is use the right-hand or left-hand sum formula, plugging in the appropriate values for $f(x)$. One thing you should watch out for is that sometimes the x-values are not evenly spaced, so make sure that you use the correct values for the widths of the rectangles. Let's do an example.

Example 4: Suppose we are given the following table of values for x and $f(x)$.

x	2	4	6	8	10	12
$f(x)$	10	13	15	14	9	3

Use a *right-hand* Riemann sum with 5 subintervals indicated by the data in the table to approximate $\int_2^{12} f(x)dx$.

Recall that the formula for finding the area under the curve using the right endpoints is $\left(\dfrac{b-a}{n}\right)[y_1 + y_2 + y_3 + \ldots + y_n]$. Here, the width of each rectangle is 2. We find the height of each rectangle by evaluating $f(x)$ at the appropriate value of x, the right endpoint of each interval on the x-axis. Here, $y_1 = 13$, $y_2 = 15$, $y_3 = 14$, $y_4 = 9$, and $y_5 = 3$. Therefore, we can approximate the integral with $\int_2^{12} f(x)dx = (2)(13) + (2)(15) + (2)(14) + (2)(9) + (2)(3) = 108$.

Let's do another example, but this time the values of x will not be evenly spaced on the x-axis.

Example 5: Given the following table of values for x and $f(x)$:

x	0	2	5	11	19	22	23
$f(x)$	4	6	16	18	22	29	50

Use a *left-hand* Riemann sum with 6 subintervals indicated by the data in the table to approximate $\int_0^{23} f(x)dx$.

Recall that the formula for finding the area under the curve using the left endpoints is $\left(\dfrac{b-a}{n}\right)[y_0 + y_1 + y_2 + \ldots + y_{n-1}]$. This formula assumes that the x-values are evenly spaced, but they aren't here, so we will replace the values of $\left(\dfrac{b-a}{n}\right)$ with the appropriate widths of each rectangle. The width of the first rectangle is $2 - 0 = 2$; the second width is $5 - 2 = 3$; the third is $11 - 5 = 6$; the fourth is $19 - 11 = 8$; the fifth is $22 - 19 = 3$; and the sixth is $23 - 22 = 1$. We find the height of each rectangle by evaluating $f(x)$ at the appropriate value of x, the left endpoint of each interval on the x-axis. Here, $y_0 = 4$, $y_1 = 6$, $y_2 = 16$, $y_3 = 18$, $y_4 = 22$, and $y_5 = 29$. Therefore, we can approximate the integral with $\int_0^{23} f(x)dx = (2)(4) + (3)(6) + (6)(16) + (8)(18) + (3)(22) + (1)(29) = 361$.

That's all there is to approximating the area under a curve using rectangles. Now let's learn how to find the area exactly. In order to evaluate this, you'll need to know. . .

The Fundamental Theorem of Calculus

Before, we said that if you create an infinite number of infinitely thin rectangles, you'll get the area under the curve, which is an integral. For the example above, the integral is

$$\int_2^3 x^3 dx$$

There is a rule for evaluating an integral like this. The rule is called the **Fundamental Theorem of Calculus**, and it says

$$\int_a^b f(x)dx = F(b) - F(a); \text{ where } F(x) \text{ is the antiderivative of } f(x).$$

Using this rule, you can find $\int_2^3 x^3 dx$ by integrating it, and we get $\dfrac{x^4}{4}$. Now all you do is plug in 3 and 2 and take the difference. We use the following notation to symbolize this:

$$\left.\frac{x^4}{4}\right|_2^3$$

Thus, we have

$$\frac{3^4}{4} - \frac{2^4}{4} = \frac{81}{4} - \frac{16}{4} = \frac{65}{4}$$

Because $\dfrac{65}{4} = 16.25$, you can see how close we were with our three earlier approximations.

Example 6: Find $\int_1^3 \left(x^2 + 2\right)dx$.

The Fundamental Theorem of Calculus yields the following:

$$\int_1^3 \left(x^2 + 2\right)dx = \left.\left(\frac{x^3}{3} + 2x\right)\right|_1^3$$

If we evaluate this, we get the following:

$$\left(\frac{3^3}{3} + 2(3)\right) - \left(\frac{1^3}{3} + 2(1)\right) = \frac{38}{3}$$

This is the first function for which we found the approximate area by using inscribed rectangles. Our final estimate, where we averaged the inscribed and circumscribed rectangles, was $\frac{51}{4}$, and as you can see, that was very close (off by $\frac{1}{12}$). When we used the midpoints, we were off by $\frac{1}{24}$.

We're going to do only a few approximations using rectangles, because it's not a big part of the AP Exam. On the other hand, definite integrals are a huge part of the rest of this book.

Example 7: $\int_1^5 \left(x^2 - x\right) dx = \left(\frac{x^3}{3} - \frac{x^2}{2}\right)\Bigg|_1^5 = \left(\frac{125}{3} - \frac{25}{2}\right) - \left(\frac{1}{3} - \frac{1}{2}\right) = \frac{88}{3}$

Example 8: $\int_0^{\frac{\pi}{2}} \sin x \, dx = (-\cos x)\Bigg|_0^{\frac{\pi}{2}} = \left(-\cos\frac{\pi}{2}\right) - (-\cos 0) = 1$

Example 9: $\int_0^{\frac{\pi}{4}} \sec^2 x \, dx = \tan x \Bigg|_0^{\frac{\pi}{4}} = \tan\frac{\pi}{4} - \tan 0 = 1$

The Trapezoid Rule

There's another approximation method that's even better than the rectangle method. Essentially, all you do is divide the region into trapezoids instead of rectangles. Let's use the problem that we did at the beginning of the chapter.

We get a picture that looks like the following:

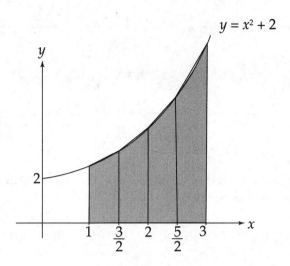

As you should recall from geometry, the formula for the area of a trapezoid is

$$\frac{1}{2}(b_1 + b_2)h$$

(Note: b_1 and b_2 are the two bases of the trapezoid.) Notice that each of the shapes is a trapezoid on its side, so the height of each trapezoid is the length of the interval $\frac{1}{2}$, and the bases are the y-values that correspond to each x-value. We found these earlier in the rectangle example; they are, in order: 3, $\frac{17}{4}$, 6, $\frac{33}{4}$, and 11. We can find the area of each trapezoid and add them up.

$$\frac{1}{2}\left(3 + \frac{17}{4}\right)\left(\frac{1}{2}\right) + \frac{1}{2}\left(\frac{17}{4} + 6\right)\left(\frac{1}{2}\right) + \frac{1}{2}\left(6 + \frac{33}{4}\right)\left(\frac{1}{2}\right) + \frac{1}{2}\left(\frac{33}{4} + 11\right)\left(\frac{1}{2}\right) = \frac{51}{4}, \text{ or } 12.75$$

Recall that the actual value of the area is $\frac{38}{3}$ or 12.67; the Trapezoid Rule gives a pretty good approximation.

Notice how each trapezoid shares a base with the trapezoid next to it, except for the end ones. This enables us to simplify the formula for using the Trapezoid Rule. Each trapezoid has a height equal to the length of the interval divided by the number of trapezoids we use. If the interval is from $x = a$ to $x = b$, and the number of trapezoids is n, then the height of each trapezoid is $\dfrac{b-a}{n}$. Then our formula is

$$\left(\frac{1}{2}\right)\left(\frac{b-a}{n}\right)\left[y_0 + 2y_1 + 2y_2 + 2y_3 \ldots + 2y_{n-2} + 2y_{n-1} + y_n\right]$$

This is all you need to know about the Trapezoid Rule. Just follow the formula, and you won't have any problems. Let's do one more example.

Example 10: Approximate the area under the curve $y = x^3$ from $x = 2$ to $x = 3$ using four inscribed trapezoids.

Following the rule, the height of each trapezoid is $\dfrac{3-2}{4} = \dfrac{1}{4}$. Thus, the approximate area is

$$\left(\frac{1}{2}\right)\left(\frac{1}{4}\right)\left[2^3 + 2\left(\frac{9}{4}\right)^3 + 2\left(\frac{5}{2}\right)^3 + 2\left(\frac{11}{4}\right)^3 + 3^3\right] = \frac{1{,}045}{64}$$

Compare this answer to the actual value we found earlier—it's pretty close!

PROBLEM 1. Approximate the area under the curve $y = 4 - x^2$ from $x = -1$ to $x = 1$ with $n = 4$ inscribed rectangles.

Answer: Draw four rectangles that look like this:

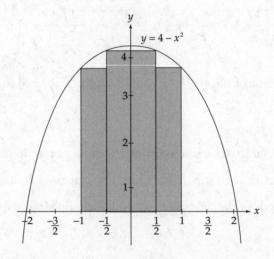

The width of each rectangle is $\frac{1}{2}$. The heights of the rectangles are found by evaluating $y = 4 - x^2$ at the appropriate endpoints.

$$\left(4-(-1)^2\right), \left(4-\left(-\frac{1}{2}\right)^2\right), \left(4-\left(\frac{1}{2}\right)^2\right), \text{ and } \left(4-(1)^2\right)$$

These can be simplified to $3, \frac{15}{4}, \frac{15}{4}$, and 3. Therefore, the area is

$$\left(\frac{1}{2}\right)(3) + \left(\frac{1}{2}\right)\left(\frac{15}{4}\right) + \left(\frac{1}{2}\right)\left(\frac{15}{4}\right) + \left(\frac{1}{2}\right)(3) = \frac{27}{4}$$

PROBLEM 2. Find the area under the curve $y = 4 - x^2$ from $x = -1$ to $x = 1$ with $n = 4$ circumscribed rectangles.

Answer: Draw four rectangles that look like the following:

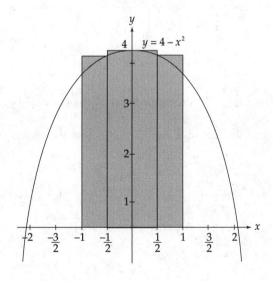

The width of each rectangle is $\frac{1}{2}$. The heights of the rectangles are found by evaluating $y = 4 - x^2$ at the appropriate endpoints.

$$\left(4-\left(-\frac{1}{2}\right)^2\right), \left(4-(0)^2\right), \left(4-(0)^2\right), \text{ and } \left(4-\left(\frac{1}{2}\right)^2\right)$$

These can be simplified to $\frac{15}{4}$, 4, 4, and $\frac{15}{4}$. Therefore, the area is

$$\left(\frac{1}{2}\right)\left(\frac{15}{4}\right)+\left(\frac{1}{2}\right)(4)+\left(\frac{1}{2}\right)(4)+\left(\frac{1}{2}\right)\left(\frac{15}{4}\right)=\frac{31}{4}$$

PROBLEM 3. Find the area under the curve $y = 4 - x^2$ from $x = -1$ to $x = 1$ using the Trapezoid Rule with $n = 4$.

Answer: Draw four trapezoids that look like the following:

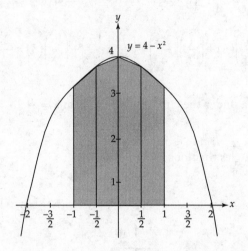

The width of each trapezoid is $\frac{1}{2}$. Evaluate the bases of the trapezoid by calculating $y = 4 - x^2$ at the appropriate endpoints. Following the rule, we get that the area is approximately

$$\left(\frac{1}{2}\right)\left(\frac{1}{2}\right)\left[\left(4-(-1)^2\right)+2\left(4-\left(-\frac{1}{2}\right)^2\right)+2\left(4-(0)^2\right)+2\left(4-\left(\frac{1}{2}\right)^2\right)+\left(4-(1)^2\right)\right]=$$
$$\left(\frac{1}{2}\right)\left(\frac{1}{2}\right)\left[3+\frac{15}{2}+8+\frac{15}{2}+3\right]=\frac{29}{4}$$

PROBLEM 4. Find the area under the curve $y = 4 - x^2$ from $x = -1$ to $x = 1$ using the Midpoint Formula with $n = 4$.

Answer: Draw four rectangles that look like the following:

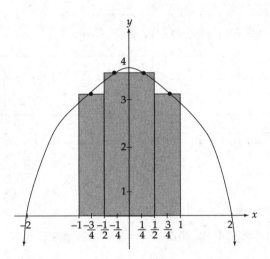

The width of each rectangle is $\dfrac{1}{2}$. The heights of the rectangles are found by evaluating $y = 4 - x^2$ at the appropriate points.

$$4 - \left(-\frac{3}{4}\right)^2 = \frac{55}{16}; \quad 4 - \left(-\frac{1}{4}\right)^2 = \frac{63}{16}; \quad 4 - \left(\frac{1}{4}\right)^2 = \frac{63}{16}; \quad 4 - \left(\frac{3}{4}\right)^2 = \frac{55}{16}$$

Multiply each height by the width of $\dfrac{1}{2}$, and add up the areas.

$$\left(\frac{1}{2}\right)\left(\frac{55}{16}\right) + \left(\frac{1}{2}\right)\left(\frac{63}{16}\right) + \left(\frac{1}{2}\right)\left(\frac{63}{16}\right) + \left(\frac{1}{2}\right)\left(\frac{55}{16}\right) = \frac{59}{8}$$

PROBLEM 5. Find the area under the curve $y = 4 - x^2$ from $x = -1$ to $x = 1$.

Answer: Now we can use the definite integral by evaluating.

$$\int_{-1}^{1} \left(4 - x^2\right) dx$$

This can be rewritten as

$$\left(4x - \frac{x^3}{3}\right)\Bigg|_{-1}^{1}$$

Follow the Fundamental Theorem of Calculus.

$$\left(4 - \frac{1}{3}\right) - \left(4(-1) - \frac{(-1)^3}{3}\right) = \frac{22}{3}$$

PROBLEM 6. Given the following table of values for t and $f(t)$:

t	0	2	4	7	11	13	14
$f(t)$	5	6	10	15	20	26	30

Use a *right-hand* Riemann sum with 6 subintervals indicated by the data in the table to approximate $\int_0^{14} f(t)\, dt$.

Answer: The width of the first rectangle is $2 - 0 = 2$; the second width is $4 - 2 = 2$; the third is $7 - 4 = 3$; the fourth is $11 - 7 = 4$; the fifth is $13 - 11 = 2$; and the sixth is $14 - 13 = 1$. We find the height of each rectangle by evaluating $f(t)$ at the appropriate value of t, the right endpoint of each interval on the t-axis. Here, $y_1 = 6$, $y_2 = 10$, $y_3 = 15$ $y_4 = 20$, $y_5 = 26$, and $y_6 = 30$. Therefore, we can approximate the integral with

$$\int_0^{14} f(t)\, dt = (2)(6) + (2)(10) + (3)(15) + (4)(20) + (2)(26) + (1)(30) = 239$$

PRACTICE PROBLEM SET 20

Here's a great opportunity to practice finding the area beneath a curve and evaluating integrals. The answers are in Chapter 19.

1. Find the area under the curve $y = 2x - x^2$ from $x = 1$ to $x = 2$ with $n = 4$ left-end-point rectangles.

2. Find the area under the curve $y = 2x - x^2$ from $x = 1$ to $x = 2$ with $n = 4$ right-end-point rectangles.

3. Find the area under the curve $y = 2x - x^2$ from $x = 1$ to $x = 2$ using the Trapezoid Rule with $n = 4$.

4. Find the area under the curve $y = 2x - x^2$ from $x = 1$ to $x = 2$ using the Midpoint Formula with $n = 4$.

5. Find the area under the curve $y = 2x - x^2$ from $x = 1$ to $x = 2$.

6. Evaluate $\int_{-\frac{\pi}{2}}^{\frac{\pi}{2}} \cos\ x\ dx$.

7. Evaluate $\int_{1}^{9} 2x\sqrt{x}\ dx$.

8. Evaluate $\int_{0}^{1} \left(x^4 - 5x^3 + 3x^2 - 4x - 6\right)dx$.

9. Evaluate $\int_{-\frac{\pi}{2}}^{\frac{\pi}{2}} \sin\ x\ dx$.

10. Suppose we are given the following table of values for x and $g(x)$:

x	0	1	3	5	9	14
$g(x)$	10	8	11	17	20	23

Use a left-hand Riemann sum with 5 subintervals indicated by the data in the table to approximate $\int_{0}^{14} g(x)dx$.

THE MEAN VALUE THEOREM FOR INTEGRALS

As you recall, we did the Mean Value Theorem once before, in Chapter 8, but this time we'll apply it to integrals, not derivatives. In fact, some books refer to it as the "Mean Value Theorem for Integrals" or MVTI. The most important aspect of the MVTI is that it enables you to find the average value of a function. In fact, the AP Exam will often ask you to find the average value of a function, which is just its way of testing your knowledge of the MVTI.

Here's the theorem.

> If $f(x)$ is continuous on a closed interval $[a, b]$, then at some point c in the interval $[a, b]$ the following is true:
>
> $$\int_a^b f(x)dx = f(c)(b-a)$$

This tells you that the area under the curve of $f(x)$ on the interval $[a, b]$ is equal to the value of the function at some value c (between a and b) times the length of the interval. If you look at this graphically, you can see that you're finding the area of a rectangle whose base is the interval and whose height is some value of $f(x)$ that creates a rectangle with the same area as the area under the curve.

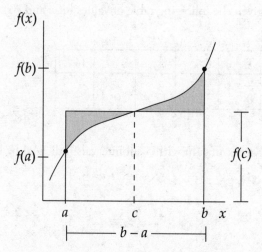

The number $f(c)$ gives us the average value of f on $[a, b]$. Thus, if we rearrange the theorem, we get the formula for finding the average value of $f(x)$ on $[a, b]$.

This is the formula for the average value.

$$f(c) = \frac{1}{b-a}\int_a^b f(x)dx$$

There's all you need to know about finding average values. Try some examples.

Example 1: Find the average value of $f(x) = x^2$ from $x = 2$ to $x = 4$.

Evaluate the integral $\dfrac{1}{4-2}\displaystyle\int_2^4 x^2\,dx$.

$$\frac{1}{4-2}\int_2^4 x^2\,dx = \left[\frac{1}{2}\ \frac{x^3}{3}\Big|_2^4\right] = \frac{1}{2}\left(\frac{64}{3}-\frac{8}{3}\right) = \frac{28}{3}$$

Example 2: Find the average value of $f(x) = \sin x$ on $[0, \pi]$.

Evaluate $\dfrac{1}{\pi-0}\displaystyle\int_0^\pi \sin x\,dx$.

$$\frac{1}{\pi-0}\int_0^\pi \sin x\,dx = \frac{1}{\pi}\left(-\cos x\right)\Big|_0^\pi = \frac{1}{\pi}\left(-\cos\pi + \cos 0\right) = \frac{2}{\pi}$$

The Second Fundamental Theorem of Calculus

As you saw in the last chapter, we've only half-learned the theorem. It has two parts, often referred to as the **First and Second Fundamental Theorems of Calculus**.

> The First Fundamental Theorem of Calculus (which you've already seen):
>
> If $f(x)$ is continuous at every point of $[a, b]$, and $F(x)$ is an antiderivative of $f(x)$ on $[a, b]$, then $\displaystyle\int_a^b f(x)dx = F(b) - F(a)$.

> The Second Fundamental Theorem of Calculus:
>
> If $f(x)$ is continuous on $[a, b]$, then the derivative of the function $F(x) = \displaystyle\int_a^x f(t)dt$ is
>
> $$\frac{dF}{dx} = \frac{d}{dx}\int_a^x f(t)dt = f(x)$$

These theorems are sometimes taught in reverse order.

We've already made use of the first theorem in evaluating definite integrals. In fact, we use the first Fundamental Theorem every time we evaluate a definite integral, so we're not going to give you any examples of that here. There is one aspect of the first Fundamental Theorem, however, that involves the area between curves (we'll discuss that in Chapter 16).

But for now, you should know the following:

If we have a point c in the interval $[a, b]$, then

$$\int_a^c f(x)\,dx + \int_c^b f(x)\,dx = \int_a^b f(x)\,dx$$

In other words, we can divide up the region into parts, add them up, and find the area of the region based on the result. We'll get back to this in the chapter on the area between two curves.

The second theorem tells us how to find the derivative of an integral.

Example 3: Find $\dfrac{d}{dx}\displaystyle\int_1^x \cos t\,dt$.

The second Fundamental Theorem says that the derivative of this integral is just $\cos x$.

Example 4: Find $\dfrac{d}{dx}\displaystyle\int_2^x \left(1 - t^3\right)dt$.

Here, the theorem says that the derivative of this integral is just $(1 - x^3)$.

Isn't this easy? Let's add a couple of nuances. First, the constant term in the limits of integration is a "dummy term." Any constant will give the same answer. For example,

$$\frac{d}{dx}\int_2^x (1 - t^3)\,dt = \frac{d}{dx}\int_{-2}^x (1 - t^3)\,dt = \frac{d}{dx}\int_\pi^x (1 - t^3)\,dt = 1 - x^3$$

In other words, all we're concerned with is the variable term.

Second, if the upper limit is a function of x, instead of just plain x, we multiply the answer by the derivative of that term. For example,

$$\frac{d}{dx}\int_2^{x^2} \left(1 - t^3\right)dt = \left[1 - \left(x^2\right)^3\right](2x) = \left(1 - x^6\right)(2x)$$

Example 5: Find $\dfrac{d}{dx}\displaystyle\int_0^{3x^4}\left(t+4t^2\right)dt=\left[3x^4+4(3x^4)^2\right]\left(12x^3\right)$.

Try these solved problems on your own. You know the drill.

PROBLEM 1. Find the average value of $f(x)=\dfrac{1}{x^2}$ on the interval [1, 3].

Answer: According to the Mean Value Theorem, the average value is found by evaluating

$$\frac{1}{3-1}\int_1^3\frac{dx}{x^2}$$

Your result should be

$$\frac{1}{2}\left(-\frac{1}{x}\right)\Bigg|_1^3=\frac{1}{2}\left(-\frac{1}{3}+1\right)=\frac{1}{3}$$

PROBLEM 2. Find the average value of $f(x)=\sin x$ on the interval $[-\pi,\pi]$.

Answer: According to the Mean Value Theorem, the average value is found by evaluating

$$\frac{1}{2\pi}\left(\int_{-\pi}^{\pi}\sin x\ dx\right)$$

Integrating, we get

$$\frac{1}{2\pi}\left(-\cos x\right)\Big|_{-\pi}^{\pi}=\frac{1}{2\pi}\left(-\cos\pi+\cos\left(-\pi\right)\right)=0$$

PROBLEM 3. Find $\dfrac{d}{dx}\displaystyle\int_1^x\frac{dt}{1-\sqrt[3]{t}}$.

Answer: According to the Second Fundamental Theorem of Calculus,

$$\frac{d}{dx}\int_1^x\frac{dt}{1-\sqrt[3]{t}}=\frac{1}{1-\sqrt[3]{x}}$$

PROBLEM 4. Find $\dfrac{d}{dx}\displaystyle\int_1^{x^2}\frac{t\ dt}{\sin t}$.

Answer: According to the Second Fundamental Theorem of Calculus,

$$\frac{d}{dx}\int_1^{x^2}\frac{t\ dt}{\sin t}=\frac{x^2}{\sin\left(x^2\right)}(2x)=\frac{2x^3}{\sin x^2}$$

Accumulation Functions

The AP Exam will also have problems that deal with **accumulation functions**. These are simply functions of the form $F(x) = \int_0^x f(t)\ dt$. These are called accumulation functions because the value of the integral is the area under the curve from the constant to the value x, and as x gets bigger, so does the area (it "accumulates"). In these functions, t is a dummy variable that is used as the variable of integration.

Let's do an example.

Example 1: Suppose we have the function $F(x) = \int_0^x t\ dt$. Let's evaluate this at different values of x. First, let's find $F(1)$. Graphically, we are looking for the area under the curve $y = t$ from $t = 0$ to $t = 1$. It looks like the following:

$F(1)$ is just the area of the triangle $A = \dfrac{1}{2}(1)(1) = \dfrac{1}{2}$. If we evaluate the integral, we get $F(1) = \int_0^1 t\ dt = \left(\dfrac{t^2}{2}\right)\Big|_0^1 = \dfrac{1}{2} - 0 = \dfrac{1}{2}$. Let's now find $F(2)$. This is the area under the curve $y = t$ from $t = 0$ to $t = 2$.

It looks like the following:

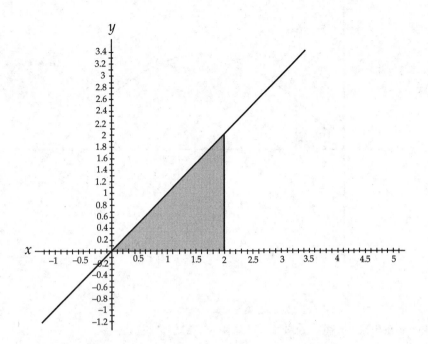

$F(2)$ is just the area of the triangle: $A = \dfrac{1}{2}(2)(2) = 2$. If we evaluate the integral, we get

$F(2) = \displaystyle\int_0^2 t\, dt = \left(\dfrac{t^2}{2}\right)\Bigg|_0^2 = 2 - 0 = 2$.

We can see that, as x increases, the function increases.

This was a fairly simple example. Let's do another one.

Example 2: Suppose we have the function $F(x) = \displaystyle\int_0^x \sin t\, dt$. Let's evaluate this as x increases from 0 to π. Obviously, $F(0) = 0$ because there is no area under the curve. So, first, let's find $F\left(\dfrac{\pi}{6}\right)$. Graphically, we are looking for the area under the curve $y = \sin t$ from $t = 0$ to $t = \dfrac{\pi}{6}$. It looks like the following:

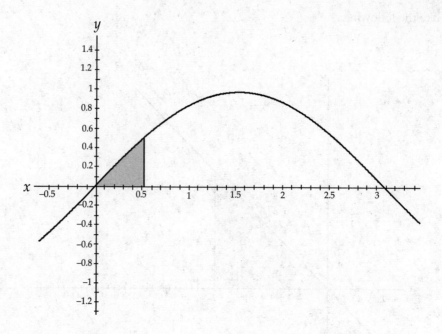

If we evaluate the integral, we get

$$F\left(\frac{\pi}{6}\right) = \int_0^{\frac{\pi}{6}} \sin t \, dt = \left(-\cos t\right)\Big|_0^{\frac{\pi}{6}} = -\cos\frac{\pi}{6} + \cos 0 = 1 - \frac{\sqrt{3}}{2} \approx 0.134$$

Now let's find $F\left(\dfrac{\pi}{4}\right)$. It looks like the following:

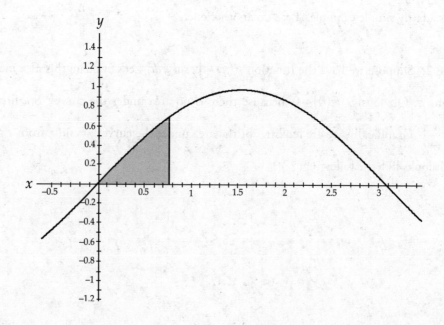

If we evaluate the integral, we get

$$F\left(\frac{\pi}{4}\right) = \int_0^{\frac{\pi}{4}} \sin t \; dt = \left(-\cos t\right)\Big|_0^{\frac{\pi}{4}} = -\cos\frac{\pi}{4} + \cos 0 = 1 - \frac{\sqrt{2}}{2} \approx 0.293$$

Let's make a table of values of the accumulation function for different values of x.

x	0	$\dfrac{\pi}{6}$	$\dfrac{\pi}{4}$	$\dfrac{\pi}{3}$	$\dfrac{\pi}{2}$	π
$F(x)$	0	$1 - \dfrac{\sqrt{3}}{2}$	$1 - \dfrac{\sqrt{2}}{2}$	$\dfrac{1}{2}$	1	2

We can see that the area is accumulating under the curve as x increases from 0 to π. Naturally, the values will shrink as we move from π to 2π because the values of $\sin x$ are negative. But, with accumulation functions, we are usually concerned only with positive areas.

Let's do one more example.

Example 3: Suppose we have the function $F(x) = \int_0^x t^2 \; dt$. Let's evaluate this as x increases from 0 to 4. First, let's find $F(1)$. Graphically, we are looking for the area under the curve $y = t^2$ from $t = 0$ to $t = 1$. It looks like the following:

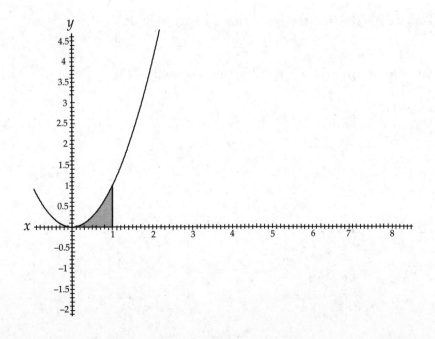

If we evaluate the integral, we get $F(1) = \int_0^1 t^2\, dt = \left(\dfrac{t^3}{3}\right)\Big|_0^1 = \dfrac{1}{3}$.

As in the previous example, let's make a table of values of the accumulation function for different values of x.

x	1	2	3	4
$F(x)$	$\dfrac{1}{3}$	$\dfrac{8}{3}$	9	$\dfrac{64}{3}$

We can see that the values of $F(x)$ will increase as x increases.

PRACTICE PROBLEM SET 21

Now try these problems. The answers are in Chapter 19.

1. Find the average value of $f(x) = 4x \cos x^2$ on the interval $\left[0, \sqrt{\dfrac{\pi}{2}}\right]$.

2. Find the average value of $f(x) = \sqrt{x}$ on the interval $[0, 16]$.

3. Find the average value of $f(x) = 2|x|$ on the interval $[-1, 1]$.

4. Find $\dfrac{d}{dx} \int_1^x \sin^2 t\, dt$.

5. Find $\dfrac{d}{dx} \int_1^{3x} \left(t^2 - t\right) dt$.

6. Find $\dfrac{d}{dx} \int_1^x -2\cos t\, dt$.

End of Chapter 14 Drill

The answers are in Chapter 20.

1. Approximate the area under the curve $y = 3 + x^2$ from $x = 1$ to $x = 5$ using $n = 4$ left-endpoint rectangles.

   (A) 21
   (B) 33
   (C) 42
   (D) 66

2. Approximate the area under the curve $y = x^3 + 4$ from $x = 0$ to $x = 8$ using $n = 4$ trapezoids.

   (A)   560
   (B) 1120
   (C) 2240
   (D) 4480

3. Approximate $\int_0^{16} \frac{3}{x} \, dx$ using the Midpoint Formula with $n = 4$.

   (A) $\dfrac{22}{35}$

   (B) $\dfrac{88}{35}$

   (C) $\dfrac{176}{35}$

   (D) $\dfrac{352}{35}$

4. Find the average value of $f(x) = \dfrac{x^2}{\left(1 - x^3\right)^2}$ on the interval [2, 4].

   (A) $\dfrac{2}{189}$

   (B) $\dfrac{4}{189}$

   (C) $\dfrac{8}{189}$

   (D) $\dfrac{16}{189}$

5. Find $\dfrac{d}{dx} \int_2^{x^2} \sin^3 t \, dt$.

   (A) $3x^2 \sin^2 x \cos x$
   (B) $6x \sin^2 x \cos x$
   (C) $x^2 \sin^3 x$
   (D) $2x \sin^3 x$

REFLECT

Respond to the following questions:

- For which topics discussed in this chapter do you feel you have achieved sufficient mastery to answer multiple-choice questions correctly?

- For which topics discussed in this chapter do you feel you have achieved sufficient mastery to answer open-ended questions correctly?

- For which topics discussed in this chapter do you feel you need more work before you can answer multiple-choice questions correctly?

- For which topics discussed in this chapter do you feel you need more work before you can answer open-ended questions correctly?

- What parts of this chapter are you going to re-review?

Chapter 15

Integrals of Exponential, Logarithmic, and Trig Functions

INVERSE TRIG FUNCTIONS

In this unit, you'll learn a broader set of integration techniques. So far, you know how to do only a few types of integrals: polynomials, some trig functions, and logs. Now we'll turn our attention to a set of integrals that result in **inverse trigonometric functions**. But first, we need to go over the derivatives of the inverse trigonometric functions. (You might want to refer back to Chapter 12 on derivatives of inverse functions.)

Suppose you have the equation $\sin y = x$. If you differentiate both sides with respect to x, you get

$$\cos y \frac{dy}{dx} = 1$$

Now divide both sides by $\cos y$.

$$\frac{dy}{dx} = \frac{1}{\cos y}$$

Because $\sin^2 y + \cos^2 y = 1$, we can replace $\cos y$ with $\sqrt{1 - \sin^2 y}$.

$$\frac{dy}{dx} = \frac{1}{\sqrt{1 - \sin^2 y}}$$

Finally, because $x = \sin y$, replace $\sin y$ with x. The derivative equals

$$\frac{1}{\sqrt{1 - x^2}}$$

Now go back to the original equation $\sin y = x$ and solve for y in terms of x: $y = \sin^{-1} x$.

If you differentiate both sides with respect to x, you get

$$\frac{dy}{dx} = \frac{d}{dx} \sin^{-1} x$$

Replace $\dfrac{dy}{dx}$, and you get the final result.

$$\frac{d}{dx} \sin^{-1} x = \frac{1}{\sqrt{1 - x^2}}$$

This is the derivative of inverse sine. By similar means, you can find the derivatives of all six inverse trig functions. They're not difficult to derive, and they're also not difficult to memorize. The choice is yours. We use u instead of x to account for the Chain Rule.

$$\frac{d}{dx}(\sin^{-1} u) = \frac{1}{\sqrt{1 - u^2}}\frac{du}{dx}; \quad -1 < u < 1 \qquad \frac{d}{dx}(\cos^{-1} u) = \frac{-1}{\sqrt{1 - u^2}}\frac{du}{dx}; \quad -1 < u < 1$$

$$\frac{d}{dx}(\tan^{-1} u) = \frac{1}{1 + u^2}\frac{du}{dx} \qquad\qquad \frac{d}{dx}(\cot^{-1} u) = \frac{-1}{1 + u^2}\frac{du}{dx}$$

$$\frac{d}{dx}(\sec^{-1} u) = \frac{1}{|u|\sqrt{u^2 - 1}}\frac{du}{dx}; \quad |u| > 1 \qquad \frac{d}{dx}(\csc^{-1} u) = \frac{-1}{|u|\sqrt{u^2 - 1}}\frac{du}{dx}; \quad |u| > 1$$

Notice the domain restrictions for inverse sine, cosine, secant, and cosecant.

Example 1: $\dfrac{d}{dx} = \sin^{-1} x^2 = \dfrac{2x}{\sqrt{1 - (x^2)^2}} = \dfrac{2x}{\sqrt{1 - x^4}}$

Example 2: $\dfrac{d}{dx} = \tan^{-1} 5x = \dfrac{5}{1 + (5x)^2} = \dfrac{5}{1 + 25x^2}$

Example 3:

$$\frac{d}{dx} = \sec^{-1}(x^2 - x) = \frac{2x - 1}{|x^2 - x|\sqrt{(x^2 - x)^2 - 1}} = \frac{2x - 1}{|x^2 - x|\sqrt{x^4 - 2x^3 + x^2 - 1}}$$

Finding the derivatives of inverse trig functions is just a matter of following the formulas. They're very rarely tested on the AP Exam in this form and are not terribly important. In addition, the AP Exam almost always tests only inverse sine and tangent, and then usually only as integrals, not as derivatives.

Therefore, it's time to learn the integrals. The derivative formulas lead directly to the integral formulas, but because the only difference between the derivatives of inverse sine, tangent, and secant and those of inverse cosine, cotangent, and cosecant is the negative sign, only the former functions are generally used.

$$\int \frac{du}{\sqrt{1 - u^2}} = \sin^{-1} u + C \qquad \text{(valid for } u^2 < 1\text{)}$$

$$\int \frac{du}{1 + u^2} = \tan^{-1} u + C$$

$$\int \frac{du}{u\sqrt{u^2 - 1}} = \sec^{-1} u + C \qquad \text{(valid for } u^2 > 1\text{)}$$

The first two always show up on the AP Exam, so you should learn to recognize the pattern. Most of the time, the integration requires some sticky algebra to get the integral into the proper form.

Example 4: $\int \frac{x \, dx}{\sqrt{1 - x^4}} =$

Don't forget about *u*-substitution. Let $u = x^2$ and $du = 2x \, dx$. Thus, $\frac{1}{2} du = x \, dx$. Substituting and integrating, we get

$$\frac{1}{2} \int \frac{du}{\sqrt{1 - u^2}} = \frac{1}{2} \sin^{-1} u + C$$

Now substitute back.

$$\frac{1}{2} \sin^{-1} (x^2) + C$$

Example 5: $\int \frac{dx}{\sqrt{4 - x^2}} =$

Anytime you see an integral where you have the square root of a constant minus a function, try to turn it into an inverse sine. You can do that here with a little simple algebra.

$$\int \frac{dx}{\sqrt{4 - x^2}} = \int \frac{dx}{\sqrt{4\left(1 - \frac{x^2}{4}\right)}} = \frac{1}{2} \int \frac{dx}{\sqrt{1 - \frac{x^2}{4}}}$$

Now use u-substitution: let $u = \dfrac{x}{2}$ and $du = \dfrac{1}{2} dx$, so $2du = dx$, and substitute.

$$\frac{1}{2}\int \frac{dx}{\sqrt{1 - \dfrac{x^2}{4}}} = \frac{1}{2}\int \frac{2\,du}{\sqrt{1 - u^2}} = \sin^{-1} u + C$$

When you substitute back, you get

$$\sin^{-1} \frac{x}{2} + C$$

Example 6: $\displaystyle\int \frac{e^x\,dx}{1 + e^{2x}} =$

Again, use u-substitution. Let $u = e^x$ and $du = e^x\,dx$, and substitute.

$$\int \frac{du}{1 + u^2} = \tan^{-1} u + C$$

Then substitute back.

$$\tan^{-1} e^x + C$$

Let's do one last type that's slightly more difficult. For this one, you need to remember how to complete the square.

Example 7: $\displaystyle\int \frac{dx}{x^2 + 4x + 5} =$

Complete the square in the denominator (your algebra teacher warned you about this).

$$x^2 + 4x + 5 = (x + 2)^2 + 1$$

Now rewrite the integral.

$$\int \frac{dx}{1 + (x + 2)^2}$$

Using u-substitution, let $u = x + 2$ and $du = dx$. Then do the substitution.

$$\int \frac{du}{1 + u^2} = \tan^{-1} u + C$$

When you put the function of x back in to the integral, it reads:

$$\tan^{-1}(x + 2) + C$$

Evaluating these integrals involves looking for a particular pattern. If it's there, all you have to do is use algebra and u-substitution to make the integrand conform to the pattern. Once that's accomplished, the rest is easy. These integrals will show up again when we do partial fractions. Otherwise, as far as the AP Exam is concerned, this is all you need to know.

Try these solved problems, and don't forget to look for those patterns. Get out your index card, and cover the answers while you work.

PROBLEM 1. Find the derivative of $y = \sin^{-1}\left(\dfrac{x}{2}\right)$.

Answer: The rule is

$$\frac{d}{dx}(\sin^{-1} u) = \frac{1}{\sqrt{1 - u^2}}\frac{du}{dx}$$

The algebra looks like the following:

$$\frac{d}{dx}\left(\sin^{-1}\left(\frac{x}{2}\right)\right) = \frac{1}{\sqrt{1 - \left(\frac{x}{2}\right)^2}}\frac{1}{2} = \frac{1}{2}\left(\frac{1}{\sqrt{1 - \frac{x^2}{4}}}\right) = \frac{1}{2}\left(\frac{1}{\sqrt{\frac{4 - x^2}{4}}}\right) = \frac{1}{2}\left(\frac{2}{\sqrt{4 - x^2}}\right) = \frac{1}{\sqrt{4 - x^2}}$$

PROBLEM 2. Evaluate $\displaystyle\int \frac{dx}{4 + x^2}$.

Answer: First, you need to do a little algebra. Factor 4 out of the denominator.

$$\int \frac{dx}{4\left(1 + \frac{x^2}{4}\right)}$$

Next, rewrite the integrand.

$$\frac{1}{4}\int \frac{dx}{\left(1 + \left(\frac{x}{2}\right)^2\right)}$$

Now you can use *u*-substitution. Let $u = \dfrac{x}{2}$ and $du = \dfrac{1}{2}\, dx$, so $2du = dx$.

$$\frac{1}{2} \int \frac{du}{1+u^2}$$

Now integrate it.

$$\frac{1}{2} \tan^{-1} u + C$$

And re-substitute.

$$\frac{1}{2} \tan^{-1} \frac{x}{2} + C$$

PROBLEM 3. Evaluate $\displaystyle\int \frac{dx}{\sqrt{-x^2 + 4x - 3}}$.

Answer: This time, you need to use some algebra by completing the square of the polynomial under the square root sign.

$$-x^2 + 4x - 3 = -(x^2 - 4x + 3) = -\left[(x-2)^2 - 1\right] = \left[1 - (x-2)^2\right]$$

Now rewrite the integrand.

$$\int \frac{dx}{\sqrt{1 - (x-2)^2}}$$

And use *u*-substitution. Let $u = x - 2$ and $du = dx$.

$$\int \frac{du}{\sqrt{1 - u^2}}$$

Now, this looks familiar. Once you integrate, you get

$$\sin^{-1} u + C$$

After you substitute back, it becomes

$$\sin^{-1}(x - 2) + C$$

PRACTICE PROBLEM SET 22

Here is some more practice work on derivatives and integrals of inverse trig functions. The answers are in Chapter 19.

1. Find the derivative of $\dfrac{1}{4}\tan^{-1}\left(\dfrac{x}{4}\right)$.

2. Find the derivative of $\sin^{-1}\left(\dfrac{1}{x}\right)$.

3. Find the derivative of $\tan^{-1}(e^x)$.

4. Evaluate $\displaystyle\int \dfrac{dx}{x\sqrt{x^2 - \pi}}$.

5. Evaluate $\displaystyle\int \dfrac{dx}{7 + x^2}$.

6. Evaluate $\displaystyle\int \dfrac{dx}{x(1 + \ln^2 x)}$.

7. Evaluate $\displaystyle\int \dfrac{\sec^2 x\, dx}{\sqrt{1 - \tan^2 x}}$.

8. Evaluate $\displaystyle\int \dfrac{dx}{\sqrt{9 - 4x^2}}$.

You've learned how to integrate polynomials and some of the trig functions (there are more of them to come), and you have the first technique of integration: *u*-substitution. Now it's time to learn how to integrate some other functions—namely, exponential and logarithmic functions. The first integral is the natural logarithm.

$$\int \frac{du}{u} = \ln|u| + C$$

Notice the absolute value in the logarithm. This ensures that you aren't taking the logarithm of a negative number. If you know that the term you're taking the log of is positive (for example, $x^2 + 1$), we can dispense with the absolute value marks. Let's do some examples.

Example 8: Find $\int \dfrac{5dx}{x+3}$.

Whenever an integrand contains a fraction, check to see if the integral is a logarithm. Usually, the process involves u-substitution. Let $u = x + 3$ and $du = dx$. Then,

$$\int \frac{5dx}{x+3} = 5 \int \frac{du}{u} = 5\ln|u| + C$$

Substituting back, the final result is

$$5 \ln |x + 3| + C$$

Example 9: Find $\int \dfrac{2x\,dx}{x^2+1}$.

Let $u = x^2 + 1$ $du = 2x\,dx$ and substitute into the integrand.

$$\int \frac{2x\,dx}{x^2+1} = \int \frac{du}{u} = \ln|u| + C$$

Then substitute back.

$$\ln (x^2 + 1) + C$$

MORE INTEGRALS OF TRIG FUNCTIONS

Remember when we started antiderivatives and we didn't do the integral of tangent, cotangent, secant, or cosecant? Well, their time has come.

Example 10: Find $\int \tan x\,dx$.

First, rewrite this integral as

$$\int \frac{\sin x}{\cos x}\,dx$$

Now, we let $u = \cos x$ and $du = -\sin x\, dx$ and substitute.

$$\int -\frac{du}{u}$$

Now, integrate and re-substitute.

$$\int -\frac{du}{u} = -\ln|u| = -\ln|\cos x| + C$$

Thus, $\int \tan x\, dx = -\ln|\cos x| + C$.

Example 11: Find $\int \cot x\, dx$.

Just as before, rewrite this integral in terms of sine and cosine.

$$\int \frac{\cos x}{\sin x}\, dx$$

Now we let $u = \sin x$ and $du = \cos x\, dx$ and substitute.

$$\int \frac{du}{u}$$

Now, integrate.

$$\ln|u| + C = \ln|\sin x| + C$$

Therefore, $\int \cot x\, dx = \ln|\sin x| + C$.

This looks a lot like the previous example, doesn't it?

Example 12: Find $\int \sec x\, dx$.

You could rewrite this integral as

$$\int \frac{1}{\cos x}\, dx$$

However, if you try u-substitution at this point, it won't work. So what should you do? You'll probably never guess, so we'll show you: multiply the sec x by $\dfrac{\sec x + \tan x}{\sec x + \tan x}$. This gives you

$$\int \sec x \left(\frac{\sec x + \tan x}{\sec x + \tan x} \right) dx = \int \frac{\sec^2 x + \sec x \tan x}{\sec x + \tan x}\, dx$$

Now you can do u-substitution. Let $u = \sec x + \tan x \; du = (\sec x \tan x + \sec^2 x)\, dx$.

Then rewrite the integral as

$$\int \frac{du}{u}$$

Pretty slick, huh?

The rest goes according to plan as you integrate.

$$\int \frac{du}{u} = \ln |u| + C = \ln |\sec x + \tan x| + C$$

> Therefore, $\displaystyle\int \sec x\, dx = \ln |\sec x + \tan x| + C.$

Example 13: Find $\displaystyle\int \csc x\, dx$.

You guessed it! Multiply $\csc x$ by $\dfrac{\csc x + \cot x}{\csc x + \cot x}\, dx$. This gives you

$$\int \csc x \left(\frac{\csc x + \cot x}{\csc x + \cot x} \right) dx = \int \frac{\csc^2 x + \csc x \cot x}{\csc x + \cot x}\, dx$$

Let $u = \csc x + \cot x$ and $du = (-\csc x \cot x - \csc^2 x)\, dx$. And, just as in Example 5, you can rewrite the integral as

$$\int - \frac{du}{u}$$

And integrate.

$$\int - \frac{du}{u} = -\ln |u| + C = -\ln |\csc x + \cot x| + C$$

Therefore, $\int \csc x \, dx = -\ln |\csc x + \cot x| + C$.

As we do more integrals, the natural log will turn up over and over. It's important that you get good at recognizing when integrating requires the use of the natural log.

INTEGRATING e^x AND a^x

Now let's learn how to find the integral of e^x. Remember that $\dfrac{d}{dx} e^x = e^x$? Well, you should be able to predict the following formula:

$$\int e^u \, du = e^u + C$$

As with the natural logarithm, most of these integrals use u-substitution.

Example 14: Find $\int e^{7x} \, dx$.

Let $u = 7x$, $du = 7dx$, and $\dfrac{1}{7} du = dx$. Then you have

$$\int e^{7x} \, dx = \frac{1}{7} \int e^u \, du = \frac{1}{7} e^u + C$$

Substituting back, you get

$$\frac{1}{7} e^{7x} + C$$

In fact, whenever you see $\int e^{kx} \, dx$, where k is a constant, the integral is

$$\int e^{kx} \, dx = \frac{1}{k} e^{kx} + C$$

Example 15: Find $\int x e^{3x^2+1}\,dx$.

Let $u = 3x^2 + 1$, $du = 6x\,dx$, and $\frac{1}{6}\,du = x\,dx$. The result is

$$\int x e^{3x^2+1}\,dx = \frac{1}{6}\int e^u\,du = \frac{1}{6}e^u + C$$

Now it's time to put the x's back in.

$$\frac{1}{6}e^{3x^2+1} + C$$

Example 16: Find $\int e^{\sin x}\cos x\,dx$.

Let $u = \sin x$ and $du = \cos x\,dx$. The substitution here couldn't be simpler.

$$\int e^u\,du = e^u + C = e^{\sin x} + C$$

As you can see, these integrals are pretty straightforward. The key is to use u-substitution to transform nasty-looking integrals into simple ones.

There's another type of exponential function whose integral you'll have to find occasionally.

$$\int a^u\,du$$

As you should recall from your rules of logarithms and exponents, the term a^u can be written as $e^{u\ln a}$. Because $\ln a$ is a constant, we can transform $\int a^u\,du$ into $\int e^{u\ln a}\,du$. If you integrate this, you'll get

$$\int e^{u\ln a}\,du = \frac{1}{\ln a}\,e^{u\ln a} + C$$

Now substituting back a^u for $e^{u\ln a}$,

$$\int a^u\,du = \frac{1}{\ln a}\,a^u + C$$

Example 17: Find $\int 5^x\, dx$.

Follow the rule we just derived.

$$\int 5^x\, dx = \frac{1}{\ln 5} 5^x + C$$

Because these integrals don't show up too often on the AP Exam, this is the last you'll see of them in this book. You should, however, be able to integrate them using the rule, or by converting them into a form of $\int e^u\, du$.

Try these on your own. Do each problem with the answers covered, and then check your answer.

PROBLEM 4. Evaluate $\int \frac{dx}{3x}$.

Answer: Move the constant term outside of the integral, like this.

$$\frac{1}{3} \int \frac{dx}{x}$$

Now you can integrate.

$$\frac{1}{3} \int \frac{dx}{x} = \frac{1}{3} \ln|x| + C$$

PROBLEM 5. Evaluate $\int \frac{3x^2}{x^3 - 1}\, dx$.

Answer: Let $u = x^3 - 1$ and $du = 3x^2\, dx$, and substitute.

$$\int \frac{du}{u}$$

Now integrate.

$$\ln|u| + C$$

And substitute back.

$$\ln|x^3 - 1| + C$$

PROBLEM 6. Evaluate $\int e^{5x}\,dx$.

Answer: Let $u = 5x$ and $du = 5dx$. Then $\frac{1}{5}\,du = dx$. Substitute in.

$$\frac{1}{5}\int e^{u}\,du$$

Integrate.

$$\frac{1}{5}e^{u} + C$$

And substitute back.

$$\frac{1}{5}e^{5x} + C$$

PROBLEM 7. Evaluate $\int 2^{3x}\,dx$.

Answer: Let $u = 3x$ and $du = 3dx$. Then $\frac{1}{3}\,du = dx$. Make the substitution.

$$\frac{1}{3}\int 2^{u}\,du$$

Integrate according to the rule. Your result should be

$$\frac{1}{3\ln 2}2^{u} + C$$

Now get back to the expression as a function of x.

$$\frac{1}{3\ln 2}2^{3x} + C = \frac{1}{\ln 8}2^{3x} + C$$

PRACTICE PROBLEM SET 23

Evaluate the following integrals. The answers are in Chapter 19.

1. $\displaystyle\int \frac{\sec^2 x}{\tan x}\, dx$

2. $\displaystyle\int \frac{\cos x}{1 - \sin x}\, dx$

3. $\displaystyle\int \frac{1}{x\ln x}\, dx$

4. $\displaystyle\int \frac{\sin x - \cos x}{\cos x}\, dx$

5. $\displaystyle\int \frac{dx}{\sqrt{x}\,(1 + 2\sqrt{x})}$

6. $\displaystyle\int \frac{e^x\, dx}{1 + e^x}$

7. $\displaystyle\int x e^{5x^2 - 1}\, dx$

8. $\displaystyle\int \frac{e^x + e^{-x}}{e^x - e^{-x}}\, dx$

9. $\displaystyle\int x 4^{-x^2}\, dx$

10. $\displaystyle\int 7^{\sin x} \cos x\, dx$

End of Chapter 15 Drill

The answers are in Chapter 20.

1. Evaluate $\int \frac{x^2}{\left(4x^3+1\right)} dx$.

   (A) $\frac{1}{12}\ln\left|4x^3+1\right|+C$

   (B) $\ln\left|4x^3+1\right|+C$

   (C) $\frac{-1}{\left(4x^3+1\right)^2}+C$

   (D) $\frac{x^3}{\left(4x^3+1\right)^2}+C$

2. Evaluate $\int x\tan x^2 \, dx$.

   (A) $\frac{1}{2}\sec^2 x^2+C$

   (B) $\sec^2 x^2+C$

   (C) $-\frac{1}{2}\ln\left|\cos x^2\right|+C$

   (D) $\frac{1}{2}\ln\left|\cos x^2\right|+C$

3. Evaluate $\int x^3 e^{1+x^4} dx$.

   (A) $\frac{x^4}{16}e^{1+x^4}+C$

   (B) $\frac{x^4}{4}e^{1+x^4}+C$

   (C) $4e^{1+x^4}+C$

   (D) $\frac{1}{4}e^{1+x^4}+C$

4. Evaluate $\int x^2 5^{2+x^3} dx$.

   (A) $\frac{1}{3\ln 5}5^{2+x^3}+C$

   (B) $\frac{1}{3}5^{2+x^3}+C$

   (C) $\frac{x^3}{3}5^{2+x^3}+C$

   (D) $\frac{x^3}{3\ln 5}5^{2+x^3}+C$

5. Evaluate $\int 4x\sec\left(x^2\right) dx$.

   (A) $\frac{1}{2}\ln\left|\sec\left(x^2\right)+\tan\left(x^2\right)\right|+C$

   (B) $2\ln\left|\sec\left(x^2\right)+\tan\left(x^2\right)\right|+C$

   (C) $2\sec\left(x^2\right)\tan\left(x^2\right)+C$

   (D) $\frac{1}{2}\sec\left(x^2\right)\tan\left(x^2\right)+C$

REFLECT

Respond to the following questions:

- For which topics discussed in this chapter do you feel you have achieved sufficient mastery to answer multiple-choice questions correctly?

- For which topics discussed in this chapter do you feel you have achieved sufficient mastery to answer open-ended questions correctly?

- For which topics discussed in this chapter do you feel you need more work before you can answer multiple-choice questions correctly?

- For which topics discussed in this chapter do you feel you need more work before you can answer open-ended questions correctly?

- What parts of this chapter are you going to re-review?

Unit 4
Integral Calculus
Applications

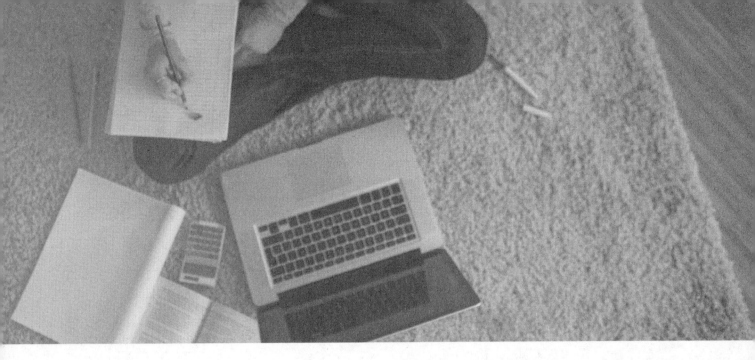

Chapter 16
The Area Between
Two Curves

These next two units discuss some of the most difficult topics you'll encounter in AP Calculus. For some reason, students have terrible trouble setting up these problems. Fortunately, the AP Exam asks only relatively simple versions of these problems on the exam.

Unfortunately, this unit and the next are always on the AP Exam. We'll try to make them as simple as possible. You've already learned that if you want to find the area under a curve, you can integrate the function of the curve by using the endpoints as limits. So far, though, we've talked only about the area between a curve and the x-axis. What if you have to find the area between two curves?

VERTICAL SLICES

Suppose you wanted to find the area between the curve $y = x$ and the curve $y = x^2$ from $x = 2$ to $x = 4$. First, sketch the curves.

You can find the area by slicing up the region vertically, into a bunch of infinitely thin strips, and adding up the areas of all the strips. The height of each strip is $x^2 - x$, and the width of each strip is dx. Add up all the strips by using the integral.

$$\int_2^4 \left(x^2 - x\right)dx$$

Then, evaluate it.

$$\left(\frac{x^3}{3} - \frac{x^2}{2}\right)\Bigg|_2^4 = \left(\frac{64}{3} - \frac{16}{2}\right) - \left(\frac{8}{3} - \frac{4}{2}\right) = \frac{38}{3}$$

That wasn't so hard, was it? Don't worry. The process gets more complicated, but the idea remains the same. Now let's generalize this and come up with a rule.

> If a region is bounded by $f(x)$ above and $g(x)$ below at all points of the interval $[a, b]$, then the area of the region is given by
>
> $$\int_a^b \left[f(x) - g(x) \right] dx$$

Example 1: Find the area of the region between the parabola $y = 1 - x^2$ and the line $y = 1 - x$.

First, make a sketch of the region.

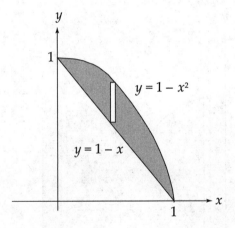

To find the points of intersection of the graphs, set the two equations equal to each other and solve for x.

$$1 - x^2 = 1 - x$$
$$x^2 - x = 0$$
$$x(x - 1) = 0$$
$$x = 0, 1$$

The left-hand edge of the region is $x = 0$ and the right-hand edge is $x = 1$, so the limits of integration are from 0 to 1.

Next, note that the top curve is always $y = 1 - x^2$, and the bottom curve is always $y = 1 - x$. (If the region has a place where the top and bottom curve switch, you need to make two integrals, one for each region. Fortunately, that's not the case here.) Thus, we need to evaluate the following:

$$\int_0^1 \left[\left(1 - x^2\right) - \left(1 - x\right)\right] dx$$

$$\int_0^1 \left[\left(1 - x^2\right) - \left(1 - x\right)\right] dx = \int_0^1 \left(-x^2 + x\right) dx = \left(-\frac{x^3}{3} + \frac{x^2}{2}\right)\Bigg|_0^1 = \frac{1}{6}$$

Sometimes you're given the endpoints of the region; sometimes you have to find them on your own.

Example 2: Find the area of the region between the curve $y = \sin x$ and the curve $y = \cos x$ from 0 to $\frac{\pi}{2}$.

First, sketch the region.

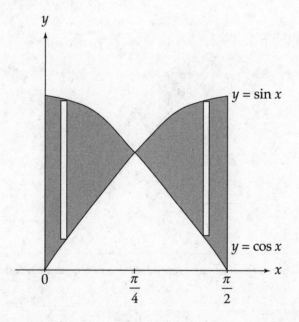

Notice that $\cos x$ is on top between 0 and $\frac{\pi}{4}$, then $\sin x$ is on top between $\frac{\pi}{4}$ and $\frac{\pi}{2}$. The point where they cross is $\frac{\pi}{4}$, so you have to divide the area into two integrals: one from 0 to $\frac{\pi}{4}$, and the other from $\frac{\pi}{4}$ to $\frac{\pi}{2}$. In the first region, $\cos x$ is above $\sin x$, so the integral to evaluate is

$$\int_0^{\frac{\pi}{4}} \left(\cos x - \sin x\right) dx$$

The integral of the second region is a little different, because sin x is above cos x.

$$\int_{\frac{\pi}{4}}^{\frac{\pi}{2}} \left(\sin x - \cos x \right) dx$$

If you add the two integrals, you'll get the area of the whole region.

$$\int_0^{\frac{\pi}{4}} \left(\cos x - \sin x \right) dx = \left(\sin x + \cos x \right) \Big|_0^{\frac{\pi}{4}} = \sqrt{2} - 1$$

$$\int_{\frac{\pi}{4}}^{\frac{\pi}{2}} \left(\sin x - \cos x \right) dx = \left(-\cos x - \sin x \right) \Big|_{\frac{\pi}{4}}^{\frac{\pi}{2}} = \sqrt{2} - 1$$

Adding these, we get that the area is $2\sqrt{2} - 2$.

HORIZONTAL SLICES

Now for the fun part. We can slice a region vertically when one function is at the top of our section and a different function is at the bottom. But what if the same function is both the top and the bottom of the slice (what we call a double-valued function)? You have to slice the region horizontally.

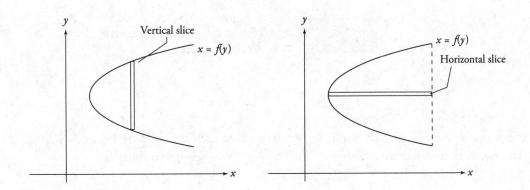

If we were to slice vertically, as in the left-hand picture, we'd have a problem. But if we were to slice horizontally, as in the right-hand picture, we don't have a problem. Instead of integrating an equation $f(x)$ with respect to x, we need to integrate an equation $f(y)$ with respect to y. As a result, our area formula changes a little.

If a region is bounded by $f(y)$ on the right and $g(y)$ on the left at all points of the interval $[c, d]$, then the area of the region is given by

$$\int_c^d [f(y) - g(y)]\, dy$$

Example 3: Find the area of the region between the curve $x = y^2$ and the curve $x = y + 6$ from $y = 0$ to $y = 3$.

First, sketch the region.

When you slice up the area horizontally, the right end of each section is the curve $x = y + 6$, and the left end of each section is always the curve $x = y^2$. Now set up our integral.

$$\int_0^3 (y + 6 - y^2)\, dy$$

Evaluating this gives us the area.

$$\int_0^3 (y + 6 - y^2)\, dy = \left(\frac{y^2}{2} + 6y - \frac{y^3}{3}\right)\Bigg|_0^3 = \frac{27}{2}$$

Example 4: Find the area between the curve $y = \sqrt{x + 3}$, the curve $y = \sqrt{3 - x}$, and the x-axis from $x = -3$ to $x = 3$.

First, sketch the curves.

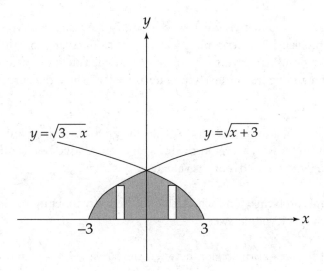

From $x = -3$ to $x = 0$, if you slice the region vertically, the curve $y = \sqrt{x + 3}$ is on top, and the x-axis is on the bottom; from $x = 0$ to $x = 3$, the curve $y = \sqrt{3 - x}$ is on top and the x-axis is on the bottom. Therefore, you can find the area by evaluating two integrals.

$$\int_{-3}^{0}(\sqrt{x + 3} - 0)\,dx \text{ and } \int_{0}^{3}(\sqrt{3 - x} - 0)\,dx$$

Your results should be

$$\frac{2}{3}(x+3)^{\frac{3}{2}}\Big|_{-3}^{0} + \left(-\frac{2}{3}(3-x)^{\frac{3}{2}}\right)\Big|_{0}^{3} = 4\sqrt{3}$$

Let's suppose you sliced the region horizontally instead. The curve $y = \sqrt{x + 3}$ is always on the left, and the curve $y = \sqrt{3 - x}$ is always on the right. If you solve each equation for x in terms of y, you save some time by using only one integral instead of two.

The two equations are $x = y^2 - 3$ and $x = 3 - y^2$. We also have to change the limits of integration from x-limits to y-limits. The two curves intersect at $y = \sqrt{3}$, so our limits of integration are from $y = 0$ to $y = \sqrt{3}$. The new integral is

$$\int_0^{\sqrt{3}} \left[(3 - y^2) - (y^2 - 3) \right] dy = \int_0^{\sqrt{3}} (6 - 2y^2) \, dy = 6y - \frac{2y^3}{3} \Big|_0^{\sqrt{3}} = 4\sqrt{3}$$

You get the same answer no matter which way you integrate (as long as you do it right!). The challenge of area problems is determining which way to integrate and then converting the equation to different terms. Unfortunately, there's no simple rule for how to do this. You have to look at the region and figure out its endpoints, as well as where the curves are with respect to each other.

Once you can do that, then the actual set-up of the integral(s) isn't that hard. Sometimes, evaluating the integrals isn't easy; however, if the integral of an AP question is difficult to evaluate, you'll be required only to set it up, not to evaluate it.

Here are some sample problems. On each, decide the best way to set up the integrals, and then evaluate them. Then check your answer.

PROBLEM 1. Find the area of the region between the curve $y = 3 - x^2$ and the line $y = 1 - x$ from $x = 0$ to $x = 2$.

Answer: First, make a sketch.

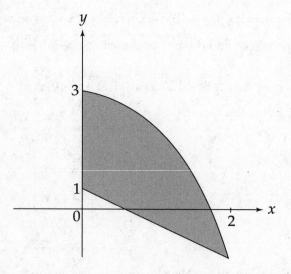

Because the curve $y = 3 - x^2$ is always above $y = 1 - x$ within the interval, you have to evaluate the following integral:

$$\int_0^2 \left[(3 - x^2) - (1 - x) \right] dx = \int_0^2 (2 + x - x^2) \, dx$$

Therefore, the area of the region is

$$\left(2x + \frac{x^2}{2} - \frac{x^3}{3}\right)\Bigg|_0^2 = \frac{10}{3}$$

PROBLEM 2. Find the area between the x-axis and the curve $y = 2 - x^2$ from $x = 0$ to $x = 2$.

Answer: First, sketch the graph over the interval.

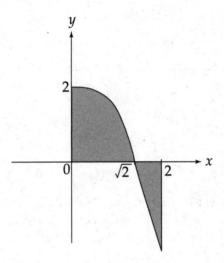

Because the curve crosses the x-axis at $\sqrt{2}$, you have to divide the region into two parts: from $x = 0$ to $x = \sqrt{2}$ and from $x = \sqrt{2}$ to $x = 2$. In the latter region, you'll need to integrate $y = -\left(2 - x^2\right) = x^2 - 2$ to adjust for the region's being below the x-axis. Therefore, we can find the area by evaluating

$$\int_0^{\sqrt{2}} (2 - x^2)\, dx + \int_{\sqrt{2}}^2 (x^2 - 2)\, dx$$

Integrating, we get

$$\left(2x + \frac{x^3}{3}\right)\Bigg|_0^{\sqrt{2}} + \left(\frac{x^3}{3} - 2x\right)\Bigg|_{\sqrt{2}}^2 = \left(2\sqrt{2} - \frac{2\sqrt{2}}{3}\right) - 0 + \left(\frac{8}{3} - 4\right) - \left(\frac{2\sqrt{2}}{3} - 2\sqrt{2}\right) = \frac{8\sqrt{2} - 4}{3}$$

PROBLEM 3. Find the area of the region between the curve $x = y^2 - 4y$ and the line $x = y$.

Answer: First, sketch the graph over the interval.

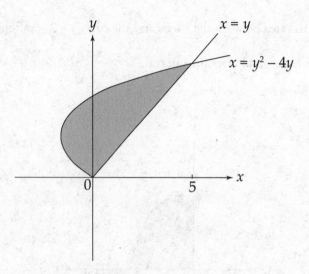

You don't have the endpoints this time, so you need to find where the two curves intersect. If you set them equal to each other, they intersect at $y = 0$ and at $y = 5$. The curve $x = y^2 - 4y$ is always to the left of $x = y$ over the interval we just found, so we can evaluate the following integral:

$$\int_0^5 \left[y - (y^2 - 4y) \right] dy = \int_0^5 (5y - y^2) \, dy$$

The result of the integration should be

$$\left(\frac{5y^2}{2} - \frac{y^3}{3} \right) \Bigg|_0^5 = \frac{125}{6}$$

PROBLEM 4. Find the area between the curve $x = y^3 - y$ and the line $x = 0$ (the y-axis).

Answer: First, sketch the graph over the interval.

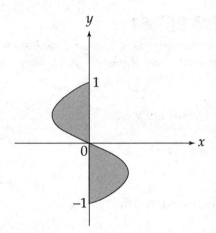

Next, find where the two curves intersect. By setting $y^3 - y = 0$, you'll find that they intersect at $y = -1$, $y = 0$, and $y = 1$. Notice that the curve is to the right of the y-axis from $y = -1$ to $y = 0$ and to the left of the y-axis from $y = 0$ to $y = 1$. Thus, the region must be divided into two parts: from $y = -1$ to $y = 0$ and from $y = 0$ to $y = 1$.

Set up the two integrals.

$$\int_{-1}^{0} (y^3 - y)\, dy + \int_{0}^{1} (y - y^3)\, dy$$

And integrate them.

$$\left(\frac{y^4}{4} - \frac{y^2}{2} \right)\Bigg|_{-1}^{0} + \left(\frac{y^2}{2} - \frac{y^4}{4} \right)\Bigg|_{0}^{1} = \frac{1}{2}$$

PRACTICE PROBLEM SET 24

Find the area of the region between the two curves in each problem, and be sure to sketch each one. (We gave you only endpoints in one of them.) The answers are in Chapter 19.

1. The curve $y = x^2 - 2$ and the line $y = 2$.

2. The curve $y = x^2$ and the curve $y = 4x - x^2$.

3. The curve $y = x^2 - 4x - 5$ and the curve $y = 2x - 5$.

4. The curve $y = x^3$ and the x-axis, from $x = -1$ to $x = 2$.

5. The curve $x = y^2$ and the line $x = y + 2$.

6. The curve $x = y^2$ and the curve $x = 3 - 2y^2$.

7. The curve $x = y^2 - 4y + 2$ and the line $x = y - 2$.

8. The curve $x = y^{\frac{2}{3}}$ and the curve $x = 2 - y^4$.

End of Chapter 16 Drill

The answers are in Chapter 20.

1. Find the area between the x-axis and $y = \cos x$ from $x = 0$ to $x = \dfrac{\pi}{2}$.

   (A) 0
   (B) 1
   (C) π
   (D) 2π

2. Find the area between $y = x^2 + 1$ and $y = 5$.

   (A) $\dfrac{16}{3}$

   (B) $\dfrac{32}{3}$

   (C) 16

   (D) 32

3. Find the area between $y = 16 - x^2$ and $y = 7$.

   (A) 18
   (B) 36
   (C) 72
   (D) 108

4. Find the area between the y-axis and $x = 25 - y^2$.

   (A) $\dfrac{5}{3}$

   (B) $\dfrac{25}{3}$

   (C) $\dfrac{125}{3}$

   (D) $\dfrac{500}{3}$

5. Find the area between the y-axis, $x = \sqrt{y}$ and $x = 4$.

   (A) $\dfrac{4}{3}$

   (B) $\dfrac{16}{3}$

   (C) $\dfrac{64}{3}$

   (D) $\dfrac{128}{3}$

REFLECT

Respond to the following questions:

- For which topics discussed in this chapter do you feel you have achieved sufficient mastery to answer multiple-choice questions correctly?

- For which topics discussed in this chapter do you feel you have achieved sufficient mastery to answer open-ended questions correctly?

- For which topics discussed in this chapter do you feel you need more work before you can answer multiple-choice questions correctly?

- For which topics discussed in this chapter do you feel you need more work before you can answer open-ended questions correctly?

- What parts of this chapter are you going to re-review?

Chapter 17
The Volume of a Solid of Revolution

Does the chapter title leave you in a cold sweat? Don't worry. You're not alone. This chapter covers a topic widely seen as one of the most difficult on the AP Exam. There is *always* a volume question on the test. The good news is that you're almost never asked to evaluate the integral— you usually only have to set it up. The difficulty with this chapter, as with Chapter 16, is that there aren't any simple rules to follow. You have to draw the picture and figure it out.

In this chapter, we're going to take the region between two curves, rotate it around a line (usually the *x*- or *y*-axis), and find the volume of the region. There are two methods of doing this: the **washers method** and the **cylindrical shells method**. Sometimes you'll hear the washers method called the **disk method**, but a disk is only a washer without a hole in the middle.

WASHERS AND DISKS

Let's look at the region between the curve $y = \sqrt{x}$ and the *x*-axis (the curve $y = 0$), from $x = 0$ to $x = 1$, and revolve it about the *x*-axis. The picture looks like the following:

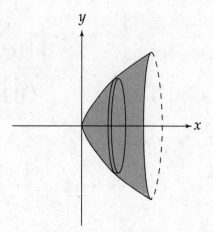

If you slice the resulting solid perpendicular to the *x*-axis, each cross-section of the solid is a circle, or disk (hence the phrase "disk method"). The radii of the disks vary from one value of *x* to the next, but you can find each radius by plugging it into the equation for *y*: Each radius is $\sqrt{x}$. Therefore, the area of each disk is

$$\pi\left(\sqrt{x}\right)^2 = \pi x$$

Each disk is infinitesimally thin, so its thickness is *dx;* if you add up the volumes of all the disks, you'll get the entire volume. The way to add these up is by using the integral, with the endpoints of the interval as the limits of integration. Therefore, to find the volume, evaluate the integral.

$$\int_0^1 \pi x \, dx = \left.\frac{\pi x^2}{2}\right|_0^1 = \frac{\pi}{2}$$

Now let's generalize this. If you have a region whose area is bounded by the curve $y = f(x)$ and the x-axis on the interval $[a, b]$, each disk has a radius of $f(x)$, and the area of each disk will be

$$\pi\left[f(x)\right]^2$$

To find the volume, evaluate the integral.

$$\pi\int_a^b \left[f(x)\right]^2 dx$$

This is the formula for finding the volume using disks.

Example 1: Find the volume of the solid that results when the region between the curve $y = x$ and the x-axis, from $x = 0$ to $x = 1$, is revolved about the x-axis.

As always, sketch the region to get a better look at the problem.

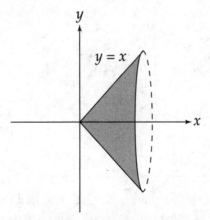

When you slice vertically, the top curve is $y = x$ and the limits of integration are from $x = 0$ to $x = 1$. Using our formula, we evaluate the integral.

$$\pi\int_0^1 x^2 \, dx$$

The result is

$$\pi \int_0^1 x^2 \, dx = \pi \frac{x^3}{3} \Big|_0^1 = \frac{\pi}{3}$$

By the way, did you notice that the solid in the problem is a cone with a height and radius of 1?

The formula for the volume of a cone is $\frac{1}{3}\pi r^2 h$, so you should expect to get $\frac{\pi}{3}$.

Now let's figure out how to find the volume of the solid that results when we revolve a region that does not touch the x-axis. Consider the region bounded above by the curve $y = x^3$ and below by the curve $y = x^2$, from $x = 2$ to $x = 4$, which is revolved about the x-axis. Sketch the region first.

If you slice this region vertically, each cross-section looks like a washer (hence the phrase "washer method").

The outer radius is $R = x^3$ and the inner radius is $r = x^2$. To find the area of the region between the two circles, take the area of the outer circle, πR^2, and subtract the area of the inner circle, πr^2.

We can simplify this to

$$\pi R^2 - \pi r^2 = \pi \left(R^2 - r^2 \right)$$

Because the outer radius is $R = x^3$ and the inner radius is $r = x^2$, the area of each region is $\pi\left(x^6 - x^4\right)$. You can sum up these regions using the integral.

$$\pi\int_2^4 \left(x^6 - x^4\right) dx = \frac{74{,}336\pi}{35}$$

Here's the general idea: in a region whose area is bounded above by the curve $y = f(x)$ and below by the curve $y = g(x)$, on the interval $[a, b]$, then each washer will have an area of

$$\pi\left[f(x)^2 - g(x)^2\right]$$

To find the volume, evaluate the integral.

$$\pi\int_a^b \left[f(x)^2 - g(x)^2\right] dx$$

This is the formula for finding the volume using washers when the region is rotated around the x-axis.

The formula using disks is just the formula for washers without the second function, $g(x)$.

Example 2: Find the volume of the solid that results when the region bounded by $y = x$ and $y = x^2$, from $x = 0$ to $x = 1$, is revolved about the x-axis.

Sketch it first.

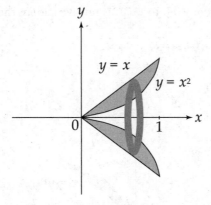

The top curve is $y = x$ and the bottom curve is $y = x^2$ throughout the region. Then our formula tells us that we evaluate the integral.

$$\pi\int_0^1 \left(x^2 - x^4\right) dx$$

The result is

$$\pi \int_0^1 \left(x^2 - x^4 \right) dx = \pi \left(\frac{x^3}{3} - \frac{x^5}{5} \right)\Bigg|_0^1 = \frac{2\pi}{15}$$

Suppose the region we're interested in is revolved around the y-axis instead of the x-axis. Now, to find the volume, you have to slice the region horizontally instead of vertically. We discussed how to do this in the previous unit on area.

Now, if you have a region whose area is bounded on the right by the curve $x = f(y)$ and on the left by the curve $x = g(y)$, on the interval $[c, d]$, then each washer has an area of

$$\pi \left[f(y)^2 - g(y)^2 \right]$$

To find the volume, evaluate the integral.

$$\pi \int_c^d \left[f(y)^2 - g(y)^2 \right] dy$$

This is the formula for finding the volume using washers when the region is rotated about the y-axis.

Example 3: Find the volume of the solid that results when the region bounded by the curve $x = y^2$ and the curve $x = y^3$, from $y = 0$ to $y = 1$, is revolved about the y-axis.

Sketch away.

Because $x = y^2$ is always on the outside and $x = y^3$ is always on the inside, you have to evaluate the integral.

$$\pi \int_0^1 \left(y^4 - y^6 \right) dy$$

You should get the following:

$$\pi \int_0^1 \left(y^4 - y^6 \right) dx = \pi \left[\frac{y^5}{5} - \frac{y^7}{7} \right]_0^1 = \frac{2\pi}{35}$$

There's only one more nuance to cover. Sometimes you'll have to revolve the region about a line instead of one of the axes. If so, this will affect the radii of the washers; you'll have to adjust the integral to reflect the shift. Once you draw a picture, it usually isn't too hard to see the difference.

Example 4: Find the volume of the solid that results when the area bounded by the curve $y = x^2$ and the curve $y = 4x$ is revolved about the line $y = -2$. <u>Set up but do not evaluate the integral</u>. (This is how the AP Exam will say it!)

You're not given the limits of integration here, so you need to find where the two curves intersect by setting the equations equal to each other.

$$x^2 = 4x$$
$$x^2 - 4x = 0$$
$$x = 0, 4$$

These will be our limits of integration. Next, sketch the curve.

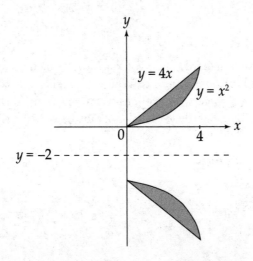

Notice that the distance from the axis of revolution is no longer found by just using each equation. Now, you need to add 2 to each equation to account for the shift in the axis. Thus, the radii are $x^2 + 2$ and $4x + 2$. This means that we need to evaluate the integral.

$$\pi \int_0^4 \left[\left(4x + 2\right)^2 - \left(x^2 + 2\right)^2 \right] dx$$

Suppose instead that the region was revolved about the line $x = -2$. Sketch the region again.

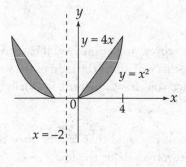

You'll have to slice the region horizontally this time; this means you're going to solve each equation for x in terms of y: $x = \sqrt{y}$ and $x = \dfrac{y}{4}$. We also need to find the y-coordinates of the intersection of the two curves: $y = 0, 16$.

Notice also that, again, each radius is going to be increased by 2 to reflect the shift in the axis of revolution. Thus, we will have to evaluate the integral.

$$\pi \int_0^{16} \left[\left(\sqrt{y} + 2\right)^2 - \left(\frac{y}{4} + 2\right)^2 \right] dy$$

Finding the volumes isn't that hard, once you've drawn a picture, figured out whether you need to slice vertically or horizontally, and determined whether the axis of revolution has been shifted. Sometimes, though, there will be times when you want to slice vertically yet revolve around the y-axis (or slice horizontally yet revolve about the x-axis). Here's the method for finding volumes in this way.

CYLINDRICAL SHELLS

Let's examine the region bounded above by the curve $y = 2 - x^2$ and below by the curve $y = x^2$, from $x = 0$ to $x = 1$. Suppose you had to revolve the region about the y-axis instead of the x-axis.

This advanced topic isn't likely to appear on the AP Exam, but it's still worth familiarizing yourself with the principles of this technique.

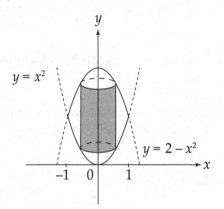

If you slice the region vertically and revolve the slice, you won't get a washer; you'll get a cylinder instead. Because each slice is an infinitesimally thin rectangle, the cylinder's "thickness" is also very, very thin, but real nonetheless. Thus, if you find the surface area of each cylinder and add them up, you'll get the volume of the region.

The formula for the surface area of a cylinder is $2\pi rh$. The height of the cylinder is the length of the vertical slice, $(2 - x^2) - x^2 = 2 - 2x^2$, and the radius of the slice is x. Thus, evaluate the integral.

$$2\pi \int_0^1 x(2 - 2x^2)\,dx$$

The math goes like the following:

$$2\pi \int_0^1 x(2 - 2x^2)\,dx = 2\pi \int_0^1 (2x - 2x^3)\,dx = 2\pi \left(x^2 - \frac{x^4}{2} \right) \bigg|_0^1 = \pi$$

Why work in the dark? Just as you spend time practicing formulas in order to memorize them, make sure you actually work on drawing these examples. If you can't visualize the problem, you won't be able to set up the integral.

Suppose you tried to slice the region horizontally and use washers. You'd have to convert each equation and find the new limits of integration. Because the region is not bounded by the same pair of curves throughout, you would have to evaluate the region using several integrals. The cylindrical shells method was invented precisely so you can avoid this.

From a general standpoint: if we have a region whose area is bounded above by the curve $y = f(x)$ and below by the curve $y = g(x)$, on the interval $[a, b]$, then each cylinder will have a height of $f(x) - g(x)$, a radius of x, and an area of $2\pi x[f(x) - g(x)]$.

To find the volume, evaluate the integral.

$$2\pi \int_a^b x[(f(x) - g(x)]\,dx$$

This is the formula for finding the volume using cylindrical shells when the region is rotated around the y-axis.

Example 5: Find the volume of the region that results when the region bounded by the curve $y = \sqrt{x}$, the x-axis, and the line $x = 9$ is revolved about the y-axis. <u>Set up but do not evaluate the integral</u>.

Your sketch should look like the following:

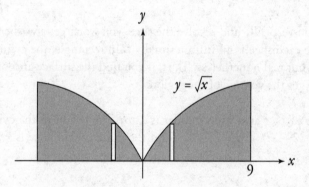

Notice that the limits of integration are from $x = 0$ to $x = 9$, and that each vertical slice is bounded from above by the curve $y = \sqrt{x}$ and from below by the x-axis ($y = 0$). We need to evaluate the integral.

$$2\pi \int_0^9 x(\sqrt{x} - 0)\,dx = 2\pi \int_0^9 x(\sqrt{x})\,dx$$

Example 6: Find the volume that results when the region in Example 5 is revolved about the line $x = -1$. <u>Set up but do not evaluate the integral</u>.

Sketch the figure.

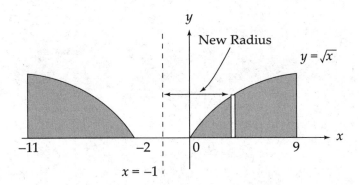

If you slice the region vertically, the height of the shell doesn't change because of the shift in axis of revolution, but you have to add 1 to each radius.

Our integral thus becomes

$$2\pi \int_0^9 (x + 1)(\sqrt{x})\, dx$$

The last formula you need to learn involves slicing the region horizontally and revolving it about the *x*-axis. As you probably guessed, you'll get a cylindrical shell.

If you have a region whose area is bounded on the right by the curve $x = f(y)$ and on the left by the curve $x = g(y)$, on the interval $[c, d]$, then each cylinder will have a height of $f(y) - g(y)$, a radius of y, and an area of $2\pi y[f(y) - g(y)]$.

To find the volume, evaluate the integral.

$$2\pi \int_c^d y[(f(y) - g(y)]\, dy$$

This is the formula for finding the volume using cylindrical shells when the region is rotated around the *x*-axis.

Example 7: Find the volume of the region that results when the region bounded by the curve $x = y^3$ and the line $x = y$, from $y = 0$ to $y = 1$, is rotated about the *x*-axis. <u>Set up but do not evaluate the integral.</u>

Let your sketch be your guide.

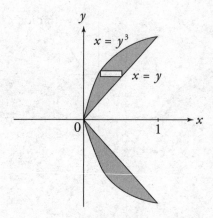

Each horizontal slice is bounded on the right by the curve $x = y$ and on the left by the line $x = y^3$. The integral to evaluate is

$$2\pi \int_0^1 y(y - y^3)\,dy$$

Suppose that you had to revolve this region about the line $y = -1$ instead. Now the region looks like the following:

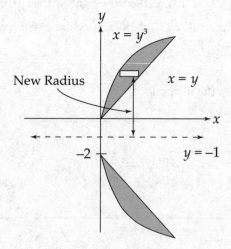

The radius of each cylinder is increased by 1 because of the shift in the axis of revolution, so the integral looks like the following:

$$2\pi \int_0^1 (y + 1)(y - y^3)\,dy$$

Wasn't this fun? Volumes of Solids of Revolution require you to sketch the region carefully and to decide whether it'll be easier to slice the region vertically or horizontally. Once you figure out the slices' boundaries and the limits of integration (and you've adjusted for an axis of revolution, if necessary), it's just a matter of plugging into the integral. Usually, you won't be asked to evaluate the integral unless it's a simple one. Once you've conquered this topic, you're ready for anything.

VOLUMES OF SOLIDS WITH KNOWN CROSS-SECTIONS

There is one other type of volume that you need to be able to find. Sometimes, you will be given an object where you know the shape of the base and where perpendicular cross-sections are all the same regular, planar geometric shape. These sound hard, but are actually quite straightforward. This is easiest to explain through an example.

Example 8:   Suppose we are asked to find the volume of a solid whose base is the circle $x^2 + y^2 = 4$, and where cross-sections perpendicular to the x-axis are all squares whose sides lie on the base of the circle. How would we find the volume?

First, make a drawing of the circle.

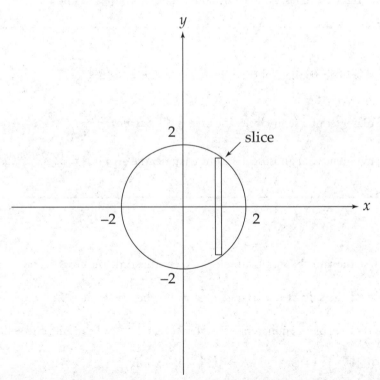

What this problem is telling us is that every time we make a vertical slice, the slice is the length of the base of a square. If we want to find the volume of the solid, all we have to do is integrate the area of the square, from one endpoint of the circle to the other.

The side of the square is the vertical slice whose length is $2y$, which we can find by solving the equation of the circle for y and multiplying by 2. We get $y = \sqrt{4 - x^2}$. Then the length of a side of the square is $2\sqrt{4 - x^2}$. Because the area of a square is *side²*, we can find the volume with $\int_{-2}^{2}(16 - 4x^2)\,dx$.

Let's perform the integration, although on some problems you will be permitted to find the answer with a calculator.

$$\int_{-2}^{2}(16 - 4x^2)\,dx = \left(16x - \frac{4x^3}{3}\right)\Bigg|_{-2}^{2} = \left(32 - \frac{32}{3}\right) - \left(32 + \frac{32}{3}\right)$$

$$= 64 - \frac{64}{3} = \frac{128}{3}$$

As you can see, the technique is very simple. First, you find the side of the cross-section in terms of y. This will involve a vertical slice. Then, you plug the side into the equation for the area of the cross-section. Then, integrate the area from one endpoint of the base to the other. On the AP Exam, cross-sections will be squares, equilateral triangles, circles, semi-circles, or maybe isosceles right triangles. So here are some handy formulas to know.

Given the side of an equilateral triangle, the area is $A = (side)^2\,\dfrac{\sqrt{3}}{4}$.

Given the diameter of a semi-circle, the area is $A = (diameter)^2\dfrac{\pi}{8}$.

Given the hypotenuse of an isosceles right triangle, the area is $A = \dfrac{(hypotenuse)^2}{4}$.

Example 9: Use the same base as Example 8, except this time the cross-sections are equilateral triangles. We find the side of the triangle just as we did above. It is $2y$, which is $2\sqrt{4 - x^2}$. Now, because the area of an equilateral triangle is $(side)^2\,\dfrac{\sqrt{3}}{4}$, we can find the volume by evaluating the integral $\dfrac{\sqrt{3}}{4}\int_{-2}^{2}(4)(4 - x^2)\,dx = \sqrt{3}\int_{-2}^{2}(4 - x^2)\,dx$.

We get $\sqrt{3}\int_{-2}^{2}(4 - x^2)\,dx = \sqrt{3}\left(4x - \dfrac{x^3}{3}\right)\Bigg|_{-2}^{2} = \dfrac{32\sqrt{3}}{3}$.

Example 10: Use the same base as Example 8, except this time the cross-sections are semi-circles whose diameters lie on the base. We find the side of the semi-circle just as we did above.

It is $2y$, which is $2\sqrt{4-x^2}$. Now, because the area of a semi-circle is $(\textit{diameter})^2\dfrac{\pi}{8}$, we can find the volume by evaluating the integral $\dfrac{\pi}{8}\displaystyle\int_{-2}^{2}(4)(4-x^2)\,dx = \dfrac{\pi}{2}\displaystyle\int_{-2}^{2}(4-x^2)\,dx$.

We get $\dfrac{\pi}{2}\displaystyle\int_{-2}^{2}(4-x^2)\,dx = \dfrac{\pi}{2}\left(4x-\dfrac{x^3}{3}\right)\Bigg|_{-2}^{2} = \dfrac{16\pi}{3}$.

Here are some solved problems. Do each problem, covering the answer first, then check your answer.

PROBLEM 1. Find the volume of the solid that results when the region bounded by the curve $y = 16 - x^2$ and the curve $y = 16 - 4x$ is rotated about the x-axis. Use the washer method and <u>set up but do not evaluate the integral</u>.

Answer: First, sketch the region.

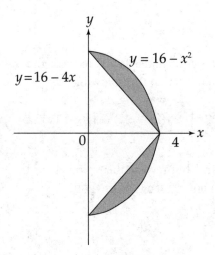

Next, find where the curves intersect by setting the two equations equal to each other.

$$16 - x^2 = 16 - 4x$$
$$x^2 = 4x$$
$$x^2 - 4x = 0$$
$$x = 0,\ 4$$

Slicing vertically, the top curve is always $y = 16 - x^2$ and the bottom is always $y = 16 - 4x$, so the integral looks like the following:

$$\pi \int_0^4 \left[(16 - x^2)^2 - (16 - 4x)^2 \right] dx$$

PROBLEM 2. Repeat Problem 1, but revolve the region about the y-axis and use the cylindrical shells method. <u>Set up but do not evaluate the integral</u>.

Answer: Sketch the situation.

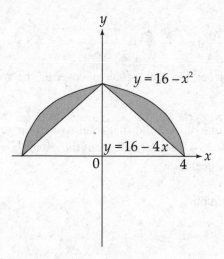

Slicing vertically, the top of each cylinder is $y = 16 - x^2$, the bottom is $y = 16 - 4x$, and the radius is x. Therefore, you should set up the following:

$$2\pi \int_0^4 x \left[(16 - x^2) - (16 - 4x) \right] dx$$

PROBLEM 3. Repeat Problem 1 but revolve the region about the x-axis and use the cylindrical shells method. <u>Set up but do not evaluate the integral</u>.

Answer: Sketch the situation.

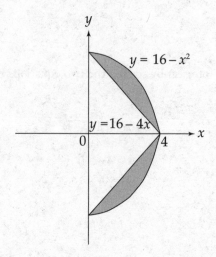

To slice horizontally, you have to solve each equation for x in terms of y and find the limits of integration with respect to y. First, solve for x in terms of y.

$$y = 16 - x^2 \text{ becomes } x = \sqrt{16 - y}$$

and

$$y = 16 - 4x \text{ becomes } x = \frac{16 - y}{4}$$

Next, determine the limits of integration: they're from $y = 0$ to $y = 16$. Slicing horizontally, the curve $x = \sqrt{16 - y}$ is always on the right and the curve $x = \dfrac{16 - y}{4}$ is always on the left. The radius is y, so we evaluate

$$2\pi \int_0^{16} y \left[\left(\sqrt{16 - y}\right) - \left(\frac{16 - y}{4}\right) \right] dy$$

Problem 4. Repeat Problem 1 but revolve the region about the y-axis and use the washers method. <u>Set up but do not evaluate the integral.</u>

Answer: Sketch.

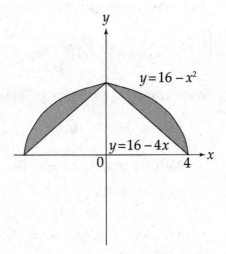

Slicing horizontally, the curve $x = \sqrt{16 - y}$ is always on the right and the curve $x = \dfrac{16 - y}{4}$ is always on the left. Your integral should look like the following:

$$\pi \int_0^{16} \left[\left(\sqrt{16 - y}\right)^2 - \left(\frac{16 - y}{4}\right)^2 \right] dy$$

PROBLEM 5. Repeat Problem 1, but revolve the region about the line $y = -3$. You may use either method. <u>Set up but do not evaluate the integral</u>.

Answer: Your sketch should resemble the one below (note that it's not drawn exactly to scale).

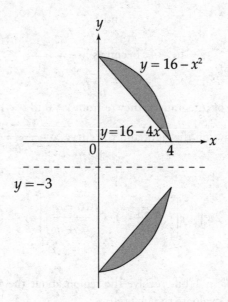

If you were to slice the region vertically, you would use washers. You'll need to add 3 to each radius to adjust for the axis of revolution. The integral to evaluate is

$$\pi \int_0^4 \left[(16 - x^2 + 3)^2 - \left(16 - 4x + 3\right)^2 \right] dx$$

To slice the region horizontally, use cylindrical shells. The radius of each shell would increase by 3, and you would evaluate

$$2\pi \int_0^{16} (y+3) \left[(\sqrt{16 - y}) - \left(\frac{16 - y}{4} \right) \right] dy$$

PROBLEM 6. Repeat Problem 1, but revolve the region about the line $x = 8$. You may use either method. <u>Set up but do not evaluate the integral</u>.

Warning! This one is tricky!

Answer: First, sketch the region.

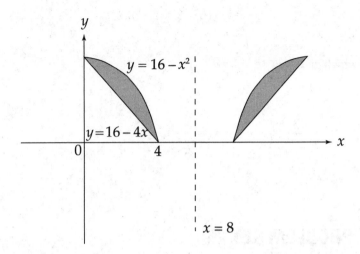

If you choose cylindrical shells, slice the region vertically; you'll need to adjust for the axis of revolution. Each radius can be found by subtracting x from 8. (Not 8 from x. That was the tricky part, in case you missed it.) The integral to evaluate is

$$2\pi \int_0^4 (8 - x)\left[(16 - x^2) - \left(16 - 4x\right)\right] dx$$

If you choose washers, slice the region horizontally. The radius of each washer is found by subtracting each equation from 8. Notice also that the curve $x = \dfrac{16 - y}{4}$ is now the outer radius of the washer, and the curve $x = \sqrt{16 - y}$ is the inner radius. The integral looks like the following:

$$\pi \int_0^{16} \left[\left(8 - \left(\frac{16 - y}{4}\right)\right)^2 - \left(8 - \left(\sqrt{16 - y}\right)\right)^2 \right] dy$$

PROBLEM 7: Find the volume of a solid whose base is the region between the x-axis and the curve $y = 4 - x^2$, and whose cross-sections perpendicular to the x-axis are equilateral triangles with a side that lies on the base.

Answer: The curve $y = 4 - x^2$ intersects the x-axis at $x = -2$ and $x = 2$. The side of the triangle is $4 - x^2$, so all that we have to do is evaluate $\dfrac{\sqrt{3}}{4} \int_{-2}^2 (4 - x^2)^2 \, dx$.

Expand the integrand to get

$$\frac{\sqrt{3}}{4}\int_{-2}^{2}(16 - 8x^2 + x^4)\,dx$$

Then integrate, which gives you

$$\frac{\sqrt{3}}{4}\left(16x - \frac{8x^3}{3} + \frac{x^5}{5}\right)\Bigg|_{-2}^{2} = \frac{128\sqrt{3}}{15} \approx 14.780$$

PRACTICE PROBLEM SET 25

Calculate the volumes below. The answers are in Chapter 19.

1. Find the volume of the solid that results when the region bounded by $y = \sqrt{9 - x^2}$ and the x-axis is revolved around the x-axis.

2. Find the volume of the solid that results when the region bounded by $y = \sec x$ and the x-axis from $x = -\dfrac{\pi}{4}$ to $x = \dfrac{\pi}{4}$ is revolved around the x-axis.

3. Find the volume of the solid that results when the region bounded by $x = 1 - y^2$ and the y-axis is revolved around the y-axis.

4. Find the volume of the solid that results when the region bounded by $x = \sqrt{5}y^2$ and the y-axis from $y = -1$ to $y = 1$ is revolved around the y-axis.

5. Use the method of cylindrical shells to find the volume of the solid that results when the region bounded by $y = x$, $x = 2$, and $y = -\dfrac{x}{2}$ is revolved around the y-axis.

6. Use the method of cylindrical shells to find the volume of the solid that results when the region bounded by $y = \sqrt{x}$, $y = 2x - 1$, and $x = 0$ is revolved around the y-axis.

7. Use the method of cylindrical shells to find the volume of the solid that results when the region bounded by $y = x^2$, $y = 4$, and $x = 0$ is revolved around the x-axis.

8. Use the method of cylindrical shells to find the volume of the solid that results when the region bounded by $y = 2\sqrt{x}$, $x = 4$, and $y = 0$ is revolved around the y-axis.

9. Find the volume of the solid whose base is the region between the semi-circle $y = \sqrt{16 - x^2}$ and the x-axis and whose cross-sections perpendicular to the x-axis are squares with a side on the base.

10. Find the volume of the solid whose base is the region between $y = x^2$ and $y = 4$ and whose perpendicular cross-sections are isosceles right triangles with the hypotenuse on the base.

End of Chapter 17 Drill

The answers are in Chapter 20.

1. Find the volume of the solid that results when the region between $y = \dfrac{1}{x}$, $x = 1$, $x = 4$, and $y = 0$ is revolved about the x-axis.

    (A) $\dfrac{3}{4}$

    (B) $\dfrac{3\pi}{4}$

    (C) $\dfrac{63}{64}$

    (D) $\dfrac{63\pi}{64}$

2. Find the volume of the solid that results when the region between $x = \sqrt{1+y}$, $x = 0$, and $y = 3$ is revolved about the y-axis.

    (A) 2π
    (B) 4π
    (C) 8π
    (D) 16π

3. Find the volume of the solid that results when the region between $y = x^2 + 1$, and $y = x + 3$ is revolved about the x-axis.

    (A) $\dfrac{9}{2}$

    (B) $\dfrac{9\pi}{2}$

    (C) $\dfrac{117}{5}$

    (D) $\dfrac{117\pi}{5}$

4. Find the volume of the solid that results when the region between $x = \csc y$, $y = \dfrac{\pi}{4}$, $y = \dfrac{3\pi}{4}$, and $x = 0$ is revolved about the y-axis.

    (A) $\dfrac{\pi}{2}$

    (B) π

    (C) 2π

    (D) 4π

5. Use cylindrical shells to find the volume of the solid that results when the region between $x^2 + y^3 = 4$, $x = 0$, $x = 4$, and $y = 0$ is revolved about the y-axis. **Set up but do not evaluate the integral.**

    (A) $2\pi \displaystyle\int_0^2 x\left(4 - x^2\right)^{\frac{1}{3}} dx - 2\pi \displaystyle\int_2^4 x\left(4 - x^2\right)^{\frac{1}{3}} dx$

    (B) $2\pi \displaystyle\int_0^2 x\left(4 - x^2\right)^{\frac{1}{3}} dx + 2\pi \displaystyle\int_2^4 x\left(4 - x^2\right)^{\frac{1}{3}} dx$

    (C) $\pi \displaystyle\int_0^2 \left(4 - x^2\right)^{\frac{2}{3}} dx - \pi \displaystyle\int_2^4 \left(4 - x^2\right)^{\frac{2}{3}} dx$

    (D) $\pi \displaystyle\int_0^2 \left(4 - x^2\right)^{\frac{2}{3}} dx + \pi \displaystyle\int_2^4 \left(4 - x^2\right)^{\frac{2}{3}} dx$

6. Find the volume of the solid whose base is the region between $y = x^2 + 4$, and $y = 3x + 8$, and whose cross-sections perpendicular to the x-axis are squares with a side on the base.

    (A) $\dfrac{125}{6}$

    (B) $\dfrac{125\pi}{6}$

    (C) $\dfrac{625}{6}$

    (D) $\dfrac{625\pi}{6}$

REFLECT

Respond to the following questions:

- For which topics discussed in this chapter do you feel you have achieved sufficient mastery to answer multiple-choice questions correctly?

- For which topics discussed in this chapter do you feel you have achieved sufficient mastery to answer open-ended questions correctly?

- For which topics discussed in this chapter do you feel you need more work before you can answer multiple-choice questions correctly?

- For which topics discussed in this chapter do you feel you need more work before you can answer open-ended questions correctly?

- What parts of this chapter are you going to re-review?

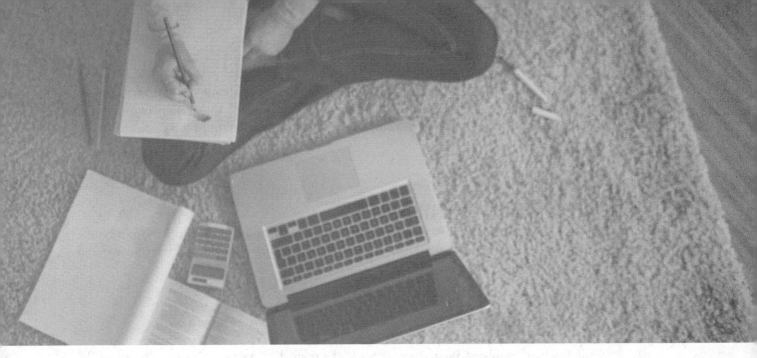

Chapter 18
Differential Equations

There are many types of differential equations, but only a very small number of them appear on the AP Exam. There are courses devoted to learning how to solve a wide variety of differential equations, but AP Calculus provides only a very basic introduction to the topic.

SEPARATION OF VARIABLES

If you're given an equation in which the derivative of a function is equal to some other function, you can determine the original function by integrating both sides of the equation and then solving for the constant term.

Example 1: If $\dfrac{dy}{dx} = \dfrac{4x}{y}$ and $y(0) = 5$, find an equation for y in terms of x.

The first step in solving these is to put all of the terms that contain y on the left side of the equals sign and all of the terms that contain x on the right side. We then have $y\,dy = 4x\,dx$. The second step is to integrate both sides.

$$\int y\,dy = \int 4x\,dx$$

And then you integrate

$$\frac{y^2}{2} = 2x^2 + C$$

You're not done yet. The final step is to solve for the constant by plugging in $x = 0$ and $y = 5$.

$$\frac{5^2}{2} = 2(0^2) + C, \text{ so } C = \frac{25}{2}$$

The solution is $\dfrac{y^2}{2} = 2x^2 + \dfrac{25}{2}$.

That's all there is to it. Separate the variables, integrate both sides, and solve for the constant. Often, the equation will involve a logarithm. Let's do an example.

Example 2: If $\dfrac{dy}{dx} = 3x^2y$ and $y(0) = 2$, find an equation for y in terms of x.

First, put the y terms on the left and the x terms on the right.

$$\frac{dy}{y} = 3x^2\,dx$$

Next, integrate both sides.

$$\int \frac{dy}{y} = \int 3x^2\,dx$$

The result is $\ln y = x^3 + C$. It's customary to solve this equation for y. You can do this by putting both sides into exponential form.

$$y = e^{x^3 + C}$$

This can be rewritten as $y = e^{x^3} \cdot e^C$ and, because e^C is a constant, the equation becomes

$$y = Ce^{x^3}$$

This is the preferred form of the equation. Now, solve for the constant. Plug in x 0 and $y = 2$, and you get $2 = Ce^0$.

Because $e^0 = 1$, $C = 2$. The solution is $y = 2e^{x^3}$.

This is the typical differential equation that you'll see on the AP Exam. Other common problem types involve position, velocity, and acceleration or exponential growths and decay (Problem 4). We did several problems of this type in Chapter 10, before you knew how to use integrals. In a sample problem, you're given the velocity and acceleration and told to find distance (the reverse of what we did before).

Example 3: If the acceleration of a particle is given by $a(t) = -32$ ft/sec², and the velocity of the particle is 64 ft/sec and the height of the particle is 32 ft at time $t = 0$, find: (a) the equation of the particle's velocity at time t; (b) the equation for the particle's height, h, at time t; and (c) the maximum height of the particle.

Part A: Because acceleration is the rate of change of velocity with respect to time, you can write that $\dfrac{dv}{dt} = -32$. Now separate the variables and integrate both sides.

$$\int dv = \int -32 \, dt$$

Integrating this expression, we get $v = -32t + C$. Now we can solve for the constant by plugging in $t = 0$ and $v = 64$. We get $64 = -32(0) + C$ and $C = 64$. Thus, velocity is $v = -32t + 64$.

Part B: Because velocity is the rate of change of displacement with respect to time, you know that

$$\frac{dh}{dt} = -32t + 64$$

Separate the variables and integrate both sides.

$$\int dh = \int \left(-32t + 64 \right) dt$$

Integrate the expression: $h = -16t^2 + 64t + C$. Now solve for the constant by plugging in $t = 0$ and $h = 32$.

$$32 = -16(0^2) + 64(0) + C \text{ and } C = 32$$

Thus, the equation for height is $h = -16t^2 + 64t + 32$.

Part C: In order to find the maximum height, you need to take the derivative of the height with respect to time and set it equal to zero. Notice that the derivative of height with respect to time is the velocity; just set the velocity equal to zero and solve for t.

$$-32t + 64 = 0, \text{ so } t = 2$$

Thus, at time $t = 2$, the height of the particle is a maximum. Now, plug $t = 2$ into the equation for height.

$$h = -16(2)^2 + 64(2) + 32 = 96$$

Therefore, the maximum height of the particle is 96 feet.

Here are some solved problems. Do each problem, covering the answer first, then check your answer.

PROBLEM 1. If $\dfrac{dy}{dx} = \dfrac{3x}{2y}$ and $y(0) = 10$, find an equation for y in terms of x.

Answer: First, separate the variables.

$$2y\, dy = 3x\, dx$$

Then, take the integral of both sides.

$$\int 2y\, dy = \int 3x\, dx$$

Next, integrate both sides.

$$y^2 = \frac{3x^2}{2} + C$$

Finally, solve for the constant.

$$10^2 = \frac{3(0)^2}{2} + C, \text{ so } C = 100$$

The solution is $y^2 = \dfrac{3x^2}{2} + 100$.

Problem 2. If $\dfrac{dy}{dx} = 4xy^2$ and $y(0) = 1$, find an equation for y in terms of x.

Answer: First, separate the variables: $\dfrac{dy}{y^2} = 4x\,dx$. Then, take the integral of both sides.

$$\int \frac{dy}{y^2} = \int 4x\,dx$$

Next, integrate both sides: $-\dfrac{1}{y} = 2x^2 + C$. You can rewrite this as $y = -\dfrac{1}{2x^2 + C}$.

Finally, solve for the constant.

$$1 = -\frac{1}{2(0)^2 + C} = \frac{-1}{C},\text{ so }C = -1$$

The solution is $y = -\dfrac{1}{2x^2 - 1}$.

Problem 3. If $\dfrac{dy}{dx} = \dfrac{y^2}{x}$ and $y(1) = \dfrac{1}{3}$, find an equation for y in terms of x.

Answer: This time, separating the variables gives us this: $\dfrac{dy}{y^2} = \dfrac{dx}{x}$.

Then, take the integral of both sides: $\int \dfrac{dy}{y^2} = \int \dfrac{dx}{x}$.

Next, integrate both sides.

$$-\frac{1}{y} = \ln x + C$$

And rearrange the equation.

$$y = \frac{-1}{\ln x + C}$$

Finally, solve for the constant. $\dfrac{1}{3} = \dfrac{-1}{C}$, so $C = -3$. The solution is $y = \dfrac{-1}{\ln x - 3}$.

Problem 4. A city had a population of 10,000 in 1980 and 13,000 in 1990. Assuming an exponential growth rate, estimate the city's population in 2000.

Answer: The phrase "exponential growth rate" means that $\dfrac{dy}{dt} = ky$, where k is a constant. Take the integral of both sides.

$$\int \frac{dy}{y} = \int k\,dt$$

Then, integrate both sides ($\ln y = kt + C$) and put them in exponential form.

$$y = e^{kt+c} = Ce^{kt}$$

Next, use the information about the population to solve for the constants. If you treat 1980 as $t = 0$ and 1990 as $t = 10$, then

$$10,000 = Ce^{k(0)} \text{ and } 13,000 = Ce^{k(10)}$$

So $C = 10,000$ and $k = \dfrac{1}{10} \ln 1.3 \approx 0.0262$.

The equation for population growth is approximately $y = 10,000e^{.0262t}$. We can estimate that the population in 2000 will be

$$y = 10,000e^{0.0262(20)} = 16,900$$

SLOPE FIELDS

The idea behind **slope fields**, also known as **direction fields**, is to make a graphical representation of the slope of a function at various points in the plane. We are given a differential equation, but not the equation itself. So how do we do this? Well, it's always easiest to start with an example.

Example 1: Given $\dfrac{dy}{dx} = x$, sketch the slope field of the function.

What does this mean? Look at the equation. It gives us the derivative of the function, which is the slope of the tangent line to the curve at any point x. In other words, the equation tells us that the slope of the curve at any point x is the x-value at that point.

For example, the slope of the curve at $x = 1$ is 1. The slope of the curve at $x = 2$ is 2. The slope of the curve at the origin is 0. The slope of the curve at $x = -1$ is -1. We will now represent these different slopes by drawing small segments of the tangent lines at those points. Let's make a sketch.

See how all of these slopes are independent of the *y*-values, so for each value of *x*, the slope is the same vertically, but is different horizontally. Compare this slope field to the next example.

Example 2: Given $\dfrac{dy}{dx} = y$, sketch the slope field of the function.

Here, the slope of the curve at *y* = 1 is 1. The slope of the curve at *y* = 2 is 2. The slope of the curve at the origin is 0. The slope of the curve at *y* = –1 is –1. Let's make a sketch.

See how all of these slopes are independent of the *x*-values, so for each value of *y*, the slope is the same horizontally, but is different vertically.

Now let's do a slightly harder example.

Example 3: Given $\dfrac{dy}{dx} = xy$, sketch the slope field of the function.

Now, we have to think about both the *x*- and *y*-values at each point. Let's calculate a few slopes.

> At (0, 0), the slope is (0)(0) = 0.
>
> At (1, 0), the slope is (1)(0) = 0.
>
> At (2, 0), the slope is (2)(0) = 0.
>
> At (0, 1), the slope is (0)(1) = 0.
>
> At (0, 2), the slope is (0)(2) = 0.

So the slope will be zero at any point on the coordinate axes.

> At (1, 1), the slope is (1)(1) = 1.
>
> At (1, 2), the slope is (1)(2) = 2.
>
> At (1, –1), the slope is (1)(–1) = –1.
>
> At (1, –2), the slope is (1)(–2) = –2.

So the slope at any point where $x = 1$ will be the y-value. Similarly, you should see that the slope at any point where $y = 1$ will be the x-value. As we move out the coordinate axes, slopes will get steeper—whether positive or negative.

Let's do one more example.

Example 4: Given $\dfrac{dy}{dx} = y - x$, sketch the slope field of the function.

We have to think about both the x- and y-values at each point. This time, let's make a table of the values of the slope at different points.

	$y = -3$	$y = -2$	$y = -1$	$y = 0$	$y = 1$	$y = 2$	$y = 3$
$x = -3$	0	1	2	3	4	5	6
$x = -2$	-1	0	1	2	3	4	5
$x = -1$	-2	-1	0	1	2	3	4
$x = 0$	-3	-2	-1	0	1	2	3
$x = 1$	-4	-3	-2	-1	0	1	2
$x = 2$	-5	-4	-3	-2	-1	0	1
$x = 3$	-6	-5	-4	-3	-2	-1	0

Now let's make a sketch of the slope field. Notice that the slopes are zero along the line $y = x$ and that the slopes get steeper as we move away from the line in either direction.

That's really all that there is to slope fields. Obviously, there are more complicated slope fields that one could come up with, but on the AP Exam, they will ask you to sketch only the simplest ones.

PRACTICE PROBLEM SET 26

Now try these problems. The answers are in Chapter 19.

1. If $\dfrac{dy}{dx} = \dfrac{7x^2}{y^3}$ and $y(3) = 2$, find an equation for y in terms of x.

2. If $\dfrac{dy}{dx} = 5x^2 y$ and $y(0) = 6$, find an equation for y in terms of x.

3. If $\dfrac{dy}{dx} = \dfrac{e^x}{y^2}$ and $y(0) = 1$, find an equation for y in terms of x.

4. If $\dfrac{dy}{dx} = \dfrac{y^2}{x^3}$ and $y(1) = 2$, find an equation for y in terms of x.

5. If $\dfrac{dy}{dx} = \dfrac{\sin x}{\cos y}$ and $y(0) = \dfrac{3\pi}{2}$, find an equation for y in terms of x.

6. A colony of bacteria grows exponentially and the colony's population is 4,000 at time $t = 0$ and 6,500 at time $t = 3$. How big is the population at time $t = 10$?

7. A rock is thrown upward with an initial velocity, $v(t)$, of 18 m/s from a height, $h(t)$, of 45 m. If the acceleration of the rock is a constant -9 m/s^2, find the height of the rock at time $t = 4$.

More Great Books
Check out The Princeton Review's test prep titles for ACT and SAT: *Cracking the ACT Premium, Cracking the SAT Premium, ACT Elite, SAT Elite*, and many more!

8. A radioactive element decays exponentially in proportion to its mass. One-half of its original amount remains after 5,750 years. If 10,000 grams of the element are present initially, how much will be left after 1,000 years?

9. Sketch the slope field for $\dfrac{dy}{dx} = 2x$.

10. Sketch the slope field for $\dfrac{dy}{dx} = -\dfrac{x}{y}$.

End of Chapter 18 Drill

The answers are in Chapter 20.

1. If $\dfrac{dy}{dx}=-4xy^2$ and $y(0)=1$, find an equation for y in terms of x.

   (A) $y=\dfrac{-1}{2x^2-1}$

   (B) $y=\dfrac{1}{2x^2-1}$

   (C) $y=\dfrac{1}{2x^2+1}$

   (D) $y=\dfrac{-1}{2x^2+1}$

2. If $\sec x\dfrac{dy}{dx}=4\tan y$ and $y(0)=\dfrac{\pi}{6}$, find an equation for y in terms of x.

   (A) $\sec^2 y=\dfrac{\sqrt{3}}{2}e^{4\sin x}$

   (B) $\sec^2 y=\dfrac{1}{2}e^{4\sin x}$

   (C) $\sin y=\dfrac{\sqrt{3}}{2}e^{4\sin x}$

   (D) $\sin y=\dfrac{1}{2}e^{4\sin x}$

3. Polonium-210 has a half-life of 140 days. It decays exponentially, where rate of decay is proportional to the amount at the time t. If we start with 100 g, how much will remain after 10 weeks?

   (A) 50
   (B) 71
   (C) 90
   (D) 280

4. Use Euler's Method, with $h = 0.25$, to estimate $y(1)$ if $\dfrac{dy}{dx}=x+y$ and $y(0)=3$.

   (A) 3.25
   (B) 7
   (C) 8.8125
   (D) 10.25

5. Moss is growing on a stone at an exponential rate, where the rate of growth is proportional to the amount at the time t. If there are 16 cm^2 of moss on June 1, and 60 cm^2 on July 1 (30 days later), to the nearest square centimeter, how much will there be on September 1 (92 days later)?

   (A) 151
   (B) 921
   (C) 3601
   (D) 9601

REFLECT

Respond to the following questions:

- For which topics discussed in this chapter do you feel you have achieved sufficient mastery to answer multiple-choice questions correctly?

- For which topics discussed in this chapter do you feel you have achieved sufficient mastery to answer open-ended questions correctly?

- For which topics discussed in this chapter do you feel you need more work before you can answer multiple-choice questions correctly?

- For which topics discussed in this chapter do you feel you need more work before you can answer open-ended questions correctly?

- What parts of this chapter are you going to re-review?

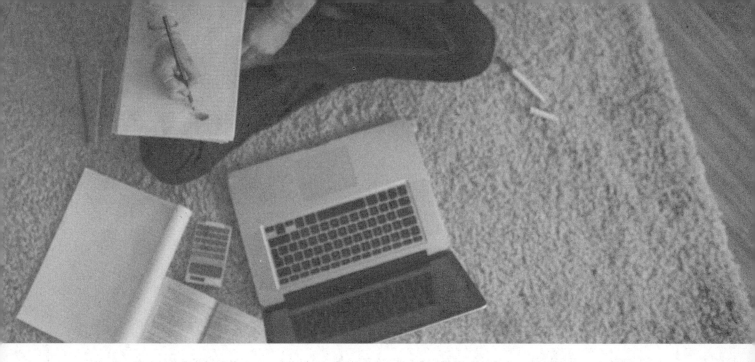

Chapter 19
Solutions to Practice Problem Sets

SOLUTIONS TO PRACTICE PROBLEM SET 1

1. 13

    To find the limit, we simply plug in 8 for x: $\lim\limits_{x \to 8}\left(x^2 - 5x - 11\right) = \left(8^2 - (5)(8) - 11\right) = 13$.

2. $\dfrac{4}{5}$

    To find the limit, we simply plug in 5 for x: $\lim\limits_{x \to 5}\left(\dfrac{x+3}{x^2 - 15}\right) = \dfrac{5+3}{5^2 - 15} = \dfrac{8}{10} = \dfrac{4}{5}$.

3. 4

    If we plug in 3 for x, we get $\dfrac{0}{0}$, which is indeterminate. When this happens, we try to factor the expression in order to get rid of the problem terms. Here we factor the top and get $\lim\limits_{x \to 3}\left(\dfrac{x^2 - 2x - 3}{x-3}\right) = \lim\limits_{x \to 3}\dfrac{(x-3)(x+1)}{x-3}$. Now we can cancel the term $x-3$ to get $\lim\limits_{x \to 3}(x+1)$. Notice that we are allowed to cancel the terms because x is not 3 but very close to 3. Now we can plug in 3 for x: $\lim\limits_{x \to 3}(x+1) = 3 + 1 = 4$.

4. $+\infty$

    Here we are finding the limit as x goes to infinity. We divide the top and bottom by the highest power of x in the expression, which is x^4: $\lim\limits_{x \to \infty}\left(\dfrac{x^4 - 8}{10x^2 + 25x + 1}\right) = \lim\limits_{x \to \infty}\left(\dfrac{\dfrac{x^4}{x^4} - \dfrac{8}{x^4}}{\dfrac{10x^2}{x^4} + \dfrac{25x}{x^4} + \dfrac{1}{x^4}}\right)$. Next, simplify the top and bottom: $\lim\limits_{x \to \infty}\left(\dfrac{1 - \dfrac{8}{x^4}}{\dfrac{10}{x^2} + \dfrac{25}{x^3} + \dfrac{1}{x^4}}\right)$. Now, if we take the limit as x goes to infinity, we get $\lim\limits_{x \to \infty}\left(\dfrac{1 - \dfrac{8}{x^4}}{\dfrac{10}{x^2} + \dfrac{25}{x^3} + \dfrac{1}{x^4}}\right) = \dfrac{1-0}{0+0+0} = \infty$.

5. $\dfrac{1}{10}$

Here we are finding the limit as x goes to infinity. We divide the top and bottom by the highest

power of x in the expression, which is x^4: $\displaystyle\lim_{x\to\infty}\left(\dfrac{x^4-8}{10x^4+25x+1}\right)=\lim_{x\to\infty}\left(\dfrac{\dfrac{x^4}{x^4}-\dfrac{8}{x^4}}{\dfrac{10x^4}{x^4}+\dfrac{25x}{x^4}+\dfrac{1}{x^4}}\right)$. Next,

simplify the top and bottom: $\displaystyle\lim_{x\to\infty}\left(\dfrac{1-\dfrac{8}{x^4}}{10+\dfrac{25}{x^3}+\dfrac{1}{x^4}}\right)$. Now, if we take the limit as x goes to infinity,

we get $\displaystyle\lim_{x\to\infty}\left(\dfrac{1-\dfrac{8}{x^4}}{10+\dfrac{25}{x^3}+\dfrac{1}{x^4}}\right)=\dfrac{1-0}{10+0+0}=\dfrac{1}{10}$.

6. $+\infty$

Here we have to think about what happens when we plug in a value that is very close to 6, but a little bit more. The top expression will approach 8. The bottom expression will approach 0, but will be a little bit bigger. Thus, the limit will be $\dfrac{8}{0^+}$, which is $+\infty$.

7. $-\infty$

Here we have to think about what happens when we plug in a value that is very close to 6, but a little bit less. The top expression will approach 8. The bottom expression will approach 0 but will be a little bit less. Thus, the limit will be $\dfrac{8}{0^-}$, which is $-\infty$.

8. The limit *Does Not Exist*.

In order to evaluate the limit as x approaches 6, we find the limit as it approaches 6^+ (from the right) and the limit as it approaches 6^- (from the left). If the two limits approach the same value, or both approach positive infinity or both approach negative infinity, then the limit is that value, or the appropriately signed infinity. If the two limits do not agree, the limit "Does Not Exist." Here, if we look at the solutions to problems 9 and 10, we find that as x approaches 6^+, the limit is $+\infty$, but as x approaches 6^-, the limit is $-\infty$. Because the two limits are *not* the same, the limit *Does Not Exist*.

9.   1

Here we have to think about what happens when we plug in a value that is very close to 0, but a little bit more. The top and bottom expressions will both be positive and the same value, so we get $\lim\limits_{x \to 0^+} \dfrac{x}{|x|} = \dfrac{0^+}{0^+} = 1$.

10. $+\infty$

Here we have to think about what happens when we plug in a value that is very close to 7, but a little bit more. The top expression will approach 7. The bottom expression will approach 0, but will be a little bit positive. Thus, the limit will be $\dfrac{7}{0^+}$, which is $+\infty$.

11.   The limit *Does Not Exist.*

In order to evaluate the limit as x approaches 7, we find the limit as it approaches 7^+ (from the right) and the limit as it approaches 7^- (from the left). If the two limits approach the same value, or both approach positive infinity or both approach negative infinity, then the limit is that value, or the appropriately signed infinity. If the two limits do not agree, the limit "Does Not Exist." Here, if we look at the solutions to Problem 10, we see that as x approaches 7^+, the limit is $+\infty$. As x approaches 7^-, the top expression will approach 7. The bottom will approach 0, but will be a little bit negative. Thus, the limit will be $\dfrac{7}{0^-}$, which is $-\infty$. Because the two limits are *not* the same, the limit *Does Not Exist.*

12.   (a) 4; (b) 5; (c) The limit *Does Not Exist.*

(a) Notice that $f(x)$ is a piecewise function, which means that we use the function $f(x) = x^2 - 5$ for all values of x less than or equal to 3. Thus, $\lim\limits_{x \to 3^-} f(x) = 3^2 - 5 = 4$.

(b) Here we use the function $f(x) = x + 2$ for all values of x greater than 3. Thus, $\lim\limits_{x \to 3^+} f(x) = 3 + 2 = 5$.

(c) In order to evaluate the limit as x approaches 3, we find the limit as it approaches 3^+ (from the right) and the limit as it approaches 3^- (from the left). If the two limits approach the same value, or both approach positive infinity or both approach negative infinity, then the limit is that value, or the appropriately signed infinity. If the two limits do not agree, the limit "Does Not Exist."

Here, if we refer to the solutions in parts (a) and (b), we see that $\lim\limits_{x\to 3^-} f(x) = 4$ and $\lim\limits_{x\to 3^+} f(x) = 5$.

Because the two limits are *not* the same, the limit *Does Not Exist*.

13.   (a) 4; (b) 4; (c) 4

(a) Notice that $f(x)$ is a piecewise function, which means that we use the function $f(x) = x^2 - 5$ for all values of x less than or equal to 3. Thus, $\lim\limits_{x\to 3^-} f(x) = 3^2 - 5 = 4$.

(b) Here we use the function $f(x) = x + 1$ for all values of x greater than 3. Thus, $\lim\limits_{x\to 3^+} f(x) = 3 + 1 = 4$.

(c) In order to evaluate the limit as x approaches 3, we find the limit as it approaches 3^+ (from the right) and the limit as it approaches 3^- (from the left). If the two limits approach the same value, or both approach positive infinity or both approach negative infinity, then the limit is that value, or the appropriately signed infinity. If the two limits do not agree, the limit "Does Not Exist." Here, if we refer to the solutions in parts (a) and (b), we see that $\lim\limits_{x\to 3^-} f(x) = 4$ and $\lim\limits_{x\to 3^+} f(x) = 4$. Because the two limits are the same, the limit is 4.

14. $\dfrac{3}{\sqrt{2}}$

Here, if we plug in $\dfrac{\pi}{4}$ for x, we get $\lim\limits_{x\to\frac{\pi}{4}} 3\cos x = 3\cos\dfrac{\pi}{4} = \dfrac{3}{\sqrt{2}}$.

15.   0

Here, if we plug in 0 for x, we get $\lim\limits_{x\to 0} 3\dfrac{x}{\cos x} = 3\dfrac{0}{\cos 0} = 3\dfrac{0}{1} = 0$.

16.   3

Remember Rule No. 1, which says that $\lim\limits_{x\to 0}\dfrac{\sin x}{x} = 1$. If we want to find the limit of its reciprocal, we can write this as $\lim\limits_{x\to 0}\dfrac{1}{\sin x} = \dfrac{1}{\lim\limits_{x\to 0}\dfrac{\sin x}{x}} = \dfrac{1}{1} = 1$. Here, if we plug in 0 for x, we get $\lim\limits_{x\to 0}\left(3\dfrac{x}{\sin x}\right) = (3)(1) = 3$.

17. $\dfrac{7}{5}$

Here, we can rewrite the expression as $\lim\limits_{x\to 0}\dfrac{\tan 7x}{\sin 5x} = \lim\limits_{x\to 0}\dfrac{\frac{\sin 7x}{\cos 7x}}{\sin 5x} = \lim\limits_{x\to 0}\dfrac{\sin 7x}{(\cos 7x)(\sin 5x)}$. Remember

Rule No. 4, which says that $\lim\limits_{x\to 0}\dfrac{\sin ax}{\sin bx} = \dfrac{a}{b}$. Here, $\lim\limits_{x\to 0}\dfrac{\sin 7x}{(\cos 7x)(\sin 5x)} = \dfrac{7}{(1)(5)} = \dfrac{7}{5}$. If we want to

evaluate the limit the long way, we first divide the numerator and the denominator of the expres-

sion by x: $\lim\limits_{x\to 0}\dfrac{\left(\frac{\sin 7x}{x}\right)}{(\cos 7x)\left(\frac{\sin 5x}{x}\right)}$. Next, we multiply the numerator and the denominator of the top

expression by 7 and the numerator and the denominator of the bottom expression by 5. We get

$\lim\limits_{x\to 0}\dfrac{\left(\frac{7\sin 7x}{7x}\right)}{(\cos 7x)\left(\frac{5\sin 5x}{5x}\right)}$. Now, we can evaluate the limit: $\lim\limits_{x\to 0}\dfrac{(7)(1)}{(1)(5)(1)} = \dfrac{7}{5}$.

18.   The limit *Does Not Exist*.

The value of sin x oscillates between –1 and 1. Thus, as x approaches infinity, sin x does not approach a specific value. Therefore, the limit *Does Not Exist*.

19.   0

Here, as x approaches infinity, $\dfrac{1}{x}$ approaches 0. Thus, $\lim\limits_{x\to\infty}\sin\dfrac{1}{x} = \sin 0 = 0$.

20. $\dfrac{49}{121}$

We can break up the limit into $\lim\limits_{x\to 0}\dfrac{\sin 7x}{\sin 11x}\dfrac{\sin 7x}{\sin 11x}$. Remember Rule No. 4, which says

that $\lim\limits_{x\to 0}\dfrac{\sin ax}{\sin bx} = \dfrac{a}{b}$. Here, $\lim\limits_{x\to 0}\dfrac{\sin 7x}{\sin 11x}\dfrac{\sin 7x}{\sin 11x} = \dfrac{7}{11}\times\dfrac{7}{11} = \dfrac{49}{121}$. If we want to evalu-

ate the limit the long way, we first divide the numerator and the denominator of the expressions by x: $\lim\limits_{x \to 0}\left(\dfrac{\dfrac{\sin 7x}{x}}{\dfrac{\sin 11x}{x}} \cdot \dfrac{\dfrac{\sin 7x}{x}}{\dfrac{\sin 11x}{x}}\right)$. Next, we multiply the numerator and the denomi-

nator of the top expression by 7 and the numerator and the denominator of the bottom

expression by 11. We get $\lim\limits_{x \to 0}\left(\dfrac{\dfrac{7\sin 7x}{7x}}{\dfrac{11\sin 11x}{11x}} \cdot \dfrac{\dfrac{7\sin 7x}{7x}}{\dfrac{11\sin 11x}{11x}}\right)$. Now, we can evaluate the limit:

$\lim\limits_{x \to 0}\left(\dfrac{\dfrac{7\sin 7x}{7x}}{\dfrac{11\sin 11x}{11x}} \cdot \dfrac{\dfrac{7\sin 7x}{7x}}{\dfrac{11\sin 11x}{11x}}\right) = \dfrac{(7)(1)}{(11)(1)}\dfrac{(7)(1)}{(11)(1)} = \left(\dfrac{7}{11}\right)\left(\dfrac{7}{11}\right) = \dfrac{49}{121}$.

21.   6

Notice that if we plug in 0 for h, we get $\dfrac{0}{0}$, which is indeterminate. If we expand the expression in

the numerator, we get $\lim\limits_{h \to 0}\dfrac{9 + 6h + h^2 - 9}{h}$. This simplifies to $\lim\limits_{h \to 0}\dfrac{6h + h^2}{h}$. Next, factor h out

of the top expression: $\lim\limits_{h \to 0}\dfrac{h(6 + h)}{h}$. Now, we can cancel the h and evaluate the limit to get

$\lim\limits_{h \to 0}\dfrac{h(6 + h)}{h} = \lim\limits_{h \to 0}(6 + h) = 6 + 0 = 6$.

22. $-\dfrac{1}{x^2}$

Notice that if we plug in 0 for h, we get $\dfrac{0}{0}$, which is indeterminate. If we combine

the two expressions on top with a common denominator, we get

$\lim\limits_{h \to 0}\dfrac{\dfrac{1}{x+h} - \dfrac{1}{x}}{h} = \lim\limits_{h \to 0}\dfrac{\dfrac{x}{x(x+h)} - \dfrac{x+h}{x(x+h)}}{h} = \lim\limits_{h \to 0}\dfrac{\dfrac{x-(x+h)}{x(x+h)}}{h}$. We can simplify the top expression, leaving us with $\lim\limits_{h \to 0} \dfrac{\dfrac{-h}{x(x+h)}}{h}$. Next, simplify the expression into $\lim\limits_{h \to 0} \dfrac{\dfrac{-h}{x(x+h)}}{h} = \lim\limits_{h \to 0} \dfrac{-h}{hx(x+h)}$.

We can cancel the h to get $\lim\limits_{h \to 0} \dfrac{-1}{x(x+h)}$. Now, if we evaluate the limit, we get

$$\lim\limits_{h \to 0} \frac{-1}{x(x+h)} = \frac{-1}{x(x)} = -\frac{1}{x^2} .$$

SOLUTIONS TO PRACTICE PROBLEM SET 2

1.   Yes. It satisfies all three conditions.

In order for a function $f(x)$ to be continuous at a point $x = c$, it must fulfill *all three* of the following conditions:

Condition 1: $f(c)$ exists.

Condition 2: $\lim\limits_{x \to c} f(x)$ exists.

Condition 3: $\lim\limits_{x \to c} f(x) = f(c)$

Let's test each condition.

$f(2) = 9$, which satisfies condition 1.

$\lim\limits_{x \to 2^-} f(x) = 9$ and $\lim\limits_{x \to 2^+} f(x) = 9$, so $\lim\limits_{x \to 2} f(x) = 9$, which satisfies condition 2.

$\lim\limits_{x \to 2} f(x) = 9 = f(2)$, which satisfies condition 3. Therefore, $f(x)$ is continuous at $x = 2$.

2.   No. It fails condition 3.

In order for a function $f(x)$ to be continuous at a point $x = c$, it must fulfill *all three* of the following conditions:

Condition 1: $f(c)$ exists.

Condition 2: $\lim\limits_{x \to c} f(x)$ exists.

Condition 3: $\lim_{x \to c} f(x) = f(c)$

Let's test each condition.

$f(3) = 29$, which satisfies condition 1.

$\lim_{x \to 3^-} f(x) = 30$ and $\lim_{x \to 3^+} f(x) = 30$, so $\lim_{x \to 3} f(x) = 30$, which satisfies condition 2.

But $\lim_{x \to 3} f(x) \neq f(3)$. Therefore, $f(x)$ is not continuous at $x = 3$ because it fails condition 3.

3.   No. It is discontinuous at any odd integral multiple of $\dfrac{\pi}{2}$.

Recall that $\sec x = \dfrac{1}{\cos x}$. This means that sec x is undefined at any value where

$\cos x = 0$, which are the odd multiples of $\dfrac{\pi}{2}$. Therefore, sec x is not continuous everywhere.

4.   No. It is discontinuous at the endpoints of the interval.

Recall that $\sec x = \dfrac{1}{\cos x}$. This means that sec x is undefined at any value where

$\cos x = 0$. Also recall that $\cos \dfrac{\pi}{2} = 0$ and $\cos\left(-\dfrac{\pi}{2}\right) = 0$. Therefore, sec x is not continuous every-

where on the interval $\left[-\dfrac{\pi}{2}, \dfrac{\pi}{2}\right]$.

5.   Yes.

Recall that $\sec x = \dfrac{1}{\cos x}$. This means that sec x is undefined at any value where

$\cos x = 0$. Also recall that $\cos \dfrac{\pi}{2} = 0$ and $\cos\left(-\dfrac{\pi}{2}\right) = 0$. Therefore, sec x is continuous everywhere

on the interval $\left(-\dfrac{\pi}{2}, \dfrac{\pi}{2}\right)$ because the interval does not include the endpoints.

6.   The function is continuous for $k = \dfrac{9}{16}$.

In order for a function $f(x)$ to be continuous at a point $x = c$, it must fulfill *all three* of the following conditions:

Condition 1: $f(c)$ exists.

Condition 2: $\lim\limits_{x \to c} f(x)$ exists.

Condition 3: $\lim\limits_{x \to c} f(x) = f(c)$

We will need to find a value, or values, of k that enables $f(x)$ to satisfy each condition.

Condition 1: $f(4) = 0$

Condition 2: $\lim\limits_{x \to 4^-} f(x) = 0$ and $\lim\limits_{x \to 4^+} f(x) = 16k - 9$. In order for the limit to exist, the two limits must be the same. If we solve $16k - 9 = 0$, we get $k = \dfrac{9}{16}$.

Condition 3: If we now let $k = \dfrac{9}{16}$, $\lim\limits_{x \to 4} f(x) = 0 = f(4)$. Therefore, the solution is $k = \dfrac{9}{16}$.

7.   The removable discontinuity is at $\left(3, \dfrac{11}{5} \right)$.

A *removable discontinuity* occurs when you have a rational expression with common factors in the numerator and denominator. Because these factors can be cancelled, the discontinuity is "removable." In practical terms, this means that the discontinuity occurs where there is a "hole" in the graph. If we factor $f(x) = \dfrac{x^2 + 5x - 24}{x^2 - x - 6}$, we get $f(x) = \dfrac{(x+8)(x-3)}{(x+2)(x-3)}$. If we cancel the common factor, we get $f(x) = \dfrac{(x+8)}{(x+2)}$. Now, if we plug in $x = 3$, we get $f(x) = \dfrac{11}{5}$. Therefore, the removable discontinuity is at $\left(3, \dfrac{11}{5} \right)$.

8. (a) 0; (b) 0; (c) 1; (d) 1; (e) $f(3)$ *Does Not Exist*; (f) a jump discontinuity at $x = -3$; a removable discontinuity at $x = 3$ and an essential discontinuity at $x = 5$.

(a) If we look at the graph, we can see that $\lim\limits_{x \to -\infty} f(x) = 0$.

(b) If we look at the graph, we can see that $\lim\limits_{x \to \infty} f(x) = 0$.

(c) If we look at the graph, we can see that $\lim\limits_{x \to 3^-} f(x) = 1$.

(d) If we look at the graph, we can see that $\lim\limits_{x \to 3^+} f(x) = 1$.

(e) $f(3)$ *Does Not Exist*.

(f) There are three discontinuities: (1) a jump discontinuity at $x = -3$; (2) a removable discontinuity at $x = 3$; and (3) an essential discontinuity at $x = 5$.

SOLUTIONS TO PRACTICE PROBLEM SET 3

1. 5

We find the derivative of a function, $f(x)$, using the definition of the derivative, which is

$f'(x) = \lim\limits_{h \to 0} \dfrac{f(x+h) - f(x)}{h}$. Here $f(x) = 5x$ and $x = 3$. This means that $f(3) = 5(3) = 15$ and

$f(3+h) = 5(3+h) = 15 + 5h$. If we now plug these into the definition of the derivative, we get

$f'(3) = \lim\limits_{h \to 0} \dfrac{f(3+h) - f(3)}{h} = \lim\limits_{h \to 0} \dfrac{15 + 5h - 15}{h}$. This simplifies to $f'(3) = \lim\limits_{h \to 0} \dfrac{5h}{h} = \lim\limits_{h \to 0} 5 = 5$.

If you noticed that the function is simply the equation of a line, then you would have seen that the derivative is simply the slope of the line, which is 5 everywhere.

2. 4

We find the derivative of a function, $f(x)$, using the definition of the derivative, which is

$f'(x) = \lim\limits_{h \to 0} \dfrac{f(x+h) - f(x)}{h}$. Here $f(x) = 4x$ and $x = -8$. This means that $f(-8) = 4(-8) = -32$

and $f(-8+h) = 4(-8+h) = -32 + 4h$. If we now plug these into the definition of the derivative, we get $f'(-8) = \lim\limits_{h \to 0} \dfrac{f(-8+h) - f(-8)}{h} = \lim\limits_{h \to 0} \dfrac{-32 + 4h + 32}{h}$. This simplifies to

$f'(-8) = \lim\limits_{h \to 0} \dfrac{4h}{h} = \lim\limits_{h \to 0} 4 = 4$.

If you noticed that the function is simply the equation of a line, then you would have seen that the derivative is simply the slope of the line, which is 4 everywhere.

3.   −10

We find the derivative of a function, $f(x)$, using the definition of the derivative, which is

$f'(x) = \lim\limits_{h \to 0} \dfrac{f(x+h) - f(x)}{h}$. Here $f(x) = 5x^2$ and $x = -1$. This means that $f(-1) = 5(-1)^2 = 5$

and $f(-1+h) = 5(-1+h)^2 = 5(1 - 2h + h^2) = 5 - 10h + 5h^2$. If we now plug these into the defi-

nition of the derivative, we get $f'(-1) = \lim\limits_{h \to 0} \dfrac{f(-1+h) - f(-1)}{h} = \lim\limits_{h \to 0} \dfrac{5 - 10h + 5h^2 - 5}{h}$. This

simplifies to $f'(-1) = \lim\limits_{h \to 0} \dfrac{-10h + 5h^2}{h}$. Now we can factor out the h from the numerator and can-

cel it with the h in the denominator: $f'(-1) = \lim\limits_{h \to 0} \dfrac{h(-10 + 5h)}{h} = \lim\limits_{h \to 0}(-10 + 5h)$. Now we take

the limit to get $f'(-1) = \lim\limits_{h \to 0}(-10 + 5h) = -10$.

4. $16x$

We find the derivative of a function, $f(x)$, using the definition of the

derivative, which is: $f'(x) = \lim\limits_{h \to 0} \dfrac{f(x+h) - f(x)}{h}$. Here $f(x) = 8x^2$ and

$f(x+h) = 8(x+h)^2 = 8(x^2 + 2xh + h^2) = 8x^2 + 16xh + 8h^2$. If we now plug these into the defi-

nition of the derivative, we get $f'(x) = \lim\limits_{h \to 0} \dfrac{f(x+h) - f(x)}{h} = \lim\limits_{h \to 0} \dfrac{8x^2 + 16xh + 8h^2 - 8x^2}{h}$. This

simplifies to $f'(x) = \lim\limits_{h \to 0} \dfrac{16xh + 8h^2}{h}$. Now we can factor out the h from the numerator and cancel it with the h in the denominator: $f'(x) = \lim\limits_{h \to 0} \dfrac{h(16x + 8h)}{h} = \lim\limits_{h \to 0}(16x + 8h)$. Now we take the

limit to get $f'(x) = \lim\limits_{h \to 0}(16x + 8h) = 16x$.

5.　　　$-20x$

We find the derivative of a function, $f(x)$, using the definition of the

derivative, which is $f'(x) = \lim\limits_{h \to 0} \dfrac{f(x + h) - f(x)}{h}$. Here $f(x) = -10x^2$ and

$f(x + h) = -10(x + h)^2 = -10(x^2 + 2xh + h^2) = -10x^2 - 20xh - 10h^2$.

If we now plug these into the definition of the derivative, we get

$f'(x) = \lim\limits_{h \to 0} \dfrac{f(x + h) - f(x)}{h} = \lim\limits_{h \to 0} \dfrac{-10x^2 - 20xh - 10h^2 + 10x^2}{h}$. This simplifies to

$f'(x) = \lim\limits_{h \to 0} \dfrac{-20xh - 10h^2}{h}$. Now we can factor out the h from the numerator and cancel it with

the h in the denominator: $f'(x) = \lim\limits_{h \to 0} \dfrac{h(-20x - 10h)}{h} = \lim\limits_{h \to 0}(-20x - 10h)$. Now we take the limit

to get $f'(x) = \lim\limits_{h \to 0}(-20x - 10h) = -20x$.

6.　　　$40a$

We find the derivative of a function, $f(x)$, using the definition of the derivative, which is

$f'(x) = \lim\limits_{h \to 0} \dfrac{f(x + h) - f(x)}{h}$. Here $f(x) = 20x^2$ and $x = a$. This means that $f(a) = 20a^2$ and

$f(a + h) = 20(a + h)^2 = 20(a^2 + 2ah + h^2) = 20a^2 + 40ah + 20h^2$. If we now plug these into the

definition of the derivative, we get $f'(a) = \lim\limits_{h \to 0} \dfrac{f(a + h) - f(a)}{h} = \lim\limits_{h \to 0} \dfrac{20a^2 + 40ah + 20h^2 - 20a^2}{h}$.

This simplifies to $f'(a) = \lim\limits_{h \to 0} \dfrac{40ah + 20h^2}{h}$. Now we can factor out the h from the numerator and cancel it with the h in the denominator: $f'(a) = \lim\limits_{h \to 0} \dfrac{h(40a + 20h)}{h} = \lim\limits_{h \to 0}(40a + 20h)$. Now we

take the limit to get $f'(a) = \lim\limits_{h \to 0}(40a + 20h) = 40a$.

7. 54

We find the derivative of a function, $f(x)$, using the defini-

tion of the derivative, which is $f'(x) = \lim\limits_{h \to 0} \dfrac{f(x+h) - f(x)}{h}$.

Here $f(x) = 2x^3$ and $x = -3$. This means that $f(-3) = 2(-3)^3 = -54$ and

$f(-3 + h) = 2(-3 + h)^3 = 2\left((-3)^3 + 3(-3)^2 h + 3(-3)h^2 + h^3\right) = -54 + 54h - 18h^2 + 2h^3$.

If we now plug these into the definition of the derivative, we get

$f'(-3) = \lim\limits_{h \to 0} \dfrac{f(-3+h) - f(-3)}{h} = \lim\limits_{h \to 0} \dfrac{-54 + 54h - 18h^2 + 2h^3 + 54}{h}$. This simplifies to

$f'(-3) = \lim\limits_{h \to 0} \dfrac{54h - 18h^2 + 2h^3}{h}$. Now we can factor out the h from the numerator and cancel it

with the h in the denominator: $f'(-3) = \lim\limits_{h \to 0} \dfrac{h(54 - 18h + 2h^2)}{h} = \lim\limits_{h \to 0}(54 - 18h + 2h^2)$. Now we

take the limit to get $f'(-3) = \lim\limits_{h \to 0}(54 - 18h + 2h^2) = 54$.

8. $-9x^2$

We find the derivative of a function, $f(x)$, using the definition of the deriva-

tive, which is $f'(x) = \lim\limits_{h \to 0} \dfrac{f(x+h) - f(x)}{h}$. Here, $f(x) = -3x^3$. This means that

$f(x + h) = -3(x + h)^3 = -3\left(x^3 + 3x^2h + 3xh^2 + h^3\right) = -3x^3 - 9x^2h - 9xh^2 - 3h^3$.

If we now plug these into the definition of the derivative, we get

$$f'(x) = \lim_{h \to 0} \frac{f(x+h) - f(x)}{h} = \lim_{h \to 0} \frac{-3x^3 - 9x^2h - 9xh^2 - 3h^3 + 3x^3}{h}.$$ This simplifies to

$$f'(x) = \lim_{h \to 0} \frac{-9x^2h - 9xh^2 - 3h^3}{h}.$$ Now we can factor out the h from the numerator and cancel it

with the h in the denominator: $f'(x) = \lim_{h \to 0} \frac{h(-9x^2 - 9xh - 3h^2)}{h} = \lim_{h \to 0}(-9x^2 - 9xh - 3h^2)$. Now

we take the limit to get $f'(x) = \lim_{h \to 0}(-9x^2 - 9xh - 3h^2) = -9x^2$.

9.　　　$5x^4$

We find the derivative of a function, $f(x)$, using the definition of the derivative, which is

$$f'(x) = \lim_{h \to 0} \frac{f(x+h) - f(x)}{h}.$$ Here $f(x) = x^5$.

This means that $f(x+h) = (x+h)^5 = (x^5 + 5x^4h + 10x^3h^2 + 10x^2h^3 + 5xh^4 + h^5)$.

If we now plug these into the definition of the derivative, we get

$$f'(x) = \lim_{h \to 0} \frac{f(x+h) - f(x)}{h} = \lim_{h \to 0} \frac{x^5 + 5x^4h + 10x^3h^2 + 10x^2h^3 + 5xh^4 + h^5 - x^5}{h}.$$

This simplifies to $f'(x) = \lim_{h \to 0} \frac{5x^4h + 10x^3h^2 + 10x^2h^3 + 5xh^4 + h^5}{h}.$ Now we can fac-

tor out the h from the numerator and cancel it with the h in the denominator:

$$f'(x) = \lim_{h \to 0} \frac{h(5x^4 + 10x^3h + 10x^2h^2 + 5xh^3 + h^4)}{h} = \lim_{h \to 0}(5x^4 + 10x^3h + 10x^2h^2 + 5xh^3 + h^4).$$

Now we take the limit to get $f'(x) = \lim_{h \to 0}(5x^4 + 10x^3h + 10x^2h^2 + 5xh^3 + h^4) = 5x^4$.

10. $\dfrac{1}{3}$

We find the derivative of a function, $f(x)$, using the definition of the derivative, which is

$f'(x) = \lim\limits_{h \to 0} \dfrac{f(x+h) - f(x)}{h}$. Here $f(x) = 2\sqrt{x}$ and $x = 9$. This means that $f(9) = 2\sqrt{9} = 6$

and $f(9+h) = 2\sqrt{9+h}$. If we now plug these into the definition of the derivative, we get

$f'(9) = \lim\limits_{h \to 0} \dfrac{f(9+h) - f(9)}{h} = \lim\limits_{h \to 0} \dfrac{2\sqrt{9+h} - 6}{h}$. Notice that if we now take the limit, we

get the indeterminate form $\dfrac{0}{0}$. With polynomials, we merely simplify the expression to elimi-

nate this problem. In any derivative of a square root, we first multiply the top and the bottom

of the expression by the conjugate of the numerator and then we can simplify. Here, the con-

jugate is $2\sqrt{9+h} + 6$. We get $f'(9) = \lim\limits_{h \to 0} \dfrac{2\sqrt{9+h} - 6}{h} \times \dfrac{2\sqrt{9+h} + 6}{2\sqrt{9+h} + 6}$. This simplifies to

$f'(9) = \lim\limits_{h \to 0} \dfrac{4(9+h) - 36}{h\left(2\sqrt{9+h} + 6\right)} = \lim\limits_{h \to 0} \dfrac{36 + 4h - 36}{h\left(2\sqrt{9+h} + 6\right)} = \lim\limits_{h \to 0} \dfrac{4h}{h\left(2\sqrt{9+h} + 6\right)}$. Now we can cancel the

h in the numerator and the denominator to get $f'(9) = \lim\limits_{h \to 0} \dfrac{4}{\left(2\sqrt{9+h} + 6\right)}$. Now we take the limit:

$f'(9) = \lim\limits_{h \to 0} \dfrac{4}{\left(2\sqrt{9+h} + 6\right)} = \dfrac{4}{\left(2\sqrt{9} + 6\right)} = \dfrac{1}{3}$.

11. $\dfrac{5}{4}$

We find the derivative of a function, $f(x)$, using the definition of the derivative, which is

$f'(x) = \lim\limits_{h \to 0} \dfrac{f(x+h) - f(x)}{h}$. Here $f(x) = 5\sqrt{2x}$ and $x = 8$. This means that $f(8) = 5\sqrt{16} = 20$

and $f(8+h) = 5\sqrt{2(8+h)} = 5\sqrt{16 + 2h}$. If we now plug these into the definition of the deriva-

tive, we get $f'(8) = \lim\limits_{h \to 0} \dfrac{f(8+h) - f(8)}{h} = \lim\limits_{h \to 0} \dfrac{5\sqrt{16 + 2h} - 20}{h}$. Notice that if we now take the limit, we get the indeterminate form $\dfrac{0}{0}$. With polynomials, we merely simplify the expression

to eliminate this problem. In any derivative of a square root, we first multiply the top and the

bottom of the expression by the conjugate of the numerator and then we simplify. Here the con-

jugate is $5\sqrt{16+2h}+20$. We get $f'(8)=\lim\limits_{h\to0}\dfrac{5\sqrt{16+2h}-20}{h}\times\dfrac{5\sqrt{16+2h}+20}{5\sqrt{16+2h}+20}$. This simplifies

to $f'(8)=\lim\limits_{h\to0}\dfrac{25(16+2h)-400}{h\left(5\sqrt{16+2h}+20\right)}=\lim\limits_{h\to0}\dfrac{400+50h-400}{h\left(5\sqrt{16+2h}+20\right)}=\lim\limits_{h\to0}\dfrac{50h}{h\left(5\sqrt{16+2h}+20\right)}$. Now we

can cancel the h in the numerator and the denominator to get $f'(8)=\lim\limits_{h\to0}\dfrac{50}{\left(5\sqrt{16+2h}+20\right)}$.

Now, we take the limit: $f'(8)=\lim\limits_{h\to0}\dfrac{50}{\left(5\sqrt{16+2h}+20\right)}=\dfrac{50}{(20+20)}=\dfrac{5}{4}$.

12. $\dfrac{1}{2}$

We find the derivative of a function, $f(x)$, using the definition of the derivative, which

is $f'(x)=\lim\limits_{h\to0}\dfrac{f(x+h)-f(x)}{h}$.   Here $f(x)=\sin x$ and $x=\dfrac{\pi}{3}$.   This   means   that

$f\left(\dfrac{\pi}{3}\right)=\sin\dfrac{\pi}{3}=\dfrac{\sqrt{3}}{2}$ and $f\left(\dfrac{\pi}{3}+h\right)=\sin\left(\dfrac{\pi}{3}+h\right)$. If we now plug these into the definition of the

derivative, we get $f'\left(\dfrac{\pi}{3}\right)=\lim\limits_{h\to0}\dfrac{f\left(\dfrac{\pi}{3}+h\right)-f\left(\dfrac{\pi}{3}\right)}{h}=\lim\limits_{h\to0}\dfrac{\sin\left(\dfrac{\pi}{3}+h\right)-\dfrac{\sqrt{3}}{2}}{h}$. Notice that if we

now take the limit, we get the indeterminate form $\dfrac{0}{0}$. We cannot eliminate this problem

merely by simplifying the expression the way that we did with a polynomial. Recall that the

trigonometric formula $\sin(A+B)=\sin A\cos B+\cos A\sin B$. Here we can rewrite the top expression as $f'\left(\dfrac{\pi}{3}\right)=\lim\limits_{h\to 0}\dfrac{\sin\left(\dfrac{\pi}{3}+h\right)-\dfrac{\sqrt{3}}{2}}{h}=\lim\limits_{h\to 0}\dfrac{\sin\dfrac{\pi}{3}\cos h+\cos\dfrac{\pi}{3}\sin h-\dfrac{\sqrt{3}}{2}}{h}$. We can

break up the limit into $\lim\limits_{h\to 0}\dfrac{\sin\dfrac{\pi}{3}\cos h-\dfrac{\sqrt{3}}{2}}{h}+\dfrac{\cos\dfrac{\pi}{3}\sin h}{h}=\lim\limits_{h\to 0}\dfrac{\dfrac{\sqrt{3}}{2}\cos h-\dfrac{\sqrt{3}}{2}}{h}+\dfrac{\left(\dfrac{1}{2}\right)\sin h}{h}$.

Next, factor $\dfrac{\sqrt{3}}{2}$ out of the top of the left-hand expression: $\lim\limits_{h\to 0}\dfrac{\dfrac{\sqrt{3}}{2}\left(\cos h-1\right)}{h}+\dfrac{\left(\dfrac{1}{2}\right)\sin h}{h}$.

Now, we can break this into separate limits: $\lim\limits_{h\to 0}\dfrac{\dfrac{\sqrt{3}}{2}\left(\cos h-1\right)}{h}+\lim\limits_{h\to 0}\dfrac{\left(\dfrac{1}{2}\right)\sin h}{h}$.

The left-hand limit is $\lim\limits_{h\to 0}\dfrac{\dfrac{\sqrt{3}}{2}\left(\cos h-1\right)}{h}=\dfrac{\sqrt{3}}{2}\lim\limits_{h\to 0}\dfrac{\left(\cos h-1\right)}{h}=\dfrac{\sqrt{3}}{2}\cdot 0=0$. The right-hand

limit is $\dfrac{1}{2}\lim\limits_{h\to 0}\dfrac{\sin h}{h}=\dfrac{1}{2}\cdot 1=\dfrac{1}{2}$. Therefore, the limit is $\dfrac{1}{2}$.

13. $2x+1$

We find the derivative of a function, $f(x)$, using the definition of the

derivative, which is $f'(x)=\lim\limits_{h\to 0}\dfrac{f(x+h)-f(x)}{h}$. Here $f(x)=x^2+x$ and

$f(x+h)=(x+h)^2+(x+h)=x^2+2xh+h^2+x+h$. If we now plug these into the definition of

the derivative, we get $f'(x)=\lim\limits_{h\to 0}\dfrac{f(x+h)-f(x)}{h}=\lim\limits_{h\to 0}\dfrac{x^2+2xh+h^2+x+h-\left(x^2+x\right)}{h}$. This

simplifies to $f'(x)=\lim\limits_{h\to 0}\dfrac{2xh+h^2+h}{h}$. Now we can factor out the h from the numerator and cancel

it with the h in the denominator: $f'(x)=\lim\limits_{h\to 0}\dfrac{2xh+h^2+h}{h}=\lim\limits_{h\to 0}\dfrac{h(2x+h+1)}{h}=\lim\limits_{h\to 0}(2x+h+1)$.

Now we take the limit to get $f'(x)=\lim\limits_{h\to 0}(2x+h+1)=2x+1$.

14. $3x^2 + 3$

We find the derivative of a function, $f(x)$, using the definition of the derivative, which is $f'(x) = \lim_{h \to 0} \dfrac{f(x+h) - f(x)}{h}$. Here $f(x) = x^3 + 3x + 2$ and

$$f(x+h) = (x+h)^3 + 3(x+h) + 2 = x^3 + 3x^2h + 3xh^2 + h^3 + 3x + 3h + 2.$$

If we now plug these into the definition of the derivative, we get

$$f'(x) = \lim_{h \to 0} \frac{f(x+h) - f(x)}{h} = \lim_{h \to 0} \frac{x^3 + 3x^2h + 3xh^2 + h^3 + 3x + 3h + 2 - (x^3 + 3x + 2)}{h}.$$

This simplifies to $f'(x) = \lim_{h \to 0} \dfrac{3x^2h + 3xh^2 + h^3 + 3h}{h}$. Now we can fac-

tor out the h from the numerator and cancel it with the h in the denominator:

$$f'(x) = \lim_{h \to 0} \frac{3x^2h + 3xh^2 + h^3 + 3h}{h} \lim_{h \to 0} \frac{h(3x^2 + 3xh + h^2 + 3)}{h} = \lim_{h \to 0}(3x^2 + 3xh + h^2 + 3). \quad \text{Now}$$

we take the limit to get $f'(x) = \lim_{h \to 0}(3x^2 + 3xh + h^2 + 3) = 3x^2 + 3$.

15. $-\dfrac{1}{x^2}$

We find the derivative of a function, $f(x)$, using the definition of the derivative, which is

$$f'(x) = \lim_{h \to 0} \frac{f(x+h) - f(x)}{h}. \quad \text{Here } f(x) = \frac{1}{x} \text{ and } f(x+h) = \frac{1}{x+h}. \text{ If we now plug these}$$

into the definition of the derivative, we get $f'(x) = \lim_{h \to 0} \dfrac{f(x+h) - f(x)}{h} = \lim_{h \to 0} \dfrac{\frac{1}{x+h} - \frac{1}{x}}{h}$.

Notice that if we now take the limit, we get the indeterminate form $\dfrac{0}{0}$. We cannot elimi-

nate this problem merely by simplifying the expression the way that we did with a poly-

nomial. Here, we combine the two terms in the numerator of the expression to get

$$f'(x) = \lim_{h \to 0} \frac{\dfrac{1}{x+h} - \dfrac{1}{x}}{h} = \lim_{h \to 0} \frac{\dfrac{x}{x(x+h)} - \dfrac{x+h}{x(x+h)}}{h} = \lim_{h \to 0} \frac{\dfrac{-h}{x(x+h)}}{h}.$$ This simplifies to

$$f'(x) = \lim_{h \to 0} \frac{\dfrac{-h}{x(x+h)}}{h} = \lim_{h \to 0} \frac{-h}{x(x+h)h}.$$ Now we can cancel the factor h in the numerator

and the denominator to get $f'(x) = \lim_{h \to 0} \dfrac{-h}{x(x+h)h} = \lim_{h \to 0} \dfrac{-1}{x(x+h)}$. Now we take the limit:

$$f'(x) = \lim_{h \to 0} \frac{-1}{x(x+h)} = -\frac{1}{x^2}.$$

16. $2ax + b$

We find the derivative of a function, $f(x)$, using the definition of the

derivative, which is $f'(x) = \lim_{h \to 0} \dfrac{f(x+h) - f(x)}{h}$. Here $f(x) = ax^2 + bx + c$ and

$$f(x+h) = a(x+h)^2 + b(x+h) + c = ax^2 + 2axh + ah^2 + bx + bh + c.$$

If we now plug these into the definition of the derivative, we get

$$f'(x) = \lim_{h \to 0} \frac{f(x+h) - f(x)}{h} = \lim_{h \to 0} \frac{ax^2 + 2axh + ah^2 + bx + bh + c - \left(ax^2 + bx + c\right)}{h}.$$ This simplifies

to $f'(x) = \lim_{h \to 0} \dfrac{2axh + ah^2 + bh}{h}$. Now we can factor out the h from the numerator and cancel it with

the h in the denominator: $f'(x) = \lim_{h \to 0} \dfrac{2axh + ah^2 + bh}{h} = \lim_{h \to 0} \dfrac{h(2ax + ah + b)}{h} = \lim_{h \to 0}(2ax + ah + b)$.

Now we take the limit to get $f'(x) = \lim_{h \to 0}(2ax + ah + b) = 2ax + b$.

SOLUTIONS TO PRACTICE PROBLEM SET 4

1. $64x^3 + 16x$

First, expand $(4x^2 + 1)^2$ to get $16x^4 + 8x^2 + 1$. Now, use the Power Rule to take the derivative of each term. The derivative of $16x^4 = 16(4x^3) = 64x^3$. The derivative of $8x^2 = 8(2x) = 16x$. The derivative of $1 = 0$ (because the derivative of a constant is zero). Therefore, the derivative is $64x^3 + 16x$.

2. $10x^9 + 36x^5 + 18x$

First, expand $(x^5 + 3x)^2$ to get $x^{10} + 6x^6 + 9x^2$. Now, use the Power Rule to take the derivative of each term. The derivative of $x^{10} = 10x^9$. The derivative of $6x^6 = 6(6x^5) = 36x^5$. The derivative of $9x^2 = 9(2x) = 18x$. Therefore, the derivative is $10x^9 + 36x^5 + 18x$.

3. $77x^6$

Simply use the Power Rule. The derivative is $11x^7 = 11(7x^6) = 77x^6$.

4. $54x^2 + 12$

Use the Power Rule to take the derivative of each term. The derivative of $18x^3 = 18(3x^2) = 54x^2$. The derivative of $12x = 12$. The derivative of $11 = 0$ (because the derivative of a constant is zero). Therefore, the derivative is $54x^2 + 12$.

5. $6x^{11}$

Use the Power Rule to take the derivative of each term. The derivative of $x^{12} = 12x^{11}$. The derivative of $17 = 0$ (because the derivative of a constant is zero). Therefore, the derivative is $\frac{1}{2}(12x^{11}) = 6x^{11}$.

6. $-3x^8 - 2x^2$

Use the Power Rule to take the derivative of each term. The derivative of $x^9 = 9x^8$. The derivative of $2x^3 = 2(3x^2) = 6x^2$. The derivative of $9 = 0$ (because the derivative of a constant is zero). Therefore, the derivative is $-\frac{1}{3}(9x^8 + 6x^2) = -3x^8 - 2x^2$.

7.   0

Don't be fooled by the power. π^5 is a constant, so the derivative is zero.

8. $64x^{-9} + \dfrac{6}{\sqrt{x}}$

Use the Power Rule to take the derivative of each term. The derivative of $-8x^{-8} = -8\left(-8x^{-9}\right) = 64x^{-9}$.

The derivative of $12\sqrt{x} = \dfrac{12}{2\sqrt{x}} = \dfrac{6}{\sqrt{x}}$ (remember the shortcut that we showed you on page 129).

Therefore, the derivative is $64x^{-9} + \dfrac{6}{\sqrt{x}}$.

9. $-42x^{-8} - \dfrac{2}{\sqrt{x}}$

Use the Power Rule to take the derivative of each term. The derivative of $6x^{-7} = 6\left(-7x^{-8}\right) = -42x^{-8}$.

The derivative of $4\sqrt{x} = \dfrac{4}{2\sqrt{x}} = \dfrac{2}{\sqrt{x}}$ (remember the shortcut that we showed you on page 129).

Therefore, the derivative is $-42x^{-8} - \dfrac{2}{\sqrt{x}}$.

10. $\dfrac{-5}{x^6} - \dfrac{8}{x^9}$

Use the Power Rule to take the derivative of each term. The derivative of $x^{-5} = -5x^{-6}$. To find the

derivative of $\dfrac{1}{x^8}$, we first rewrite it as x^{-8}. The derivative of $x^{-8} = -8x^{-9}$. Therefore, the derivative

is $-5x^{-6} - 8x^{-9} = \dfrac{-5}{x^6} - \dfrac{8}{x^9}$.

11. $216x^2 - 48x + 36$

First, expand $\left(6x^2 + 3\right)\left(12x - 4\right)$ to get $72x^3 - 24x^2 + 36x - 12$. Now, use the Power Rule to take the derivative of each term. The derivative of $72x^3 = 72\left(3x^2\right) = 216x^2$. The derivative of $24x^2 = 24\left(2x\right) = 48x$. The derivative of $36x = 36$. The derivative of 12 = 0 (because the derivative of a constant is zero). Therefore, the derivative is $216x^2 - 48x + 36$.

12. $-6 - 36x^2 + 12x^3 - 5x^4 - 14x^6$

First, expand $\left(3 - x - 2x^3\right)\left(6 + x^4\right)$ to get $18 - 6x - 12x^3 + 3x^4 - x^5 - 2x^7$. Now, use the Power Rule to take the derivative of each term. The derivative of $18 = 0$ (because the derivative of a constant is zero). The derivative of $6x = 6$. The derivative of $12x^3 = 12\left(3x^2\right) = 36x^2$. The derivative of $3x^4 = 3\left(4x^3\right) = 12x^3$. The derivative of $x^5 = 5x^4$. The derivative of $2x^7 = 2\left(7x^6\right) = 14x^6$. Therefore, the derivative is $-6 - 36x^2 + 12x^3 - 5x^4 - 14x^6$.

13. 0

Don't be fooled by the powers. Each term is a constant so the derivative is zero.

14. $-\dfrac{16}{x^5} + \dfrac{10}{x^6} + \dfrac{36}{x^7}$

First, expand $\left(\dfrac{1}{x} + \dfrac{1}{x^2}\right)\left(\dfrac{4}{x^3} - \dfrac{6}{x^4}\right)$ to get $\dfrac{4}{x^4} - \dfrac{6}{x^5} + \dfrac{4}{x^5} - \dfrac{6}{x^6} = \dfrac{4}{x^4} - \dfrac{2}{x^5} - \dfrac{6}{x^6}$. Next, rewrite the terms as: $4x^{-4} - 2x^{-5} - 6x^{-6}$. Now, use the Power Rule to take the derivative of each term. The derivative of $4x^{-4} = 4\left(-4x^{-5}\right) = -16x^{-5}$. The derivative of $2x^{-5} = 2\left(-5x^{-6}\right) = -10x^{-6}$. The derivative of $6x^{-6} = -36x^{-7}$. Therefore, the derivative is $-16x^{-5} + 10x^{-6} + 36x^{-7} = -\dfrac{16}{x^5} + \dfrac{10}{x^6} + \dfrac{36}{x^7}$.

15. $\dfrac{1}{2\sqrt{x}}$

Use the Power Rule to take the derivative of each term. The derivative of $\sqrt{x} = \dfrac{1}{2\sqrt{x}}$ (remember the shortcut that we showed you on page 129). The derivative of $\dfrac{1}{\sqrt{3}} = 0$ (because the derivative of a constant is zero). Therefore, the derivative is $\dfrac{1}{2\sqrt{x}}$.

16. 0

The derivative of a constant is zero.

17.　　　$3x^2 + 6x + 3$

First, expand $(x+1)^3$ to get $x^3 + 3x^2 + 3x + 1$. Now, use the Power Rule to take the derivative of each term. The derivative of $x^3 = 3x^2$. The derivative of $3x^2 = 3(2x) = 6x$. The derivative of $3x = 3$. The derivative of $1 = 0$ (because the derivative of a constant is zero). Therefore, the derivative is $3x^2 + 6x + 3$.

18.　　　$\dfrac{1}{2\sqrt{x}} + \dfrac{1}{3\sqrt[3]{x^2}} + \dfrac{2}{3\sqrt[3]{x}}$

Use the Power Rule to take the derivative of each term. The derivative of $\sqrt{x} = \dfrac{1}{2\sqrt{x}}$ (remember the shortcut that we showed you on page 129). Rewrite $\sqrt[3]{x}$ as $x^{\frac{1}{3}}$ and $\sqrt[3]{x^2}$ as $x^{\frac{2}{3}}$.

The derivative of $x^{\frac{1}{3}} = \dfrac{1}{3}x^{-\frac{2}{3}}$. The derivative of $x^{\frac{2}{3}} = \dfrac{2}{3}x^{-\frac{1}{3}}$. Therefore, the derivative is

$$\dfrac{1}{2\sqrt{x}} + \dfrac{1}{3}x^{-\frac{2}{3}} + \dfrac{2}{3}x^{-\frac{1}{3}} = \dfrac{1}{2\sqrt{x}} + \dfrac{1}{3\sqrt[3]{x^2}} + \dfrac{2}{3\sqrt[3]{x}}.$$

19.　　　$6x^2 + 6x - 14$

First, expand $x(2x+7)(x-2)$ to get $x(2x^2 + 3x - 14) = 2x^3 + 3x^2 - 14x$. Now, use the Power Rule to take the derivative of each term. The derivative of $2x^3 = 2(3x^2) = 6x^2$. The derivative of $3x^2 = 3(2x) = 6x$. The derivative of $14x = 14$. Therefore, the derivative is $6x^2 + 6x - 14$.

20.　　　$\dfrac{5}{6\sqrt[6]{x}} + \dfrac{7}{10\sqrt[10]{x^3}}$

First, rewrite the terms as $x^{\frac{1}{2}}\left(x^{\frac{1}{3}} + x^{\frac{1}{5}}\right)$. Next, distribute to get $x^{\frac{1}{2}}\left(x^{\frac{1}{3}} + x^{\frac{1}{5}}\right) = x^{\frac{5}{6}} + x^{\frac{7}{10}}$. Now, use the Power Rule to take the derivative of each term. The derivative of $x^{\frac{5}{6}} = \dfrac{5}{6}x^{-\frac{1}{6}}$. The derivative of $x^{\frac{7}{10}} = \dfrac{7}{10}x^{-\frac{3}{10}}$. Therefore, the derivative is $\dfrac{5}{6}x^{-\frac{1}{6}} + \dfrac{7}{10}x^{-\frac{3}{10}} = \dfrac{5}{6\sqrt[6]{x}} + \dfrac{7}{10\sqrt[10]{x^3}}$.

SOLUTIONS TO PRACTICE PROBLEM SET 5

1. $\dfrac{-80x^9 + 75x^8 + 12x^2 - 6x}{\left(5x^7 + 1\right)^2}$

We find the derivative using the Quotient Rule, which says that if $f(x) = \dfrac{u}{v}$, then

$f'(x) = \dfrac{v\dfrac{du}{dx} - u\dfrac{dv}{dx}}{v^2}$. Here $f(x) = \dfrac{4x^3 - 3x^2}{5x^7 + 1}$, so $u = 4x^3 - 3x^2$ and $v = 5x^7 + 1$. Using the Quo-

tient Rule, we get $f'(x) = \dfrac{\left(5x^7 + 1\right)\left(12x^2 - 6x\right) - \left(4x^3 - 3x^2\right)\left(35x^6\right)}{\left(5x^7 + 1\right)^2}$. This can be simplified to

$f'(x) = \dfrac{-80x^9 + 75x^8 + 12x^2 - 6x}{\left(5x^7 + 1\right)^2}$.

2. $3x^2 - 6x - 1$

We find the derivative using the Product Rule, which says that if $f(x) = uv$, then

$f'(x) = u\dfrac{dv}{dx} + v\dfrac{du}{dx}$. Here $f(x) = \left(x^2 - 4x + 3\right)(x + 1)$, so $u = x^2 - 4x + 3$ and $v = x + 1$.

Using the Product Rule, we get $f'(x) = \left(x^2 - 4x + 3\right)(1) + (x + 1)(2x - 4)$. This can be simplified

to $f'(x) = 3x^2 - 6x - 1$.

3. $\dfrac{16x^2 - 32}{\sqrt{\left(x^2 - 4\right)}}$

We find the derivative using the Chain Rule, which says that if $y = f(g(x))$,

then $y' = \left(\dfrac{df(g(x))}{dg}\right)\left(\dfrac{dg}{dx}\right)$. Here $f(x) = 8\sqrt{\left(x^4 - 4x^2\right)}$, which can be written as

$f(x) = 8\left(x^4 - 4x^2\right)^{\frac{1}{2}}$. Using the Chain Rule, we get $f'(x) = 8\left(\dfrac{1}{2}\right)\left(x^4 - 4x^2\right)^{-\frac{1}{2}}\left(4x^3 - 8x\right)$.

This can be simplified to $f'(x) = \dfrac{16x^2 - 32}{\sqrt{\left(x^2 - 4\right)}}$.

4. $\dfrac{3x^2 - 3x^4}{\left(x^2+1\right)^4}$

Here we will find the derivative using the Chain Rule. We will also need the Quotient Rule

to take the derivative of the expression inside the parentheses. The Chain Rule says that if

$y = f\left(g(x)\right)$, then $y' = \left(\dfrac{df\left(g(x)\right)}{dg}\right)\left(\dfrac{dg}{dx}\right)$, and the Quotient Rule says that if $f(x) = \dfrac{u}{v}$, then

$f'(x) = \dfrac{v\dfrac{du}{dx} - u\dfrac{dv}{dx}}{v^2}$. We get $f'(x) = 3\left(\dfrac{x}{x^2+1}\right)^2\left(\dfrac{\left(x^2+1\right)(1) - x(2x)}{\left(x^2+1\right)^2}\right)$. This can be simplified to

$f'(x) = \dfrac{3x^2 - 3x^4}{\left(x^2+1\right)^4}$.

5. $\dfrac{1}{4}\left(\dfrac{2x-5}{5x+2}\right)^{-\frac{3}{4}}\left(\dfrac{29}{\left(5x+2\right)^2}\right)$

Here we will find the derivative using the Chain Rule. We will also need the Quotient Rule to

take the derivative of the expression inside the parentheses. The Chain Rule says

that if $y = f\left(g(x)\right)$, then $y' = \left(\dfrac{df\left(g(x)\right)}{dg}\right)\left(\dfrac{dg}{dx}\right)$, and the Quotient Rule says that if $f(x) = \dfrac{u}{v}$,

then $f'(x) = \dfrac{v\dfrac{du}{dx} - u\dfrac{dv}{dx}}{v^2}$. We get $f'(x) = \dfrac{1}{4}\left(\dfrac{2x-5}{5x+2}\right)^{-\frac{3}{4}}\left(\dfrac{(5x+2)(2) - (2x-5)(5)}{\left(5x+2\right)^2}\right)$. This can

be simplified to $f'(x) = \dfrac{1}{4}\left(\dfrac{2x-5}{5x+2}\right)^{-\frac{3}{4}}\left(\dfrac{29}{\left(5x+2\right)^2}\right)$.

6. $3x^2 + 1 + \dfrac{1}{x^2} + \dfrac{3}{x^4}$

We have two ways that we could solve this. We could expand the expression first and then take the derivative of each term, or we could find the derivative using the Product Rule. Let's do both methods just to see that they both give us the same answer. First, let's use the Product Rule, which says that if $f(x) = uv$, then $f'(x) = u\dfrac{dv}{dx} + v\dfrac{du}{dx}$. Here

$f(x) = \left(x + \dfrac{1}{x}\right)\left(x^2 - \dfrac{1}{x^2}\right)$, so $u = \left(x + \dfrac{1}{x}\right)$ and $v = \left(x^2 - \dfrac{1}{x^2}\right)$. Using the Product Rule, we get

$f'(x) = \left(x + \dfrac{1}{x}\right)\left(2x - (-2)x^{-3}\right) + \left(x^2 - \dfrac{1}{x^2}\right)\left(1 - \dfrac{1}{x^2}\right) = \left(x + \dfrac{1}{x}\right)\left(2x + \dfrac{2}{x^3}\right) + \left(x^2 - \dfrac{1}{x^2}\right)\left(1 - \dfrac{1}{x^2}\right)$.

This can be simplified to $f'(x) = \left(2x^2 + \dfrac{2}{x^2} + 2 + \dfrac{2}{x^4}\right) + \left(x^2 - 1 - \dfrac{1}{x^2} + \dfrac{1}{x^4}\right) = 3x^2 + 1 + \dfrac{1}{x^2} + \dfrac{3}{x^4}$.

The other way we could find the derivative is to expand the expression first and then take the derivative. We get $f(x) = \left(x + \dfrac{1}{x}\right)\left(x^2 - \dfrac{1}{x^2}\right) = x^3 - \dfrac{1}{x} + x - \dfrac{1}{x^3}$. Now we can take the derivative of each term. We get $f'(x) = 3x^2 + \dfrac{1}{x^2} + 1 - (-3)x^{-4} = 3x^2 + \dfrac{1}{x^2} + 1 + \dfrac{3}{x^4}$. As we can see, the second method is a little quicker, and they both give the same result.

7. $$\frac{4x^3}{(x+1)^5}$$

Here we will find the derivative using the Chain Rule. We will also need the Quotient Rule

to take the derivative of the expression inside the parentheses. The Chain Rule says that if

$y = f(g(x))$, then $y' = \left(\dfrac{df(g(x))}{dg}\right)\left(\dfrac{dg}{dx}\right)$, and the Quotient Rule says that if $f(x) = \dfrac{u}{v}$, then

$f'(x) = \dfrac{v\dfrac{du}{dx} - u\dfrac{dv}{dx}}{v^2}$. We get $f'(x) = 4\left(\dfrac{x}{x+1}\right)^3\left(\dfrac{(x+1)(1) - (x)(1)}{(x+1)^2}\right)$. This can be simplified

to $f'(x) = 4\left(\dfrac{x}{x+1}\right)^3\left(\dfrac{1}{(x+1)^2}\right) = \dfrac{4x^3}{(x+1)^5}$.

8. $$100(x^2 + x)^{99}(2x+1)$$

Here we will find the derivative using the Chain Rule. The Chain Rule says that if $y = f(g(x))$,

then $y' = \left(\dfrac{df(g(x))}{dg}\right)\left(\dfrac{dg}{dx}\right)$. We get $f'(x) = 100(x^2 + x)^{99}(2x+1)$.

9. $$\frac{-2x}{(x^2+1)^{\frac{1}{2}}(x^2-1)^{\frac{3}{2}}}$$

Here we will find the derivative using the Chain Rule. We will also need the Quotient Rule

to take the derivative of the expression inside the parentheses. The Chain Rule says that if

$y = f(g(x))$, then $y' = \left(\dfrac{df(g(x))}{dg}\right)\left(\dfrac{dg}{dx}\right)$, and the Quotient Rule says that if $f(x) = \dfrac{u}{v}$, then

$f'(x) = \dfrac{v\dfrac{du}{dx} - u\dfrac{dv}{dx}}{v^2}$. We get $f'(x) = \dfrac{1}{2}\left(\dfrac{x^2+1}{x^2-1}\right)^{-\frac{1}{2}}\left(\dfrac{(x^2-1)(2x)-(x^2+1)(2x)}{(x^2-1)^2}\right)$. This can be

simplified to $f'(x) = \dfrac{1}{2}\left(\dfrac{x^2+1}{x^2-1}\right)^{-\frac{1}{2}}\left(\dfrac{-4x}{(x-1)^2}\right) = \dfrac{-2x}{(x^2+1)^{\frac{1}{2}}(x^2-1)^{\frac{3}{2}}}$.

10. $\dfrac{9}{64}$

We find the derivative using the Quotient Rule, which says that if $f(x) = \dfrac{u}{v}$, then

$f'(x) = \dfrac{v\dfrac{du}{dx} - u\dfrac{dv}{dx}}{v^2}$. Here $f(x) = \dfrac{(x+4)(x-8)}{(x+6)(x-6)}$, so $u = (x+4)(x-8)$ and $v = (x+6)(x-6)$.

Before we take the derivative, we can simplify the numerator and denominator of the

expression: $f(x) = \dfrac{(x+4)(x-8)}{(x+6)(x-6)} = \dfrac{x^2-4x-32}{x^2-36}$. Now, using the Quotient Rule, we get

$f(x) = \dfrac{(x^2-36)(2x-4)-(x^2-4x-32)(2x)}{(x^2-36)^2}$. Next, we don't simplify. We simply plug in $x = 2$

to get $f(x) = \dfrac{((2)^2-36)(2(2)-4)-((2)^2-4(2)-32)(2(2))}{((2)^2-36)^2} = \dfrac{(-32)(0)-(-36)(4)}{(-32)^2} = \dfrac{9}{64}$.

11.   106

We find the derivative using the Quotient Rule, which says that if $f(x) = \dfrac{u}{v}$,

then $f'(x) = \dfrac{v\dfrac{du}{dx} - u\dfrac{dv}{dx}}{v^2}$. We will also need the Chain Rule to take the derivative

of the expression in the denominator. The Chain Rule says that if $y = f(g(x))$, then

$y' = \left(\dfrac{df\left(g(x)\right)}{dg} \right)\left(\dfrac{dg}{dx} \right)$, here $f(x) = \dfrac{x^6 + 4x^3 + 6}{\left(x^4 - 2\right)^2}$, so $u = x^6 + 4x^3 + 6$ and $v = \left(x^4 - 2\right)^2$. We

get $f(x) = \dfrac{\left(x^4 - 2\right)^2 \left(6x^5 + 12x^2\right) - \left(x^6 + 4x^3 + 6\right)2\left(x^4 - 2\right)\left(4x^3\right)}{\left(x^4 - 2\right)^4}$. Now we don't simplify.

We simply plug in $x = 1$ to get

$$f(x) = \dfrac{\left((1)^4 - 2\right)^2\left(6(1)^5 + 12(1)^2\right) - \left((1)^6 + 4(1)^3 + 6\right)2\left((1)^4 - 2\right)\left(4(1)^3\right)}{\left((1)^4 - 2\right)^4} =$$

$$\dfrac{(1)(18) - (11)2(-1)(4)}{(-1)^4} = 106$$

12. $\dfrac{x^2 - 6x + 3}{(x-3)^2}$

We find the derivative using the Quotient Rule, which says that if $f(x) = \dfrac{u}{v}$, then

$f'(x) = \dfrac{v\dfrac{du}{dx} - u\dfrac{dv}{dx}}{v^2}$. Here $f(x) = \dfrac{x^2 - 3}{x - 3}$, so $u = x^2 - 3$ and $v = x - 3$. Using the Quotient Rule,

we get $f'(x) = \dfrac{(x-3)(2x) - \left(x^2 - 3\right)(1)}{(x-3)^2}$. This can be simplified to $f'(x) = \dfrac{x^2 - 6x + 3}{(x-3)^2}$.

13. 6

We find the derivative using the Product Rule, which says that if $f(x) = uv$, then

$f'(x) = u\dfrac{dv}{dx} + v\dfrac{du}{dx}$. Here $f(x) = \left(x^4 - x^2\right)\left(2x^3 + x\right)$, so $u = x^4 - x^2$ and $v = 2x^3 + x$. Using the

Product Rule, we get $f'(x) = \left(x^4 - x^2\right)\left(6x^2 + 1\right) + \left(2x^3 + x\right)\left(4x^3 - 2x\right)$. Now we don't simplify.

We simply plug in $x = 1$ to get $f'(x) = \left((1)^4 - (1)^2\right)\left(6(1)^2 + 1\right) + \left(2(1)^3 + (1)\right)\left(4(1)^3 - 2(1)\right) = 6$.

14. $-\dfrac{7}{4}$

We find the derivative using the Quotient Rule, which says that if $f(x)=\dfrac{u}{v}$,

then $f'(x)=\dfrac{v\dfrac{du}{dx}-u\dfrac{dv}{dx}}{v^2}$. Here $f(x)=\dfrac{x^2+2x}{x^4-x^3}$, so $u=x^2+2x$ and $v=x^4-x^3$.

Using the Quotient Rule, we get $f'(x)=\dfrac{\left(x^4-x^3\right)(2x+2)-\left(x^2+2x\right)\left(4x^3-3x^2\right)}{\left(x^4-x^3\right)^2}$.

Now, we don't simplify. We simply plug in $x=2$ to get

$$f'(x)=\dfrac{\left((2)^4-(2)^3\right)(2(2)+2)-\left((2)^2+2(2)\right)\left(4(2)^3-3(2)^2\right)}{\left((2)^4-(2)^3\right)^2}=-\dfrac{7}{4}.$$

15. $\dfrac{2x^2+1}{\sqrt{x^2+1}}$

We find the derivative using the Chain Rule, which says that if $y=f\big(g(x)\big)$, then

$y'=\left(\dfrac{df\big(g(x)\big)}{dg}\right)\left(\dfrac{dg}{dx}\right)$. Here $f(x)=\sqrt{x^4+x^2}=\left(x^4+x^2\right)^{\frac{1}{2}}$. Using the Chain Rule, we get

$f'(x)=\dfrac{1}{2}\left(x^4+x^2\right)^{-\frac{1}{2}}\left(4x^3+2x\right)$. This can be simplified to $f'(x)=\dfrac{2x^2+1}{\sqrt{x^2+1}}$.

16. $-\dfrac{2}{(x-1)^3}$

We find the derivative using the Chain Rule, which says that if $y=y(v)$ and $v=v(x)$, then

$\dfrac{dy}{dx}=\dfrac{dy}{dv}\dfrac{dv}{dx}$. Here $\dfrac{dy}{du}=2u$ and $\dfrac{du}{dx}=-1(x-1)^{-2}=-\dfrac{1}{(x-1)^2}$. Thus, $\dfrac{dy}{dx}=(2u)\dfrac{-1}{(x-1)^2}$ and

because $u=\dfrac{1}{x-1}$, $\dfrac{dy}{dx}=\left(\dfrac{2}{x-1}\right)\left(\dfrac{-1}{(x-1)^2}\right)=-\dfrac{2}{(x-1)^3}$.

17. -24

We find the derivative using the Chain Rule, which says that if $y=y(v)$ and

$v=v(x)$, then $\dfrac{dy}{dx}=\dfrac{dy}{dv}\dfrac{dv}{dx}$. Here $\dfrac{dy}{dt}=\dfrac{\left(t^2-2\right)(2t)-\left(t^2+2\right)(2t)}{\left(t^2-2\right)^2}$ and $\dfrac{dt}{dx}=3x^2$.

Now we plug $x = 1$ into the derivative. Note that where $x = 1$, $t = (1)^3 = 1$.

We get $\dfrac{dy}{dt} = \dfrac{\left((1)^2 - 2\right)(2(1)) - \left((1)^2 + 2\right)(2(1))}{\left((1)^2 - 2\right)^2} = -8$ and $\dfrac{dt}{dx} = 3(1)^2 = 3$. Thus,

$\dfrac{dy}{dx} = \dfrac{dy}{dt}\dfrac{dt}{dx} = (-8)(3) = -24.$

18. $\left(t^3 - 6t^{\frac{5}{2}}\right)\left(5 + \dfrac{1}{2\sqrt{t}}\right) + \left(5t + \sqrt{t}\right)\left(3t^2 - 15t^{\frac{3}{2}}\right)$

Here the solution will be much simpler if we first substitute $x = \sqrt{t}$ into the expression for y.

We get $y = \left(t^3 - 6t^{\frac{5}{2}}\right)\left(5t + \sqrt{t}\right)$. Now we find the derivative using the Product Rule, which says

that if $f(x) = uv$, then $f'(x) = u\dfrac{dv}{dx} + v\dfrac{du}{dx}$. Here $y = \left(t^3 - 6t^{\frac{5}{2}}\right)\left(5t + \sqrt{t}\right)$, so $u = t^3 - 6t^{\frac{5}{2}}$ and

$v = 5t + \sqrt{t}$. Using the Product Rule, we get $\dfrac{dy}{dt} = \left(t^3 - 6t^{\frac{5}{2}}\right)\left(5 + \dfrac{1}{2\sqrt{t}}\right) + \left(5t + \sqrt{t}\right)\left(3t^2 - 15t^{\frac{3}{2}}\right)$.

19. 2

We find the derivative using the Chain Rule, which says that if $y = y(v)$ and $v = v(x)$, then

$\dfrac{dy}{dx} = \dfrac{dy}{dv}\dfrac{dv}{dx}$. Here $\dfrac{dy}{du} = \dfrac{\left(1 + u^2\right)(1) - (1 + u)(2u)}{\left(1 + u^2\right)^2}$ and $\dfrac{du}{dx} = 2x$. Now, we plug $x = 1$ into the

derivative. Note that, where $x = 1$, $u = 0$. We get $\dfrac{dy}{du} = \dfrac{(1 + 0)(1) - (1 + 0)(0)}{(1 + 0)^2} = 1$ and $\dfrac{du}{dx} = 2$, so

$\dfrac{dy}{dx} = (1)(2) = 2.$

20. $\dfrac{48v^5}{\left(v^2+8\right)^4}$

We find the derivative using the Chain Rule, which says that if $y=y(v)$ and

$v=v(x)$, then $\dfrac{dy}{dx}=\dfrac{dy}{dv}\dfrac{dv}{dx}$, although, in this case, we have $u(y)$, $y(x)$, and $x(v)$,

so we will find $\dfrac{du}{dv}$ by $\dfrac{du}{dv}=\dfrac{du}{dy}\dfrac{dy}{dx}\dfrac{dx}{dv}$. Here $\dfrac{du}{dy}=3y^2$, $\dfrac{dy}{dx}=\dfrac{(x+8)(1)-(x)(1)}{(x+8)^2}=\dfrac{8}{(x+8)^2}$,

and $\dfrac{dx}{dv}=2v$. Next, $\dfrac{du}{dv}=\dfrac{du}{dy}\dfrac{dy}{dx}\dfrac{dx}{dv}=\left(3y^2\right)\left(\dfrac{8}{(x+8)^2}\right)(2v)$. Now because $x=v^2$ and

$y=\dfrac{x}{x+8}=\dfrac{v^2}{v^2+8}$, we get $\dfrac{du}{dv}=\left(3\dfrac{\left(v^2\right)^2}{\left(v^2+8\right)^2}\right)\left(\dfrac{8}{\left(v^2+8\right)^2}\right)(2v)=\dfrac{48v^5}{\left(v^2+8\right)^4}$.

SOLUTIONS TO PRACTICE PROBLEM SET 6

1. $2\sin x\cos x$ or $\sin 2x$

Recall that $\dfrac{d}{dx}(\sin x)=\cos x$. Here we use the Chain Rule to find the derivative:

$\dfrac{dy}{dx}=2(\sin x)(\cos x)$. If you recall your trigonometric identities, $2\sin x\cos x=\sin 2x$. Either

answer is acceptable.

2. $-2x\sin\left(x^2\right)$

Recall that $\dfrac{d}{dx}(\cos x)=-\sin x$. Here, we use the Chain Rule to find the derivative:

$\dfrac{dy}{dx}=\left(-\sin\left(x^2\right)\right)(2x)=-2x\sin\left(x^2\right)$.

3. $2\sec^3 x - \sec x$

Recall that $\dfrac{d}{dx}(\tan x) = \sec^2 x$ and that $\dfrac{d}{dx}(\sec x) = \sec x \tan x$. Using the Product Rule, we get $\dfrac{dy}{dx} = (\tan x)(\sec x \tan x) + (\sec x)(\sec^2 x)$. This can be simplified to $\sec^3 x + \sec x \tan^2 x = 2\sec^3 x - \sec x$.

4. $\dfrac{3\cos 3x}{2\sqrt{\sin 3x}}$

Recall that $\dfrac{d}{dx}(\sin x) = \cos x$. Here we use the Chain Rule to find the derivative: $\dfrac{dy}{dx} = \dfrac{1}{2}(\sin 3x)^{-\frac{1}{2}}(\cos 3x)(3)$. This can be simplified to $\dfrac{dy}{dx} = \dfrac{3\cos 3x}{2\sqrt{\sin 3x}}$.

5. $\dfrac{2\cos x}{(1-\sin x)^2}$

Recall that $\dfrac{d}{dx}(\sin x) = \cos x$. Here we use the Quotient Rule to find the derivative: $\dfrac{dy}{dx} = \dfrac{(1-\sin x)(\cos x) - (1+\sin x)(-\cos x)}{(1-\sin x)^2}$. This can be simplified to $\dfrac{2\cos x}{(1-\sin x)^2}$.

6. $-4x\csc^2(x^2)\cot(x^2)$

Recall that $\dfrac{d}{dx}(\csc x) = -\csc x \cot x$. Here we use the Chain Rule to find the derivative: $\dfrac{dy}{dx} = (2\csc x^2)(-\csc x^2 \cot x^2)(2x) = -4x\csc^2(x^2)\cot(x^2)$.

7. $16\sin 2x$

Recall that $\dfrac{d}{dx}(\sin x) = \cos x$ and that $\dfrac{d}{dx}(\cos x) = -\sin x$. Here we will use the Chain Rule four times to find the fourth derivative.

The first derivative is $\dfrac{dy}{dx} = (\cos 2x)(2) = 2\cos 2x$.

The second derivative is $\dfrac{d^2 y}{dx^2} = 2(-\sin 2x)(2) = -4\sin 2x$.

The third derivative is $\dfrac{d^3 y}{dx^3} = -4(\cos 2x)(2) = -8\cos 2x$.

And the fourth derivative is $\dfrac{d^4 y}{dx^4} = -8(-\sin 2x)(2) = 16\sin 2x$.

8. $\left[\cos(1+\cos^2 x) + \sin(1+\cos^2 x)\right](-2\sin x \cos x)$

Recall that $\dfrac{d}{dx}(\sin x) = \cos x$ and that $\dfrac{d}{dx}(\cos x) = -\sin x$. Here we will use the Chain Rule to find the derivative: $\dfrac{dy}{dt} = \cos t - (-\sin t) = \cos t + \sin t$ and

$\dfrac{dt}{dx} = 2(\cos x)(-\sin x) = -2\sin x \cos x$. Next, because $\dfrac{dy}{dx} = \dfrac{dy}{dt}\dfrac{dt}{dx}$ and $t = 1 + \cos^2 x$, we get

$\dfrac{dy}{dx} = (\cos t + \sin t)(-2\sin x \cos x) = \left[\cos(1+\cos^2 x) + \sin(1+\cos^2 x)\right](-2\sin x \cos x)$.

9. $\dfrac{2\tan x \sec^2 x}{(1-\tan x)^3}$

Recall that $\dfrac{d}{dx}(\tan x) = \sec^2 x$. Using the Quotient Rule and the Chain Rule, we get

$\dfrac{dy}{dx} = 2\left(\dfrac{\tan x}{1-\tan x}\right)\dfrac{(1-\tan x)(\sec^2 x) - (\tan x)(-\sec^2 x)}{(1-\tan x)^2}$. This simplifies to $\dfrac{2\tan x \sec^2 x}{(1-\tan x)^3}$.

10. $-\left[\sin(1+\sin\theta)\right](\cos\theta)$

Recall that $\dfrac{d}{dx}(\sin x) = \cos x$ and that $\dfrac{d}{dx}(\cos x) = -\sin x$. Using the Chain Rule, we get

$\dfrac{dr}{d\theta} = -\left[\sin(1+\sin\theta)\right](\cos\theta)$.

11. $$\frac{\sec\theta\left(\tan\theta-1\right)}{\left(1+\tan\theta\right)^{2}}$$

Recall that $\frac{d}{dx}(\tan x)=\sec^{2}x$ and that $\frac{d}{dx}(\sec x)=\sec x\tan x$. Using the Quotient Rule, we

get $\frac{dr}{d\theta}=\frac{(1+\tan\theta)(\sec\theta\tan\theta)-(\sec\theta)(\sec^{2}\theta)}{(1+\tan\theta)^{2}}$. This can be simplified (using trigonometric

identities) to $\frac{\sec\theta\left(\tan\theta-1\right)}{\left(1+\tan\theta\right)^{2}}$.

12. $$-\frac{\left(\dfrac{4}{x^{2}}\right)\left(\csc^{2}\left(\dfrac{2}{x}\right)\right)}{\left(1+\cot\left(\dfrac{2}{x}\right)\right)^{3}}$$

Recall that $\frac{d}{dx}(\cot x)=-\csc^{2}x$. Here we use the Chain Rule to find the derivative:

$$\frac{dy}{dx}=-2\left(1+\cot\left(\frac{2}{x}\right)\right)^{-3}\left(-\csc^{2}\left(\frac{2}{x}\right)\right)\left(\frac{-2}{x^{2}}\right)=-\frac{\left(\dfrac{4}{x^{2}}\right)\left(\csc^{2}\left(\dfrac{2}{x}\right)\right)}{\left(1+\cot\left(\dfrac{2}{x}\right)\right)^{3}}.$$

SOLUTIONS TO PRACTICE PROBLEM SET 7

1. $$\frac{3x^{2}}{1+3y^{2}}$$

We take the derivative of each term with respect to x: $(3x^{2})\left(\dfrac{dx}{dx}\right)-(3y^{2})\left(\dfrac{dy}{dx}\right)=(1)\left(\dfrac{dy}{dx}\right)$.

Next, because $\frac{dx}{dx}=1$, we can eliminate that term and get $(3x^{2})-(3y^{2})\left(\dfrac{dy}{dx}\right)=\left(\dfrac{dy}{dx}\right)$.

Next, group the terms containing $\frac{dy}{dx}$: $(3x^{2})=\left(\dfrac{dy}{dx}\right)+(3y^{2})\left(\dfrac{dy}{dx}\right)$.

Factor out the term $\frac{dy}{dx}$: $(3x^{2})=\left(\dfrac{dy}{dx}\right)(1+3y^{2})$. Now, we can isolate $\frac{dy}{dx}$: $\dfrac{dy}{dx}=\dfrac{3x^{2}}{1+3y^{2}}$.

2. $\dfrac{8y - x}{y - 8x}$

We take the derivative of each term with respect to x:

$$(2x)\left(\frac{dx}{dx}\right) - 16\left[(x)\left(\frac{dy}{dx}\right) + (y)\left(\frac{dx}{dx}\right)\right] + (2y)\left(\frac{dy}{dx}\right) = 0.$$

Next, because $\dfrac{dx}{dx} = 1$, we can eliminate that term and we can distribute the -16 to get

$$2x - 16x\left(\frac{dy}{dx}\right) - 16y + 2y\left(\frac{dy}{dx}\right) = 0.$$

Next, group the terms containing $\dfrac{dy}{dx}$ on one side of the equals sign and the other terms on the

other side: $-16x\left(\dfrac{dy}{dx}\right) + 2y\left(\dfrac{dy}{dx}\right) = 16y - 2x$.

Factor out the term $\dfrac{dy}{dx} : \left(\dfrac{dy}{dx}\right)(2y - 16x) = 16y - 2x$. Now we can isolate $\dfrac{dy}{dx} : \dfrac{dy}{dx} = \dfrac{16y - 2x}{2y - 16x}$,

which can be reduced to $\dfrac{dy}{dx} = \dfrac{8y - x}{y - 8x}$.

3. $\dfrac{1}{2}$

First, cross-multiply so that we don't have to use the Quotient Rule: $x + y = 3x - 3y$. Next, simplify

$4y = 2x$, which reduces to $y = \dfrac{1}{2}x$. Now, we can take the derivative: $\dfrac{dy}{dx} = \dfrac{1}{2}$. Note that just

because a problem has the x's and y's mixed together doesn't mean that we need to use implicit

differentiation to solve it!

4. $\dfrac{8}{7}$

We take the derivative of each term with respect to x:

$$(32x)\left(\frac{dx}{dx}\right) - 16\left[(x)\left(\frac{dy}{dx}\right) + (y)\left(\frac{dx}{dx}\right)\right] + (2y)\left(\frac{dy}{dx}\right) = 0.$$

Next, because $\dfrac{dx}{dx} = 1$, we can eliminate that term to get $32x - 16x\left(\dfrac{dy}{dx}\right) - 16y + 2y\left(\dfrac{dy}{dx}\right) = 0$.

Next, don't simplify. Plug in (1, 1) for x and y: $32(1) - 16(1)\left(\dfrac{dy}{dx}\right) - 16(1) + 2(1)\left(\dfrac{dy}{dx}\right) = 0$,

which simplifies to $16 - 14\left(\dfrac{dy}{dx}\right) = 0$.

Finally, we can solve for $\dfrac{dy}{dx}$: $\dfrac{dy}{dx} = \dfrac{8}{7}$.

5. -1

We take the derivative of each term with respect to x:

$$(x\cos y)\left(\dfrac{dy}{dx}\right) + (\sin y)\left(\dfrac{dx}{dx}\right) + (y\cos x)\left(\dfrac{dx}{dx}\right) + (\sin x)\left(\dfrac{dy}{dx}\right) = 0.$$

Next, because $\dfrac{dx}{dx} = 1$, we can eliminate that term to get

$(x\cos y)\left(\dfrac{dy}{dx}\right) + \sin y + y\cos x + (\sin x)\left(\dfrac{dy}{dx}\right) = 0$. Next, don't simplify. Plug in

$\left(\dfrac{\pi}{4}, \dfrac{\pi}{4}\right)$ for x and y: $\left(\dfrac{\pi}{4}\cos\dfrac{\pi}{4}\right)\left(\dfrac{dy}{dx}\right) + \sin\dfrac{\pi}{4} + \dfrac{\pi}{4}\cos\dfrac{\pi}{4} + \left(\sin\dfrac{\pi}{4}\right)\left(\dfrac{dy}{dx}\right) = 0$, which simpli-

fies to $\left(\dfrac{\pi}{4}\dfrac{1}{\sqrt{2}}\right)\left(\dfrac{dy}{dx}\right) + \dfrac{1}{\sqrt{2}} + \dfrac{\pi}{4}\dfrac{1}{\sqrt{2}} + \dfrac{1}{\sqrt{2}}\left(\dfrac{dy}{dx}\right) = 0$. If we multiply through by $\sqrt{2}$, we get

$\dfrac{\pi}{4}\dfrac{dy}{dx} + 1 + \dfrac{\pi}{4} + \dfrac{dy}{dx} = 0$.

Finally, we can solve for $\dfrac{dy}{dx}$: $\dfrac{dy}{dx} = -1$.

6. $-\dfrac{1}{16y^3}$

We take the derivative of each term with respect to x: $(2x)\left(\dfrac{dx}{dx}\right) + (8y)\left(\dfrac{dy}{dx}\right) = 0$.

Next, because $\dfrac{dx}{dx} = 1$, we can eliminate that term to get $2x + (8y)\left(\dfrac{dy}{dx}\right) = 0$. Next, we can iso-

late $\dfrac{dy}{dx}$: $\dfrac{dy}{dx} = -\dfrac{x}{4y}$. Now, we take the derivative again: $\dfrac{d^2y}{dx^2} = -\dfrac{(4y)\left(\dfrac{dx}{dx}\right) - (x)\left(4\dfrac{dy}{dx}\right)}{16y^2}$. Next, because $\dfrac{dx}{dx}=1$ and $\dfrac{dy}{dx}=-\dfrac{x}{4y}$, we get $\dfrac{d^2y}{dx^2}=-\dfrac{4y-4x\left(-\dfrac{x}{4y}\right)}{16y^2}$. This can be simplified to

$$\dfrac{d^2y}{dx^2}=-\dfrac{4y+\dfrac{x^2}{y}}{16y^2}=-\dfrac{4y^2+x^2}{16y^3}=-\dfrac{1}{16y^3}.$$

7. $\dfrac{\sin x\sin^2 y-\cos y\cos^2 x}{\sin^3 y}$

We take the derivative of each term with respect to x: $(\cos x)\left(\dfrac{dx}{dx}\right)=(-\sin y)\left(\dfrac{dy}{dx}\right)$.

Next, because $\dfrac{dx}{dx}=1$, we can eliminate that term to get

$\cos x=(-\sin y)\left(\dfrac{dy}{dx}\right)$. Next, we can isolate $\dfrac{dy}{dx}$: $\dfrac{dy}{dx}=-\dfrac{\cos x}{\sin y}$. Now, we take the deriva-

tive again: $\dfrac{d^2y}{dx^2}=\dfrac{(\sin y)(\sin x)\left(\dfrac{dx}{dx}\right)+(\cos x)(\cos y)\left(\dfrac{dy}{dx}\right)}{\sin^2 y}$. Next, because $\dfrac{dx}{dx}=1$ and

$\dfrac{dy}{dx}=-\dfrac{\cos x}{\sin y}$, we get $\dfrac{d^2y}{dx^2}=\dfrac{(\sin y)(\sin x)+(\cos x)(\cos y)\left(-\dfrac{\cos x}{\sin y}\right)}{\sin^2 y}$. This can be simplified

to $\dfrac{d^2y}{dx^2}=\dfrac{\sin x\sin^2 y-\cos y\cos^2 x}{\sin^3 y}$.

8. 1

We can easily isolate y in this equation: $y=\dfrac{1}{2}x^2-2x+1$. We take the derivative: $\dfrac{dy}{dx}=x-2$. And

we take the derivative again: $\dfrac{d^2y}{dx^2}=1$. Note that just because a problem has the x's and y's mixed

together doesn't mean that we need to use implicit differentiation to solve it!

SOLUTIONS TO PRACTICE PROBLEM SET 8

1. $y - 2 = 5(x - 1)$

Remember that the equation of a line through a point (x_1, y_1) with slope m is $y - y_1 = m(x - x_1)$.

We find the y-coordinate by plugging $x = 1$ into the equation $y = 3x^2 - x$, and we find the slope by plugging $x = 1$ into the derivative of the equation.

First, we find the y-coordinate, $y_1 : y = 3(1)^2 - 1 = 2$. This means that the line passes through the point $(1, 2)$.

Next, we take the derivative: $\dfrac{dy}{dx} = 6x - 1$. Now, we can find the slope, m: $\dfrac{dy}{dx}\bigg|_{x=1} = 6(1) - 1 = 5$.
Finally, we plug in the point $(1, 2)$ and the slope $m = 5$ to get the equation of the tangent line:
$y - 2 = 5(x - 1)$.

2. $y - 18 = 24(x - 3)$

Remember that the equation of a line through a point (x_1, y_1) with slope m is $y - y_1 = m(x - x_1)$.

We find the y-coordinate by plugging $x = 3$ into the equation $y = x^3 - 3x$, and we find the slope by

plugging $x = 3$ into the derivative of the equation.

First, we find the y-coordinate, $y_1 : y = (3)^3 - 3(3) = 18$. This means that the line passes through the point $(3, 18)$.

Next, we take the derivative: $\dfrac{dy}{dx} = 3x^2 - 3$. Now, we can find the slope, m: $\dfrac{dy}{dx}\bigg|_{x=3} = 3(3)^2 - 3 = 24$.
Finally, we plug in the point $(3, 18)$ and the slope $m = 24$ to get the equation of the tangent line:
$y - 18 = 24(x - 3)$.

3. $y - \dfrac{1}{4} = -\dfrac{3}{64}(x - 3)$

Remember that the equation of a line through a point (x_1, y_1) with slope m is $y - y_1 = m(x - x_1)$.

We find the y-coordinate by plugging $x = 3$ into the equation $y = \dfrac{1}{\sqrt{x^2 + 7}}$, and we find the slope

by plugging $x = 3$ into the derivative of the equation.

First, we find the y-coordinate, y_1: $y = \dfrac{1}{\sqrt{(3)^2 + 7}} = \dfrac{1}{4}$. This means that the line passes through the

point $\left(3, \dfrac{1}{4}\right)$.

Next, we take the derivative: $\dfrac{dy}{dx} = -\dfrac{1}{2}\left(x^2 + 7\right)^{-\frac{3}{2}}(2x) = -\dfrac{x}{\left(\sqrt{x^2 + 7}\right)^3}$. Now, we can find the

slope, m: $\dfrac{dy}{dx}\bigg|_{x=3} = -\dfrac{3}{\left(\sqrt{(3)^2 + 7}\right)^3} = -\dfrac{3}{64}$. Finally, we plug in the point $\left(3, \dfrac{1}{4}\right)$ and the slope

$m = -\dfrac{3}{64}$ to get the equation of the tangent line: $y - \dfrac{1}{4} = -\dfrac{3}{64}(x - 3)$.

4. $y - 7 = \dfrac{1}{6}(x - 4)$

Remember that the equation of a line through a point (x_1, y_1) with slope m is $y - y_1 = m(x - x_1)$.

We find the y-coordinate by plugging $x = 4$ into the equation $y = \dfrac{x + 3}{x - 3}$, and we find the slope by

plugging $x = 4$ into the derivative of the equation.

First, we find the y-coordinate, y_1: $y = \dfrac{4 + 3}{4 - 3} = 7$. This means that the line passes through the

point $(4, 7)$.

Next, we take the derivative: $\dfrac{dy}{dx} = \dfrac{(x - 3)(1) - (x + 3)(1)}{(x - 3)^2} = -\dfrac{6}{(x - 3)^2}$. Now, we can find the

slope, m: $\dfrac{dy}{dx}\bigg|_{x=4} = -\dfrac{6}{(4 - 3)^2} = -6$. However, this is the slope of the *tangent* line. The *normal* line

is perpendicular to the tangent line, so its slope will be the negative reciprocal of the tangent line's slope. In this case, the slope of the normal line is $\dfrac{-1}{-6} = \dfrac{1}{6}$. Finally, we plug in the point $(4, 7)$ and

the slope $m = \dfrac{1}{6}$ to get the equation of the normal line: $y - 7 = \dfrac{1}{6}(x - 4)$.

5. $y = 0$

Remember that the equation of a line through a point (x_1, y_1) with slope m is $y - y_1 = m(x - x_1)$.

We find the y-coordinate by plugging $x = 2$ into the equation $y = 2x^3 - 3x^2 - 12x + 20$, and we

find the slope by plugging $x = 2$ into the derivative of the equation.

First, we find the y-coordinate, y_1: $y = 2(2)^3 - 3(2)^2 - 12(2) + 20 = 0$. This means that the line

passes through the point $(2, 0)$.

Next, we take the derivative: $\dfrac{dy}{dx} = 6x^2 - 6x - 12$. Now, we can find the slope,

m: $\dfrac{dy}{dx}\Big|_{x=2} = 6(2)^2 - 6(2) - 12 = 0$. Finally, we plug in the point $(2, 0)$ and the slope $m = 0$ to

get the equation of the tangent line: $y - 0 = 0(x - 2)$ or $y = 0$.

6. $y + 29 = -39(x - 5)$

Remember that the equation of a line through a point (x_1, y_1) with slope m is $y - y_1 = m(x - x_1)$.

We find the y-coordinate by plugging $x = 5$ into the equation $y = \dfrac{x^2 + 4}{x - 6}$, and we find the slope by

plugging $x = 5$ into the derivative of the equation.

First, we find the y-coordinate, y_1: $y = \dfrac{(5)^2 + 4}{(5) - 6} = -29$. This means that the line passes through

the point $(5, -29)$.

Next, we take the derivative: $\dfrac{dy}{dx} = \dfrac{(x - 6)(2x) - (x^2 + 4)(1)}{(x - 6)^2} = \dfrac{x^2 - 12x - 4}{(x - 6)^2}$. Now, we can find

the slope, m: $\dfrac{dy}{dx}\Big|_{x=5} = \dfrac{(5)^2 - 12(5) - 4}{(5 - 6)^2} = -39$. Finally, we plug in the point $(5, -29)$ and the slope $m = -39$ to get the equation of the tangent line: $y + 29 = -39(x - 5)$.

7. $y - 7 = \dfrac{24}{7}(x - 4)$

Remember that the equation of a line through a point (x_1, y_1) with slope m is $y - y_1 = m(x - x_1)$.

We find the slope by plugging $x = 4$ into the derivative of the equation $y = \sqrt{x^3 - 15}$. First,

we take the derivative: $\dfrac{dy}{dx} = \dfrac{1}{2}(x^3 - 15)^{-\frac{1}{2}}(3x^2) = \dfrac{3x^2}{2\sqrt{x^3 - 15}}$. Now, we can find the slope,

m: $\dfrac{dy}{dx}\bigg|_{x=4} = \dfrac{3(4)^2}{2\sqrt{(4)^3 - 15}} = \dfrac{24}{7}$. Finally, we plug in the point $(4, 7)$ and the slope $m = \dfrac{24}{7}$ to get

the equation of the tangent line: $y - 7 = \dfrac{24}{7}(x - 4)$.

8. $x = \pm\sqrt{\dfrac{3}{2}}$

The slope of the line $y = x$ is 1, so we want to know where the slope of the tangent line is equal

to 1. We find the slope of the tangent line by taking the derivative: $\dfrac{dy}{dx} = 6x^2 - 8$. Now we set the

derivative equal to 1: $6x^2 - 8 = 1$. If we solve for x, we get $x = \pm\sqrt{\dfrac{3}{2}}$.

9. $y - 7 = \dfrac{1}{2}(x - 3)$

Remember that the equation of a line through a point (x_1, y_1) with slope m is $y - y_1 = m(x - x_1)$.

We find the y-coordinate by plugging $x = 3$ into the equation $y = \dfrac{3x + 5}{x - 1}$, and we find the slope by

plugging $x = 3$ into the derivative of the equation.

First, we find the y-coordinate, y_1: $y = \dfrac{3(3) + 5}{(3) - 1} = 7$. This means that the line passes through the

point $(3, 7)$.

Next, we take the derivative: $\dfrac{dy}{dx} = \dfrac{(x-1)(3)-(3x+5)(1)}{(x-1)^2} = \dfrac{-8}{(x-1)^2}$. Now, we can find the slope, m: $\dfrac{dy}{dx}\bigg|_{x=3} = \dfrac{-8}{(3-1)^2} = -2$. However, this is the slope of the *tangent* line. The *normal* line is perpendicular to the tangent line, so its slope will be the negative reciprocal of the tangent line's slope. In this case, the slope of the normal line is $\dfrac{-1}{-2} = \dfrac{1}{2}$. Finally, we plug in the point $(3, 7)$ and the slope $m = \dfrac{1}{2}$ to get the equation of the normal line: $y - 7 = \dfrac{1}{2}(x-3)$.

10. $x = 9$

A line that is parallel to the y-axis has an infinite (or undefined) slope. In order to find where the normal line has an infinite slope, we first take the derivative to find the slope of the tangent line: $\dfrac{dy}{dx} = 2(x-9)(1) = 2x - 18$. Next, because the normal line is perpendicular to the tangent line, the slope of the normal line is the negative reciprocal of the slope of the tangent line: $m = \dfrac{-1}{2x-18}$. Now, we need to find where the slope is infinite. This is simply where the denominator of the slope is zero: $x = 9$.

11. $\left(-\dfrac{3}{2}, \dfrac{41}{4}\right)$

A line that is parallel to the x-axis has a zero slope. In order to find where the tangent line has a zero slope, we first take the derivative: $\dfrac{dy}{dx} = -3 - 2x$. Now we need to find where the slope is zero. The derivative $-3 - 2x = 0$ at $x = -\dfrac{3}{2}$. Now, we need to find the y-coordinate, which we get by plugging $x = -\dfrac{3}{2}$ into the equation for y: $8 - 3\left(-\dfrac{3}{2}\right) - \left(-\dfrac{3}{2}\right)^2 = \dfrac{41}{4}$. Therefore, the answer is $\left(-\dfrac{3}{2}, \dfrac{41}{4}\right)$.

12. $a = 1$, $b = 0$, and $c = 1$.

The two equations will have a common tangent line where they have the same slope, which we find by taking the derivative of each equation. The derivative of the first equation is: $\dfrac{dy}{dx} = 2x + a$. The derivative of the second equation is $\dfrac{dy}{dx} = c + 2x$. Setting the two derivatives equal to each other, we get $a = c$. Each equation will pass through the point $\left(-1, 0\right)$. If we plug $\left(-1, 0\right)$ into the first equation, we get $0 = \left(-1\right)^2 + a\left(-1\right) + b$, which simplifies to: $a - b = 1$. If we plug $\left(-1, 0\right)$ into the second equation, we get $0 = c\left(-1\right) + \left(-1\right)^2$, which simplifies to $c = 1$. Now we can find the values for a, b, and c. We get $a = 1$, $b = 0$, and $c = 1$.

SOLUTIONS TO PRACTICE PROBLEM SET 9

1. $c = 0$

The Mean Value Theorem says that if $f(x)$ is continuous on the interval $[a, b]$ and is differentiable everywhere on the interval (a, b), then there exists at least one number c on the interval (a, b) such that $f'(c) = \dfrac{f(b) - f(a)}{b - a}$. Here the function is $f(x) = 3x^2 + 5x - 2$ and the interval is $[-1, 1]$. Thus, the Mean Value Theorem says that $f'(c) = \dfrac{\left(3(1)^2 + 5(1) - 2\right) - \left(3(-1)^2 + 5(-1) - 2\right)}{(1 + 1)}$. This simplifies to $f'(c) = 5$. Next, we need to find $f'(c)$. The derivative of $f(x)$ is $f'(x) = 6x + 5$, so $f'(c) = 6c + 5$.

Now, we can solve for c: $6c + 5 = 5$ and $c = 0$. Note that 0 is in the interval $(-1, 1)$, just as we expected.

2. $c = \dfrac{4}{\sqrt{3}}$

The Mean Value Theorem says that if $f(x)$ is continuous on the interval $[a, b]$ and is differentiable everywhere on the interval (a, b), then there exists at least one number c on the interval (a, b) such that $f'(c) = \dfrac{f(b) - f(a)}{b - a}$. Here the function is $f(x) = x^3 + 24x - 16$ and the interval is [0, 4]. Thus, the

Mean Value Theorem says that $f'(c) = \dfrac{\left((4)^3 + 24(4) - 16\right) - \left((0)^3 + 24(0) - 16\right)}{(4 - 0)}$. This simplifies to

$f'(c) = 40$. Next, we need to find $f'(c)$ from the equation. The derivative of $f(x)$ is $f'(x) = 3x^2 + 24$, so

$f'(c) = 3c^2 + 24$. Now, we can solve for c: $3c^2 + 24 = 40$ and $c = \pm\dfrac{4}{\sqrt{3}}$. Note that $\dfrac{4}{\sqrt{3}}$ is in the

interval (0, 4), but $-\dfrac{4}{\sqrt{3}}$ is *not* in the interval. Thus, the answer is only $c = \dfrac{4}{\sqrt{3}}$. It's *very*

important to check that the answers you get for c fall in the given interval when doing Mean Value

Theorem problems.

3. $c = \sqrt{2}$

The Mean Value Theorem says that if $f(x)$ is continuous on the interval [a, b] and is differentiable

everywhere on the interval (a, b), then there exists at least one number c on the interval (a, b) such

that $f'(c) = \dfrac{f(b) - f(a)}{b - a}$. Here the function is $f(x) = \dfrac{6}{x} - 3$ and the interval is [1, 2]. Thus, the

Mean Value Theorem says that $f'(c) = \dfrac{\left(\dfrac{6}{2} - 3\right) - \left(\dfrac{6}{1} - 3\right)}{(2 - 1)}$. This simplifies to $f'(c) = -3$. Next, we

need to find $f'(c)$ from the equation. The derivative of $f(x)$ is $f'(x) = -\dfrac{6}{x^2}$, so $f'(c) = -\dfrac{6}{c^2}$. Now,

we can solve for c: $-\dfrac{6}{c^2} = -3$ and $c = \pm\sqrt{2}$. Note that $c = \sqrt{2}$ is in the interval (1, 2), but $-\sqrt{2}$ is

not in the interval. Thus, the answer is only $c = \sqrt{2}$. It's *very* important to check that the answers

you get for c fall in the given interval when doing Mean Value Theorem problems.

4.   *No Solution.*

The Mean Value Theorem says that if $f(x)$ is continuous on the interval $[a, b]$ and is differentiable everywhere on the interval (a, b), then there exists at least one number c on the interval (a, b) such that $f'(c) = \dfrac{f(b) - f(a)}{b - a}$. Here the function is $f(x) = \dfrac{6}{x} - 3$ and the interval is $[-1, 2]$. Note that the function is *not* continuous on the interval. It has an essential discontinuity (vertical asymptote) at $x = 0$. Thus, the Mean Value Theorem does not apply on the interval, and there is no solution.

Suppose that we were to apply the theorem anyway. We would get $f'(c) = \dfrac{\left(\dfrac{6}{2} - 3\right) - \left(\dfrac{6}{-1} - 3\right)}{(2 + 1)}$.

This simplifies to $f'(c) = 3$. Next, we need to find $f'(c)$ from the equation. The derivative of $f(x)$ is $f'(x) = -\dfrac{6}{x^2}$, so $f'(c) = -\dfrac{6}{c^2}$. Now, we can solve for c: $-\dfrac{6}{c^2} = 3$. This has no real solution.

Therefore, remember that it's *very* important to check that the function is continuous and differentiable everywhere on the given interval (it does not have to be differentiable at the endpoints) when doing Mean Value Theorem problems. If it is not, then the theorem does not apply.

5. $c = 4$

Rolle's Theorem says that if $f(x)$ is continuous on the interval $[a, b]$ and is differentiable everywhere on the interval (a, b), and if $f(a) = f(b) = 0$, then there exists at least one number c on the interval (a, b) such that $f'(c) = 0$. Here the function is $f(x) = x^2 - 8x + 12$ and the interval is $[2, 6]$. First, we check if the function is equal to zero at both of the endpoints: $f(6) = (6)^2 - 8(6) + 12 = 0$ and $f(2) = (2)^2 - 8(2) + 12 = 0$. Next, we take the derivative to find $f'(c)$: $f'(x) = 2x - 8$, so $f'(c) = 2c - 8$. Now, we can solve for c: $2c - 8 = 0$ and $c = 4$. Note that 4 is in the interval $(2, 6)$, just as we expected.

6. $c = \dfrac{1}{2}$

Rolle's Theorem says that if $f(x)$ is continuous on the interval $[a, b]$ and is differentiable everywhere on the interval (a, b), and if $f(a) = f(b) = 0$, then there exists at least one number c on the interval

(a, b) such that $f'(c) = 0$. Here the function is $f(x) = x(1 - x)$ and the interval is $[0, 1]$. First, we check if the function is equal to zero at both of the endpoints: $f(0) = (0)(1 - 0) = 0$ and $f(1) = (1)(1 - 1) = 0$. Next, we take the derivative to find $f'(c)$: $f'(x) = 1 - 2x$, so $f'(c) = 1 - 2c$. Now, we can solve for c: $1 - 2c = 0$ and $c = \frac{1}{2}$. Note that $\frac{1}{2}$ is in the interval $(0, 1)$, just as we expected.

7. *No Solution.*

Rolle's Theorem says that if $f(x)$ is continuous on the interval $[a, b]$ and is differentiable everywhere on the interval (a, b), and if $f(a) = f(b) = 0$, then there exists at least one number c on the interval (a, b) such that $f'(c) = 0$. Here the function is $f(x) = 1 - \frac{1}{x^2}$, and the interval is $[-1, 1]$. Note that the function is *not* continuous on the interval. It has an essential discontinuity (vertical asymptote) at $x = 0$. Thus, Rolle's Theorem does not apply on the interval, and there is no solution.

Suppose we were to apply the theorem anyway. First, we check if the function is equal to zero at both of the endpoints: $f(1) = 1 - \frac{1}{(1)^2} = 0$ and $f(-1) = 1 - \frac{1}{(-1)^2} = 0$. Next, we take the derivative to find $f'(c)$: $f'(x) = \frac{2}{x^3}$, so $f'(c) = \frac{2}{c^3}$. This has no solution.

Therefore, remember that it's *very* important to check that the function is continuous and differentiable everywhere on the given interval (it does not have to be differentiable at the endpoints) when doing Rolle's Theorem problems. If it is not, then the theorem does not apply.

8. $c = \frac{1}{8}$

Rolle's Theorem says that if $f(x)$ is continuous on the interval $[a, b]$ and is differentiable everywhere on the interval (a, b), and if $f(a) = f(b) = 0$, then there exists at least one number c on the interval (a, b) such that $f'(c) = 0$. Here the function is $f(x) = x^{\frac{2}{3}} - x^{\frac{1}{3}}$ and the interval is $[0, 1]$. First, we check if the function is equal to zero at both of the endpoints: $f(0) = (0)^{\frac{2}{3}} - (0)^{\frac{1}{3}} = 0$ and $f(1) = (1)^{\frac{2}{3}} - (1)^{\frac{1}{3}} = 0$. Next, we take the derivative to find $f'(c)$ $f'(x) = \frac{2}{3}x^{-\frac{1}{3}} - \frac{1}{3}x^{-\frac{2}{3}} = \frac{2}{3\sqrt[3]{x}} - \frac{1}{3\sqrt[3]{x^2}}$, so $f'(c) = \frac{2}{3\sqrt[3]{c}} - \frac{1}{3\sqrt[3]{c^2}}$. Now, we can solve for c: $\frac{2}{3\sqrt[3]{c}} - \frac{1}{3\sqrt[3]{c^2}} = 0$ and $c = \frac{1}{8}$. Note that $\frac{1}{8}$ is in the interval $(0, 1)$, just as we expected.

SOLUTIONS TO PRACTICE PROBLEM SET 10

1.   The area is 32.

First, let's draw a picture.

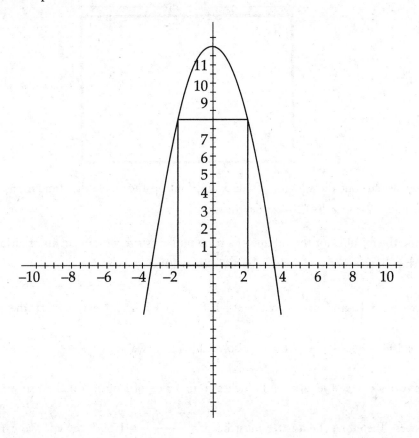

The rectangle can be expressed as a function of x, where the height is $12 - x^2$ and the base is $2x$.

The area is: $A = 2x (12 - x^2) = 24x - 2x^3$. Now, we take the derivative: $\dfrac{dA}{dx} = 24 - 6x^2$. Next, we

set the derivative equal to zero: $24 - 6x^2 = 0$. If we solve this for x, we get $x = \pm 2$. A negative

answer doesn't make any sense in the case, so we use the solution $x = 2$. We can then find the

area by plugging in $x = 2$ to get $A = 24(2) - 2(2)^3 = 32$. We can verify that this is a maximum by

taking the second derivative: $\dfrac{d^2 A}{dx^2} = -12x$. Next, we plug in $x = 2$: $\dfrac{d^2 A}{dx^2} = -12(2) = -24$. Because

the value of the second derivative is negative, according to the second derivative test (see page 187),

the area is a maximum at $x = 2$.

2. $x = \dfrac{7 - \sqrt{13}}{2} \approx 1.697$ inches

First, let's draw a picture.

After we cut out the squares of side x and fold up the sides, the dimensions of the box will be: width: $9 - 2x$; length: $12 - 2x$; depth: x.

Using the formula for the volume of a rectangular prism, we can get an equation for the volume of the box in terms of x: $V = x(9 - 2x)(12 - 2x) = 108x - 42x^2 + 4x^3$.

Now, we take the derivative: $\dfrac{dV}{dx} = 108 - 84x + 12x^2$. Next, we set the derivative equal to zero: $108 - 84x + 12x^2 = 0$. If we solve this for x, we get $x = \dfrac{7 \pm \sqrt{13}}{2} \approx 5.303, 1.697$. We can't cut two squares of length 5.303 inches from a side of length 9 inches, so we can get rid of that answer. Therefore, the answer must be $x = \dfrac{7 - \sqrt{13}}{2} \approx 1.697$ inches. We can verify that this is a maximum by taking the second derivative: $\dfrac{d^2V}{dx^2} = -84 + 24x$. Next, we plug in $x = 1.697$ to get approximately $\dfrac{d^2V}{dx^2} = -84 + 24(1.697) = -43.272$. Because the value of the second derivative is negative, according to the second derivative test (see page 187), the volume is a maximum at $x = \dfrac{7 - \sqrt{13}}{2} \approx 1.697$ inches.

3.   16 meters by 24 meters

First, let's draw a picture.

If we call the length of the plot y and the width x, the area of the plot is $A = xy = 384$. The perimeter is $P = 3x + 2y$. So, if we want to minimize the length of the fence, we need to minimize the perimeter of the plot. If we solve the area equation for y, we get $y = \dfrac{384}{x}$. Now we can substitute this for y in the perimeter equation: $P = 3x + 2\left(\dfrac{384}{x}\right) = 3x + \dfrac{768}{x}$. Now we take the derivative of P: $\dfrac{dP}{dx} = 3 - \dfrac{768}{x^2}$. If we solve this for x, we get $x = \pm 16$. A negative answer doesn't make any sense in the case, so we use the solution $x = 16$ meters. Now we can solve for y:

$$y = \frac{384}{16} = 24 \text{ meters.}$$

We can verify that this is a minimum by taking the second derivative: $\dfrac{d^2 P}{dx^2} = \dfrac{1536}{x^3}$. Next, we plug in $x = 16$ to get $\dfrac{d^2 P}{dx^2} = \dfrac{1536}{16^3} = \dfrac{3}{8}$. Because the value of the second derivative is positive, according to the second derivative test (see page 187), the perimeter is a minimum at $x = 16$.

4.   Radius is $\sqrt[3]{\dfrac{256}{\pi}}$ inches.

First, let's draw a picture.

The volume of a cylinder is $V = \pi r^2 h = 512$. The material for the can is the surface area of the cylinder (don't forget the ends!) $S = 2\pi rh + 2\pi r^2$. If we solve the volume equation for h, we get $h = \dfrac{512}{\pi r^2}$.

Now we can substitute this for h in the surface area equation: $S = 2\pi r\left(\dfrac{512}{\pi r^2}\right) + 2\pi r^2 = \dfrac{1024}{r} + 2\pi r^2$.

Now we take the derivative of S: $\dfrac{dS}{dr} = -\dfrac{1024}{r^2} + 4\pi r$. If we solve this for r, we get $r = \sqrt[3]{\dfrac{256}{\pi}}$

inches. We can verify that this is a minimum by taking the second derivative: $\dfrac{d^2S}{dx^2} = \dfrac{2048}{r^3} + 4\pi$.

Next, we plug in $\sqrt[3]{\dfrac{256}{\pi}}$ and we can see that the value of the second derivative is positive. Therefore,

according to the second derivative test (see page 187), the perimeter is a minimum at $r = \sqrt[3]{\dfrac{256}{\pi}}$.

5.   1,352.786 meters

Let's think about the situation. If the swimmer swims the whole distance to the cottage, she will be traveling the entire time at her slowest speed. If she swims straight to shore first, minimizing her swimming distance, she will be maximizing her running distance. Therefore, there should be a point, somewhere between the cottage and the point on the shore directly opposite her, where the swimmer should come on land to switch from swimming to running to get to the cottage in the shortest time.

Let's draw a picture.

Let x be the distance from the point on the shore directly opposite the swimmer to the point where she comes on land. We have two distances to consider. The first is the diagonal distance that the swimmer swims. This distance, which we'll call D_1, is $D_1 = \sqrt{500^2 + x^2} = \sqrt{250{,}000 + x^2}$.

The second distance, which we'll call D_2, is simply $D_2 = 1{,}800 - x$. Remember that *rate × time = distance*? We'll use this formula to find the total time that the swimmer needs.

The time for the swimmer to travel D_1 is $T_1 = \dfrac{D_1}{4} = \dfrac{\sqrt{250{,}000 + x^2}}{4}$ (because she swims at 4 m/s), and the time for the swimmer to travel D_2 is $T_2 = \dfrac{D_2}{6} = \dfrac{1{,}800 - x}{6} = 300 - \dfrac{x}{6}$.

Therefore, the total time is $T = \dfrac{\sqrt{250,000 + x^2}}{4} + 300 - \dfrac{x}{6}$. Now we simply take the derivative:

$\dfrac{dT}{dx} = \dfrac{1}{4}\left(\dfrac{1}{2}\right)\left(250,000 + x^2\right)^{-\frac{1}{2}}(2x) - \dfrac{1}{6} = \dfrac{x}{4\sqrt{250,000 + x^2}} - \dfrac{1}{6}$. Next, we set this equal to zero and

solve: $\dfrac{x}{4\sqrt{250,000 + x^2}} - \dfrac{1}{6} = 0$. The best way to solve this is to move the $\dfrac{1}{6}$ to the other side of

the equals sign and cross-multiply:

$$\dfrac{x}{4\sqrt{250,000 + x^2}} = \dfrac{1}{6}$$

$$6x = 4\sqrt{250,000 + x^2}$$

Next, we square both sides: $36x^2 = 16(250,000 + x^2)$.

Simplify: $20x^2 = 4,000,000$.

Solve for x: $x = \pm 447.214$ m. (We can ignore the negative answer.) Therefore, she should land $1,800 - 447.214 = 1,352.786$ meters from the cottage. We could verify that this is a minimum by taking the second derivative but that would be messy. It is simpler to use the calculator to check the sign of the derivative at a point on either side of the answer, or to graph the equation for the time.

6. $\left(\dfrac{2}{\sqrt{5}}, \dfrac{1}{\sqrt{5}}\right)$

First, let's draw a picture.

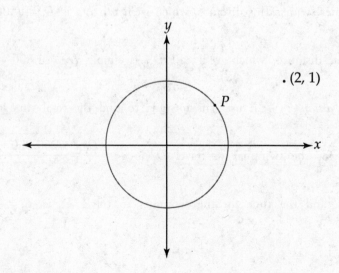

We need to find an expression for the distance from the point P to the point (2, 1) and then minimize the distance. If we call the coordinates of P (x, y), then we can find the distance to (2, 1) using the distance formula: $D^2 = (x - 2)^2 + (y - 1)^2$. Next, just as we did in sample problem 4 (page 196), we can let $L = D^2$ and minimize L: $L = (x - 2)^2 + (y - 1)^2 = x^2 - 4x + 4 + y^2 - 2y + 1$.

Because $x^2 + y^2 = 1$, we can substitute for y to get $L = x^2 - 4x + 4 + (1 - x^2) - 2\sqrt{1 - x^2} + 1$, which simplifies to $L = -4x + 6 - 2\sqrt{1 - x^2}$. Next, we take the derivative:

$\dfrac{dL}{dx} = -4 - 2\left(\dfrac{1}{2}\right)(1 - x^2)^{-\frac{1}{2}}(-2x) = -4 + \dfrac{2x}{\sqrt{1 - x^2}}$. Next, we set the derivative equal to zero:

$-4 + \dfrac{2x}{\sqrt{1 - x^2}} = 0$. The best way to solve this is to move the 4 to the other side of the equals sign and cross-multiply.

$$\frac{2x}{\sqrt{1 - x^2}} = 4$$

$$2x = 4\sqrt{1 - x^2}$$

Next, we can simplify and square both sides: $x^2 = 4(1 - x^2)$. Now, we can solve this easily. We get

$x = \pm\dfrac{2}{\sqrt{5}}$. Next, we find the y-coordinate: $y = \pm\dfrac{1}{\sqrt{5}}$. There are thus four possible answers but, if we look at the picture, the answer is obviously the point $\left(\dfrac{2}{\sqrt{5}}, \dfrac{1}{\sqrt{5}}\right)$. We could verify that this is a minimum by taking the second derivative, but that will be messy. It is simpler to use the calculator to check the sign of the derivative at a point on either side of the answer, or to graph the equation for the distance.

7. $r = \dfrac{288}{4 + \pi}$ inches

First, let's draw a picture.

Call the width of the window $2r$. Notice that this is the diameter of the semicircle. Call the height of the window h. The area of the rectangular portion of the window is $2rh$, and the perimeter is $2r + 2h$. The area of the semicircular portion of the window is $\dfrac{\pi r^2}{2}$, and the perimeter is $\dfrac{2\pi r}{2} = \pi r$. Therefore, the area of the window is $A = 2rh + \dfrac{\pi r^2}{2}$, and the perimeter is $2r + 2h + \pi r = 288$. We can use the equation for the perimeter to eliminate a variable from the equation for the area. Let's isolate h: $h = 144 - r - \dfrac{\pi r}{2}$. Now we can substitute for h in the equation for the area: $A = 2r\left(144 - r - \dfrac{\pi r}{2}\right) + \dfrac{\pi r^2}{2} = 288r - 2r^2 - \dfrac{\pi r^2}{2}$. Next, we can take the derivative: $\dfrac{dA}{dr} = 288 - 4r - \pi r$. If we set this equal to zero and solve for r, we get $r = \dfrac{288}{4 + \pi}$ inches. We can verify that this is a maximum by taking the second derivative: $\dfrac{d^2 A}{dr^2} = -4 - \pi$. We can see that the value of the second derivative is negative. Therefore, according to the second derivative test (see page 187), the area is a maximum at $r = \dfrac{288}{4 + \pi}$ inches.

8. $\theta = \dfrac{\pi}{4}$ radians (or 45 degrees)

We simply take the derivative: $\dfrac{dR}{d\theta} = \dfrac{v_0^{\,2}}{g}\left(2\cos 2\theta\right)$. Note that v_0 and g are constants, so the only variable we need to take the derivative with respect to is θ. Now we set the derivative equal to zero: $\dfrac{v_0^{\,2}}{g}\left(2\cos 2\theta\right) = 0$. Although this has an infinite number of solutions, we are interested only in values of θ between 0 and $\dfrac{\pi}{2}$ radians. The value of θ that makes the derivative zero is $\theta = \dfrac{\pi}{4}$ radians (or 45 degrees). We can verify that this is a maximum by taking the second derivative: $\dfrac{d^2 R}{d\theta^2} = \dfrac{v_0^{\,2}}{g}\left(-4\sin 2\theta\right)$. We can see that the value of the second derivative is negative at $\theta = \dfrac{\pi}{4}$. Therefore, according to the second derivative test (see page 187), the range is a maximum at $\theta = \dfrac{\pi}{4}$ radians.

9.   10 feet by 10 feet

Let's label the sides of the box. The base's length and width are the same (because it's a square) so let's call them both x. Let's call the height of the box y. This makes the volume of the box $V = x^2 y$. The surface area is the area of the bottom, x^2, plus the areas of the four sides, each of which is xy. There is no top, so the surface area is $A = x^2 + 4xy$.

We are given that the volume is 500 ft^3, so $x^2 y = 500$. Let's isolate y and then plug the expression into the area formula. First, isolate $y = \dfrac{500}{x^2}$, then plug in: $A = x^2 + 4x\left(\dfrac{500}{x^2}\right) = x^2 + \dfrac{2000}{x}$.

Now we take the derivative $A' = 2x - \dfrac{2000}{x^2}$.

Set the derivative equal to zero and solve:

$$2x - \dfrac{2000}{x^2} = 0$$

$$2x^3 - 2000 = 0$$

$$x^3 = 1000$$

$$x = 10 \text{ ft}$$

The dimensions of the base are 10 feet by 10 feet.

10. $\left(-\sqrt{\dfrac{1}{3}}, \dfrac{3}{4}\right)$

First, we need to find the slope of the tangent line to $y = \dfrac{1}{1+x^2}$. This is the derivative:

$y' = \dfrac{-2x}{\left(1+x^2\right)^2}$. Now, to find where the slope is the greatest, we take the derivative and set it

equal to zero. We get $y'' = \dfrac{\left(1+x^2\right)^2(-2) + 2x(2)\left(1+x^2\right)(2x)}{\left(1+x^2\right)^4} = 0$, which can be simplified to

$\dfrac{\left(1+x^2\right)(-2) + 2x(2)(2x)}{\left(1+x^2\right)^3} = 0$.

Note that we just need to solve for where the numerator is 0. Let's simplify the numerator:

$$-2 - 2x^2 + 8x^2 = 0$$
$$-2 + 6x^2 = 0$$

We can solve for x: $x = \pm\sqrt{\dfrac{1}{3}}$. If we plug these two values into the equation for the slope, we

find that at $x = \sqrt{\dfrac{1}{3}}$, $y' = \dfrac{-2\sqrt{\dfrac{1}{3}}}{\left(1+\dfrac{1}{3}\right)^2}$ and at $x = -\sqrt{\dfrac{1}{3}}$ $y' = \dfrac{2\sqrt{\dfrac{1}{3}}}{\left(1+\dfrac{1}{3}\right)^2}$. This is the larger of the two,

so the maximum slope is at $x = -\sqrt{\dfrac{1}{3}}$. Now we just have to find the y-coordinate, which we get

by plugging x into the original equation: $y = \dfrac{1}{1+\dfrac{1}{3}} = \dfrac{3}{4}$. Therefore, the maximum slope is at the

point $\left(-\sqrt{\dfrac{1}{3}}, \dfrac{3}{4}\right)$.

SOLUTIONS TO PRACTICE PROBLEM SET 11

1. Minimum at $\left(\sqrt{3}, -6\sqrt{3} - 6\right)$; Maximum at $\left(-\sqrt{3}, 6\sqrt{3} - 6\right)$; Point of inflection at $(0, -6)$.

First, let's find the y-intercept. We set $x = 0$ to get $y = (0)^3 - 9(0) - 6 = -6$. Therefore, the y-intercept is $(0, -6)$. Next, we find any critical points using the first derivative. The derivative is $\dfrac{dy}{dx} = 3x^2 - 9$. If we set this equal to zero and solve for x, we get $x = \pm\sqrt{3}$. Plug $x = \sqrt{3}$ and $x = -\sqrt{3}$ into the original equation to find the y-coordinates of the critical points. When $x = \sqrt{3}$, $y = \left(\sqrt{3}\right)^3 - 9\left(\sqrt{3}\right) - 6 = -6\sqrt{3} - 6$. When $x = -\sqrt{3}$, $y = \left(-\sqrt{3}\right)^3 - 9\left(-\sqrt{3}\right) - 6 = 6\sqrt{3} - 6$. Thus, we have critical points at $\left(\sqrt{3}, -6\sqrt{3} - 6\right)$ and $\left(-\sqrt{3}, 6\sqrt{3} - 6\right)$. Next, we take the second derivative to find any points of inflection. The second derivative is $\dfrac{d^2y}{dx^2} = 6x$, which is equal to zero at $x = 0$. Note that this is the y-intercept $(0, -6)$, which we already found, so there is a point of inflection at $(0, -6)$. Next, we need to determine if each critical point is maximum, minimum, or something else. If we plug $x = \sqrt{3}$ into the second derivative, the value is obviously positive, so $\left(\sqrt{3}, -6\sqrt{3} - 6\right)$ is a minimum. If we plug $x = -\sqrt{3}$ into the second derivative, the value is obviously negative, so $\left(-\sqrt{3}, 6\sqrt{3} - 6\right)$ is a maximum. Now, we can draw the curve. It looks like the following:

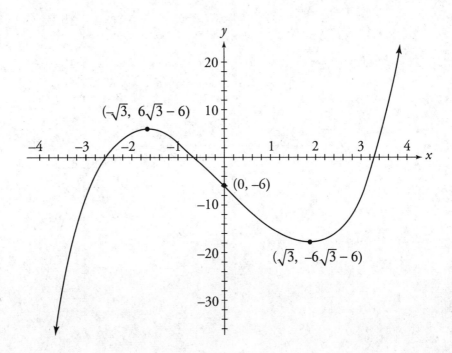

2. Minimum at (–3, –4); Maximum at (–1, 0); Point of inflection at (–2, –2).

First, let's find the y-intercept. We set $x = 0$ to get: $y = -(0)^3 - 6(0)^2 - 9(0) - 4 = -4$. Therefore, the y-intercept is (0, –4). Next, we find any critical points using the first derivative. The derivative is $\frac{dy}{dx} = -3x^2 - 12x - 9$. If we set this equal to zero and solve for x, we get $x = -1$ and $x = -3$. Plug $x = -1$ and $x = -3$ into the original equation to find the y-coordinates of the critical points: When $x = -1$, $y = -(-1)^3 - 6(-1)^2 - 9(-1) - 4 = 0$. When $x = -3$, $y = -(-3)^3 - 6(-3)^2 - 9(-3) - 4 = -4$. Thus, we have critical points at (–1, 0) and (–3, –4). Next, we take the second derivative to find any points of inflection. The second derivative is $\frac{d^2 y}{dx^2} = -6x - 12$, which is equal to zero at $x = -2$. Plug $x = -2$ into the original equation to find the y-coordinate: $y = -(-2)^3 - 6(-2)^2 - 9(-2) - 4 = -2$, so there is a point of inflection at (–2, –2). Next, we need to determine if each critical point is maximum, minimum, or something else. If we plug $x = -1$ into the second derivative, the value is negative, so (–1, 0) is a maximum. If we plug $x = -3$ into the second derivative, the value is positive, so (–3, –4) is a minimum. Now, we can draw the curve. It looks like the following:

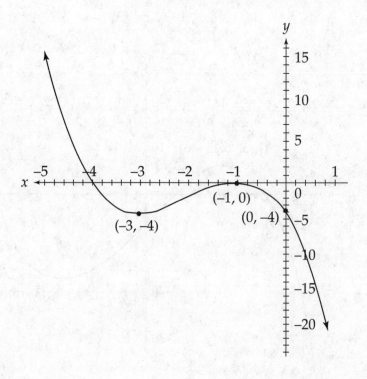

3. Minimum at $(0, -36)$; Maxima at $\left(\sqrt{\dfrac{13}{2}}, \dfrac{25}{4}\right)$ and $\left(-\sqrt{\dfrac{13}{2}}, \dfrac{25}{4}\right)$; points of inflection at $\left(\sqrt{\dfrac{13}{6}}, -\dfrac{451}{36}\right)$ and $\left(-\sqrt{\dfrac{13}{6}}, -\dfrac{451}{36}\right)$.

First, we can easily see that the graph has x-intercepts at $x = \pm 2$ and $x = \pm 3$. Next, before we take the derivative, let's multiply the two terms so we don't have to use the Product Rule. We get

$y = -x^4 + 13x^2 - 36$. Now we can take the derivative: $\dfrac{dy}{dx} = -4x^3 + 26x$. Next, we set the derivative equal to zero to find the critical points. There are three solutions: $x = 0$, $x = \sqrt{\dfrac{13}{2}}$, and $x = -\sqrt{\dfrac{13}{2}}$.

We plug these values into the original equation to find the y-coordinates of the critical points: When $x = 0$, $y = -(0)^4 + 13(0)^2 - 36 = -36$. When $x = \sqrt{\dfrac{13}{2}}$, $y = -\left(\sqrt{\dfrac{13}{2}}\right)^4 + 13\left(\sqrt{\dfrac{13}{2}}\right)^2 - 36 = \dfrac{25}{4}$.

When $x = -\sqrt{\dfrac{13}{2}}$, $y = -\left(-\sqrt{\dfrac{13}{2}}\right)^4 + 13\left(-\sqrt{\dfrac{13}{2}}\right)^2 - 36 = \dfrac{25}{4}$.

Thus, we have critical points at $(0, -36)$, $\left(\sqrt{\dfrac{13}{2}}, \dfrac{25}{4}\right)$, and $\left(-\sqrt{\dfrac{13}{2}}, \dfrac{25}{4}\right)$.

Next, we take the second derivative to find any points of inflection. The second derivative is $\dfrac{d^2y}{dx^2} = -12x^2 + 26$, which is equal to zero at $x = \sqrt{\dfrac{13}{6}}$ and $x = -\sqrt{\dfrac{13}{6}}$. We plug these values into the original equation to find the y-coordinates.

When $x = \sqrt{\dfrac{13}{6}}$, then $y = -\left(\sqrt{\dfrac{13}{6}}\right)^4 + 13\left(\sqrt{\dfrac{13}{6}}\right)^2 - 36 = -\dfrac{451}{36}$.

When $x = -\sqrt{\dfrac{13}{6}}$, then $y = -\left(-\sqrt{\dfrac{13}{6}}\right)^4 + 13\left(-\sqrt{\dfrac{13}{6}}\right)^2 - 36 = -\dfrac{451}{36}$. So there are points of inflection at $\left(\sqrt{\dfrac{13}{6}}, -\dfrac{451}{36}\right)$ and $\left(-\sqrt{\dfrac{13}{6}}, -\dfrac{451}{36}\right)$. Next, we need to determine if each critical

point is maximum, minimum, or something else. If we plug $x = 0$ into the second derivative, the

value is positive, so $(0, -36)$ is a minimum. If we plug $x = \sqrt{\dfrac{13}{2}}$ into the second derivative, the

value is negative, so $\left(\sqrt{\dfrac{13}{2}}, \dfrac{25}{4}\right)$ is a maximum. If we plug $x = -\sqrt{\dfrac{13}{2}}$ into the second derivative,

the value is negative, so $\left(-\sqrt{\dfrac{13}{2}}, \dfrac{25}{4}\right)$ is a maximum. Now, we can draw the curve. It looks like

the following:

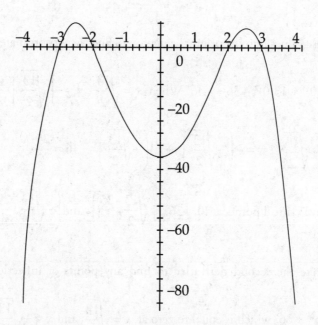

4.   Vertical asymptote at $x = -8$; Horizontal asymptote at $y = 1$; No maxima, minima, or points of inflection.

First, notice that there is an x-intercept at $x = 3$ and that the y-intercept is $\left(0, -\dfrac{3}{8}\right)$. Next, we take

the derivative: $\dfrac{dy}{dx} = \dfrac{(x+8)(1)-(x-3)(1)}{(x+8)^2} = \dfrac{11}{(x+8)^2}$. Next, we set the derivative equal to zero to

find the critical points. There is no solution. Next, we take the second derivative: $\dfrac{d^2y}{dx^2} = -\dfrac{22}{(x+8)^3}$.

If we set this equal to zero, there is also no solution. Therefore, there are no maxima, minima, or points of inflection. Note that the first derivative is always positive. This means that the curve is always increasing. Also notice that the second derivative changes sign from positive to negative at $x = -8$. This means that the curve is concave up for values of x less than $x = -8$ and concave down for values of x greater than $x = -8$.

Now, we can draw the curve. It looks like the following:

5. Vertical asymptote at $x = 3$; oblique asymptote of $y = x + 3$; maximum at $\left(3 + \sqrt{5},\ 6 + 2\sqrt{5}\right)$; minimum at $\left(3 - \sqrt{5},\ 6 - 2\sqrt{5}\right)$; no points of inflection.

First, notice that there are x-intercepts at $x = \pm 2$ and that the y-intercept is $\left(0, \dfrac{4}{3}\right)$. There is a vertical asymptote at $x = 3$. There is no horizontal asymptote, but notice that the degree of the numerator of the function is 1 greater than the denominator. This means that there is an oblique (slant) asymptote. We find this by dividing the denominator into the numerator and looking at the quotient. We get

$$x - 3 \overline{\smash{\big)}\ x^2 + 0x - 4} \quad = \quad x + 3 + \frac{5}{x - 3}$$

This means that as $x \to \pm\infty$, the function will behave like the function $y = x + 3$. This means that there is an oblique asymptote of $y = x + 3$. Next, we take the derivative:

$\dfrac{dy}{dx} = \dfrac{(x-3)(2x) - (x^2 - 4)(1)}{(x-3)^2} = \dfrac{x^2 - 6x + 4}{(x-3)^2} = 1 - \dfrac{5}{(x-3)^2}$. Next, we set the derivative equal to

zero to find the critical points. There are two solutions: $x = 3 + \sqrt{5}$ and $x = 3 - \sqrt{5}$. We plug these values into the original equation to find the y-coordinates of the critical points: When $x = 3 + \sqrt{5}$,

$y = \dfrac{\left(3 + \sqrt{5}\right)^2 - 4}{\left(3 + \sqrt{5}\right) - 3} = 6 + 2\sqrt{5}$. When $x = 3 - \sqrt{5}$, $y = \dfrac{\left(3 - \sqrt{5}\right)^2 - 4}{\left(3 - \sqrt{5}\right) - 3} = 6 - 2\sqrt{5}$. Thus, we have

critical points at $\left(3 + \sqrt{5}, 6 + 2\sqrt{5}\right)$ and $\left(3 - \sqrt{5}, 6 - 2\sqrt{5}\right)$. Next, we take the second derivative:

$\dfrac{d^2 y}{dx^2} = \dfrac{10}{(x-3)^3}$. If we set this equal to zero, there is no solution. Therefore, there is no point of

inflection. But, notice that the second derivative changes sign from negative to positive at $x = 3$.

This means that the curve is concave down for values of x less than $x = 3$ and concave up for

values of x greater than $x = 3$. Next, we need to determine if each critical point is maximum, mini-

mum, or something else. If we plug $x = 3 + \sqrt{5}$ into the second derivative, the value is positive, so

$\left(3 + \sqrt{5}, 6 + 2\sqrt{5}\right)$ is a minimum. If we plug $x = 3 - \sqrt{5}$ into the second derivative, the value is

positive, so $\left(3 - \sqrt{5}, 6 - 2\sqrt{5}\right)$ is a maximum.

Now, we can draw the curve. It looks like the following:

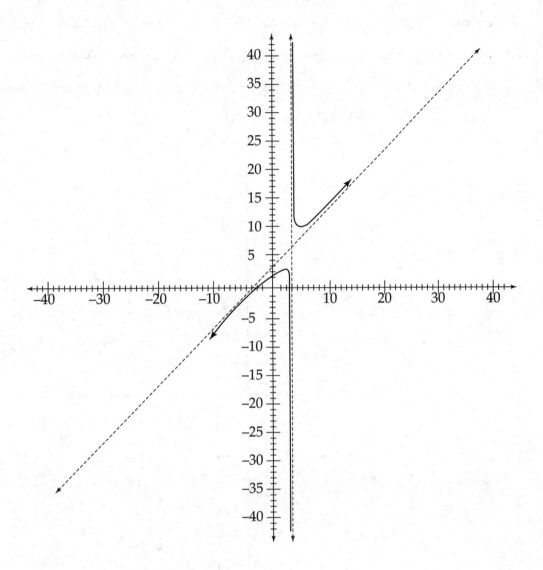

6.   No maxima, minima, or points of inflection; cusp at (0, 3).

First, notice that the function is always positive. There is a y-intercept at (0, 3). There is no x-intercept. Next, we take the derivative: $\dfrac{dy}{dx} = \dfrac{2}{3} x^{-\frac{1}{3}}$. If we set the derivative equal to zero, there is no solution. But, notice that the derivative is not defined at $x = 0$. This means that the function has either a vertical tangent or a cusp at $x = 0$. We'll be able to determine which after we take the second derivative. Notice also that the derivative is negative for $x < 0$ and positive for $x > 0$.

Therefore, the curve is decreasing for $x < 0$ and increasing for $x > 0$. Next, we take the second derivative: $\dfrac{dy}{dx} = -\dfrac{2}{9}x^{-\frac{4}{3}}$. If we set this equal to zero, there is no solution. The second derivative is always negative, which tells us that the curve is always concave down and that the curve has a cusp at $x = 0$. Note that if it had switched concavity there, then $x = 0$ would be a vertical tangent. Now, we can draw the curve. It looks like the following:

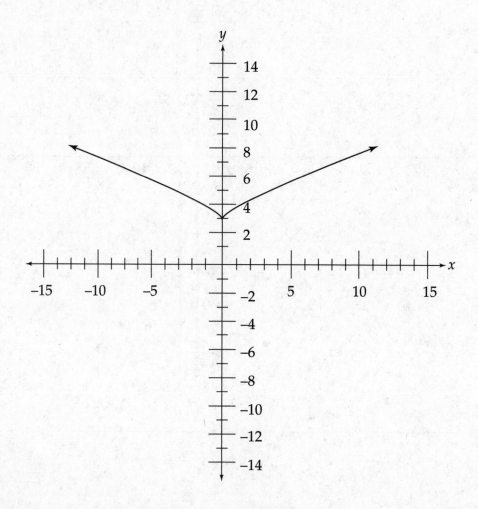

7.   Maximum at $(1, 1)$; no point of inflection; cusp at $(0, 0)$.

First, we notice that the curve has x-intercepts at $x = 0$ and $x = \dfrac{27}{8}$ and that there is a y-intercept at $(0, 0)$. Before we take the derivative, let's expand the expression. That way, we won't have to use the Product Rule. We get $y = 3x^{\frac{2}{3}} - 2x$. Next, we take the derivative: $\dfrac{dy}{dx} = 2x^{-\frac{1}{3}} - 2$. Next, we set the derivative equal to zero to find the critical points. There is one solution: $x = 1$. We plug this value into the original equation to find the y-coordinate of the critical point: When $x = 1$,

$y = 3(1)^{\frac{2}{3}} - 2(1) = 1$. Thus, we have a critical point at (1, 1). But, notice that the derivative is not defined at $x = 0$. This means that the function has either a vertical tangent or a cusp at $x = 0$. We'll be able to determine which after we take the second derivative. Notice also that the derivative is negative for $x < 0$ and for $x > 1$ and positive for $0 < x < 1$. Therefore, the curve is decreasing for $x < 0$ and for $x > 1$, and increasing for $0 < x < 1$. Next, we take the second derivative: $\dfrac{d^2 y}{dx^2} = -\dfrac{2}{3} x^{-\frac{4}{3}}$. If we set this equal to zero, there is no solution. Therefore, there is no point of inflection. The second derivative is always negative, which tells us that the curve is always concave down and that the curve has a cusp at $x = 0$. Note that if it had switched concavity there, then $x = 0$ would be a vertical tangent. Next, we need to determine if each critical point is maximum, minimum, or something else. If we plug $x = 1$ into the second derivative, the value is negative, so (1, 1) is a maximum. Now, we can draw the curve. It looks like the following:

SOLUTIONS TO PRACTICE PROBLEM SET 12

1.   2,000 ft²/s

We are given the rate at which the circumference is increasing, $\dfrac{dC}{dt} = 40$ and are looking for the rate at which the area is increasing, $\dfrac{dA}{dt}$. Thus, we need to find a way to relate the area of a circle to its circumference. Recall that the circumference of a circle is $C = 2\pi r$, and the area is $A = \pi r^2$. We could find C in terms of r and then plug it into the equation for A, or we could work with the equations separately and then relate them. Let's do both and compare.

Method 1: First, we find C in terms of r: $r = \dfrac{C}{2\pi}$. Now, we plug this in for r in the equation for A: $A = \pi \left(\dfrac{C}{2\pi}\right)^2 = \dfrac{C^2}{4\pi}$. Next, we take the derivative of the equation with respect to t: $\dfrac{dA}{dt} = \dfrac{1}{4\pi}(2C)\dfrac{dC}{dt} = \left(\dfrac{C}{2\pi}\right)\dfrac{dC}{dt}$. Next, we plug in $C = 100\pi$ and $\dfrac{dC}{dt} = 40$ and solve: $\dfrac{dA}{dt} = \left(\dfrac{100\pi}{2\pi}\right)(40) = 2{,}000$. Therefore, the answer is 2,000 ft²/s.

Method 2: First, we take the derivative of C with respect to t: $\dfrac{dC}{dt} = 2\pi\left(\dfrac{dr}{dt}\right)$. Next, we plug in $\dfrac{dC}{dt} = 40$ and solve for $\dfrac{dr}{dt}$: $40 = 2\pi\left(\dfrac{dr}{dt}\right)$, so $\dfrac{dr}{dt} = \dfrac{20}{\pi}$. Next, we take the derivative of A with respect to t: $\dfrac{dA}{dt} = 2\pi r \dfrac{dr}{dt}$. Now, we can plug in for $\dfrac{dr}{dt}$ and r and solve for $\dfrac{dA}{dt}$. Note that when the circumference is 100π, $r = 50$: $\dfrac{dA}{dt} = 2\pi(50)\left(\dfrac{20}{\pi}\right) = 2{,}000$. Therefore, the answer is 2,000 ft²/s.

Which method is better? In this case they are about the same. Method 1 is going to be more efficient if it is easy to solve for one variable in terms of the other, and it is also easy to take the derivative of the resulting expression. Otherwise, we will prefer to use Method 2 (See Example 3 on page 218).

2. $\frac{3}{4}$ in./s

We are given the rate at which the volume is increasing, $\frac{dV}{dt} = 27\pi$ and are looking for the rate at which the radius is increasing, $\frac{dr}{dt}$. Thus, we need to find a way to relate the volume of a sphere to its radius. Recall that the volume of a sphere is $V = \frac{4}{3}\pi r^3$. All we have to do is take the derivative of the equation with respect to t: $\frac{dV}{dt} = \frac{4}{3}\pi\left(3r^2\right)\left(\frac{dr}{dt}\right) = 4\pi r^2\frac{dr}{dt}$. Now we substitute $\frac{dV}{dt} = 27\pi$ and $r = 3$: $27\pi = 4\pi(3)^2\frac{dr}{dt}$. If we solve for $\frac{dr}{dt}$, we get $\frac{dr}{dt} = \frac{3}{4}$ in./s.

3. 100 km/hr

We are given the rate at which Car A is moving south, $\frac{dA}{dt} = 80$, and the rate at which Car B is moving west, $\frac{dB}{dt} = 60$, and we are looking for the rate at which the distance between them is increasing, which we'll call $\frac{dC}{dt}$. Note that the directions south and west are at right angles to each other. Thus, the distance that Car A is from the starting point, which we'll call A, and the distance that Car B is from the starting point, which we'll call B, are the legs of a right triangle, with C as the hypotenuse. We can relate the three distances using the Pythagorean Theorem. Here, because A and B are the legs and C is the hypotenuse, $A^2 + B^2 = C^2$. Now, we take the derivative of the equation with respect to t: $2A\frac{dA}{dt} + 2B\frac{dB}{dt} = 2C\frac{dC}{dt}$. This simplifies to $A\frac{dA}{dt} + B\frac{dB}{dt} = C\frac{dC}{dt}$. We know that Car A has been driving for 3 hours at 80 km/hr and Car B has been driving for 3 hours at 60 km/hr, so $A = 240$ and $B = 180$. Using the Pythagorean Theorem, $240^2 + 180^2 = C^2$, so $C = 300$. Now we can substitute into our derivative equation: $(240)(80) + (180)(60) = (300)\frac{dC}{dt}$. If we solve for C, we get $C = 100$ km/hr.

4. $243\sqrt{3}$ in.²/s

We are given the rate at which the sides of the triangle are increasing, $\frac{ds}{dt} = 27$, and are looking for the rate at which the area is increasing, $\frac{dA}{dt}$. Thus, we need to find a way to relate the area of an equilateral triangle to the length of a side. We know that the area of an equilateral triangle, in terms of its sides, is $A = \dfrac{s^2\sqrt{3}}{4}$. (If you don't know this formula, memorize it! It will come in very handy in future math problems.) Now we take the derivative of this equation with respect to t: $\dfrac{dA}{dt} = \dfrac{\sqrt{3}}{4}(2s)\left(\dfrac{ds}{dt}\right)$. Next, we plug $\dfrac{ds}{dt} = 27$ and $s = 18$ into the derivative and we get $\dfrac{dA}{dt} = \dfrac{\sqrt{3}}{4}(36)(27) = 243\sqrt{3}$ in.²/s.

5. $-\dfrac{5}{7}$ in./s

We are given the rate at which the water is flowing out of the container, $\dfrac{dV}{dt} = -35\pi$ (Why is it negative?), and are looking for the rate at which the depth of the water is dropping, $\dfrac{dh}{dt}$. Thus, we need to find a way to relate the volume of a cone to its height. We know that the volume of a cone is $V = \dfrac{1}{3}\pi r^2 h$. Notice that we have a problem. We have a third variable, r, in the equation. We cannot treat it as a constant the way we did in problem 4 because as the volume of a cone changes, both its height and radius change. But, we also know that in any cone, the ratio of the radius to the height is a constant. Here, when the radius is 21 (because the diameter is 42), the height is 15. Thus, $\dfrac{r}{h} = \dfrac{21}{15} = \dfrac{7}{5}$. We can now isolate r in this equation: $r = \dfrac{7h}{5}$. Now we can plug it into the volume formula to get rid of r: $V = \dfrac{1}{3}\pi\left(\dfrac{7h}{5}\right)^2 h$. This simplifies to $V = \dfrac{49\pi}{75}h^3$. Now we can take the derivative of this equation with respect to t: $\dfrac{dV}{dt} = \dfrac{49\pi}{75}(3h^2)\dfrac{dh}{dt}$. Next, we plug $\dfrac{dV}{dt} = -35\pi$ and $h = 5$ into the derivative and we get $-35\pi = \dfrac{49\pi}{75}(3(5)^2)\dfrac{dh}{dt}$. Now we can solve for $\dfrac{dh}{dt}$: $\dfrac{dh}{dt} = -\dfrac{5}{7}$ in./s.

6. $-\dfrac{25}{6}$ ft/s

We are given the rate at which the length of the rope, R, is changing, $\dfrac{dR}{dt} = -4$, and are looking for

the rate at which the boat, B, is approaching the dock, $\dfrac{dB}{dt}$. The key to this problem is to realize

that the vertical distance from the dock to the bow, the distance from the boat to the dock, and the

length of the rope form a right triangle.

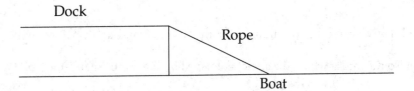

Dock

Rope

Boat

The vertical distance from the dock to the bow is always 7, so, using the Pythagorean Theorem, we

get $7^2 + B^2 = R^2$. Now we can take the derivative of this equation with respect to t: $2B\dfrac{dB}{dt} = 2R\dfrac{dR}{dt}$,

which simplifies to $B\dfrac{dB}{dt} = R\dfrac{dR}{dt}$. We know that $R = 25$ and can use the Pythagorean Theorem to

find B: $7^2 + B^2 = 25^2$, so $B = 24$. Now we plug $\dfrac{dR}{dt} = -4$, $R = 25$, and $B = 24$ into the derivative and

we get $24\dfrac{dB}{dt} = 25(-4)$. Now we can solve for $\dfrac{dB}{dt}$: $\dfrac{dB}{dt} = -\dfrac{25}{6}$ ft/s.

7. $\dfrac{4}{3}$ ft/s

We are given the rate at which the woman is walking away from the street lamp and are looking for

the rate at which the length of her shadow is changing. It helps to draw a picture of the situation.

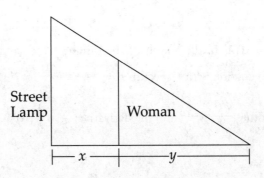

Street
Lamp

Woman

$\longmapsto x \longmapsto \longmapsto y \longmapsto$

If we label the distance between the woman and the street lamp, *x*, and the length of the woman's shadow, *y*, we can use similar triangles to get $\dfrac{y}{6} = \dfrac{x+y}{24}$. We can cross-multiply and simplify.

$$24y = 6x + 6y$$
$$3y = x$$

Next, we take the derivative of the equation with respect to *t*: $3\dfrac{dy}{dt} = \dfrac{dx}{dt}$. Now, we can plug $\dfrac{dx}{dt} = 4$ into the derivative and solve: $\dfrac{dy}{dt} = \dfrac{4}{3}\,\text{ft/s}$.

8. $\dfrac{3\pi}{5}$ in.²/min

We know that it takes 60 minutes for the minute hand of a clock to make one complete revolution, so the rate at which the angle, θ, formed by the minute hand and noon increasing, in terms of radians/min, is $\dfrac{d\theta}{dt} = \dfrac{2\pi}{60} = \dfrac{\pi}{30}$. Next, we know that the area of the sector of a circle is proportional to its central angle, so $\dfrac{\theta}{2\pi} = \dfrac{S}{\pi r^2}$, which can be simplified to $S = \dfrac{1}{2}r^2\theta$.

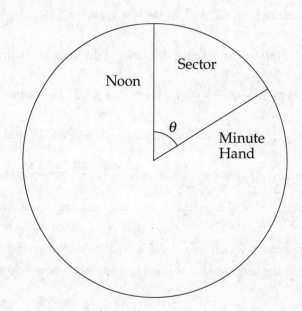

Note that the radius is 6 (the length of the minute hand) and is a constant. Next, we take the derivative of the equation with respect to *t*: $\dfrac{dS}{dt} = \dfrac{1}{2}r^2\dfrac{d\theta}{dt}$.

Finally, we substitute $\dfrac{d\theta}{dt} = \dfrac{\pi}{30}$ into the derivative: $\dfrac{dS}{dt} = \dfrac{1}{2}(6)^2\left(\dfrac{\pi}{30}\right) = \dfrac{3\pi}{5}\,\text{in.}^2/\text{min}$.

SOLUTIONS TO PRACTICE PROBLEM SET 13

1. $v(t) = 3t^2 - 18t + 24$; $a(t) = 6t - 18$

    In order to find the velocity function of the particle, we simply take the derivative of the position function with respect to t: $\dfrac{dx}{dt} = v(t) = 3t^2 - 18t + 24$. In order to find the acceleration function of the particle, we simply take the derivative of the velocity function with respect to t: $\dfrac{dv}{dt} = 6t - 18$.

2. $v(t) = 2\cos(2t) - \sin(t)$; $a(t) = -4\sin(2t) - \cos(t)$

    In order to find the velocity function of the particle, we simply take the derivative of the position function with respect to t: $\dfrac{dx}{dt} = v(t) = 2\cos(2t) - \sin(t)$. In order to find the acceleration function of the particle, we simply take the derivative of the velocity function with respect to t: $\dfrac{dv}{dt} = a(t) = -4\sin(2t) - \cos(t)$.

3. $t = \pi$ and $t = 3\pi$

    In order to find where the particle is changing direction, we need to find where the velocity of the particle changes sign. The velocity function of the particle is the derivative of the position function: $\dfrac{dx}{dt} = v(t) = \dfrac{1}{2}\cos\left(\dfrac{t}{2}\right)$. Next, we set the velocity equal to zero. The solutions are: $t = \pi$ and $t = 3\pi$. Actually, there are an infinite number of solutions, but remember that we are restricted to $0 < t < 4\pi$. Next, we check the sign of the velocity on the intervals $0 < t < \pi$, $\pi < t < 3\pi$, and $3\pi < t < 4\pi$. When $0 < t < \pi$, the velocity is positive, so the particle is moving to the right. When $\pi < t < 3\pi$, the velocity is negative, so the particle is moving to the left. When $3\pi < t < 4\pi$, the velocity is positive, so the particle is moving to the right. Therefore, the particle is changing direction at $t = \pi$ and $t = 3\pi$.

4. The distance is 69.

    In order to find the distance that the particle travels, we need to look at the position of the particle at $t = 2$ and at $t = 5$. We also need to see if the particle changes direction anywhere on the interval between the two times. If so, we will need to look at the particle's position at those "turning points" as well. The way to find out if the particle is changing direction is to look at the velocity of the particle, which we find by taking the derivative of the position function. We get $\dfrac{dx}{dt} = v(t) = 6t + 2$. If we set the velocity equal to zero, we get $t = -\dfrac{1}{3}$, which is not in the time interval. This means that the velocity doesn't change signs, and thus the particle does not change direction. Now we look at the position of the particle on the interval. At $t = 2$, the particle's position is: $x = 3(2)^2 + 2(2) + 4 = 20$. At $t = 5$, the particle's position is: $x = 3(5)^2 + 2(5) + 4 = 89$. Therefore, the particle travels a distance of 69.

5. The distance is 48.

In order to find the distance that the particle travels, we need to look at the position of the particle at $t = 0$ and at $t = 4$. We also need to see if the particle changes direction anywhere on the interval between the two times. If so, we will need to look at the particle's position at those "turning points" as well. The way to find out if the particle is changing direction is to look at the velocity of the particle, which we find by taking the derivative of the position function. We get $\dfrac{dx}{dt} = v(t) = 2t + 8$. If we set the velocity equal to zero, we get $t = -4$, which is not in the time interval. This means that the velocity doesn't change signs, and thus the particle does not change direction. Now we look at the position of the particle on the interval. At $t = 0$, the particle's position is $x = (0)^2 + 8(0) = 0$. At $t = 4$, the particle's position is $x = (4)^2 + 8(4) = 48$. Therefore, the particle travels a distance of 48.

6. Velocity is 0; acceleration is 0.

This should not be a surprise because $2\sin^2 t + 2\cos^2 t = 2$, so the position is a constant. This means that the particle is not moving and thus has a velocity and acceleration of 0.

7. $t = \dfrac{-8 + \sqrt{70}}{3} \approx 0.122$

In order to find where the particle is changing direction, we need to find where the velocity of the particle changes sign. The velocity function of the particle is the derivative of the position function: $\dfrac{dx}{dt} = v(t) = 3t^2 + 16t - 2$. Next, we set the velocity equal to zero. The solutions are $t = \dfrac{-8 + \sqrt{70}}{3}$ and $t = \dfrac{-8 - \sqrt{70}}{3}$. We can eliminate the second solution because it is negative and we are restricted to $t > 0$. Next, we check the sign of the velocity on the intervals $0 < t < \dfrac{-8 + \sqrt{70}}{3}$ and $t > \dfrac{-8 + \sqrt{70}}{3}$. When $0 < t < \dfrac{-8 + \sqrt{70}}{3}$, the velocity is negative, so the particle is moving to the left. When $t > \dfrac{-8 + \sqrt{70}}{3}$, the velocity is positive, so the particle is moving to the right. Therefore, the particle is changing direction at $t = \dfrac{-8 + \sqrt{70}}{3}$.

8. The velocity is never 0, which means that it never changes signs and thus the particle does not change direction.

In order to find where the particle is changing direction, we need to find where the velocity of the particle changes signs. The velocity function of the particle is the derivative of the position function: $\dfrac{dx}{dt} = v(t) = 6t^2 - 12t + 12$. Next, we set the velocity equal to zero. There are no real solutions. If we try a few values, we can see that the velocity is always positive. Therefore, the particle does not change direction.

SOLUTIONS TO PRACTICE PROBLEM SET 14

1. $f'(x) = \dfrac{4x^3}{x^4 + 8}$

The rule for finding the derivative of $y = \ln u$ is $\dfrac{dy}{dx} = \dfrac{1}{u}\dfrac{du}{dx}$, where u is a function of x. Here $u = x^4 + 8$. Therefore, the derivative is $f'(x) = \dfrac{1}{x^4 + 8}\left(4x^3\right) = \dfrac{4x^3}{x^4 + 8}$.

2. $f'(x) = \dfrac{1}{x} + \dfrac{1}{6 + 2x}$

The rule for finding the derivative of $y = \ln u$ is $\dfrac{dy}{dx} = \dfrac{1}{u}\dfrac{du}{dx}$, where u is a function of x. Before we find the derivative, we can use the laws of logarithms to expand the logarithm. This way, we won't have to use the Product Rule. We get $\ln\left(3x\sqrt{3 + x}\right) = \ln 3 + \ln x + \ln\sqrt{3 + x} = \ln 3 + \ln x + \dfrac{1}{2}\ln(3 + x)$.

Now we can find the derivative: $f'(x) = 0 + \dfrac{1}{x} + \dfrac{1}{2}\dfrac{1}{3 + x} = \dfrac{1}{x} + \dfrac{1}{6 + 2x}$.

3. $f'(x) = \csc x$

The rule for finding the derivative of $y = \ln u$ is $\dfrac{dy}{dx} = \dfrac{1}{u}\dfrac{du}{dx}$, where u is a function of x.

Here $u = \cot x - \csc x$. Therefore, the derivative is $f'(x) = \dfrac{1}{\cot x - \csc x}\left(-\csc^2 x + \csc x \cot x\right)$.

This can be simplified to $f'(x) = \dfrac{\left(-\csc x + \cot x\right)\left(\csc x\right)}{\cot x - \csc x} = \csc x$.

4. $f'(x) = \dfrac{2}{x} - \dfrac{x}{5+x^2}$

The rule for finding the derivative of $y = \ln u$ is $\dfrac{dy}{dx} = \dfrac{1}{u}\dfrac{du}{dx}$, where u is a function of x. Before we find the derivative, we can use the laws of logarithms to expand the logarithm. This way, we won't have to use the Product Rule or the Quotient Rule. We get

$\ln\left(\dfrac{5x^2}{\sqrt{5+x^2}}\right) = \ln 5 + \ln x^2 - \ln\sqrt{5+x^2} = \ln 5 + 2\ln x - \dfrac{1}{2}\ln(5+x^2)$. Now we can find the

derivative: $f'(x) = 0 + 2\dfrac{1}{x} - \dfrac{1}{2}\dfrac{1}{5+x^2}(2x) = \dfrac{2}{x} - \dfrac{x}{5+x^2}$.

5. $f'(x) = e^{x\cos x}\left(\cos x - x\sin x\right)$

The rule for finding the derivative of $y = e^u$ is $\dfrac{dy}{dx} = e^u\dfrac{du}{dx}$, where u is a function of x. Here we will use the Product Rule to find the derivative of the exponent: $f'(x) = e^{x\cos x}\left(\cos x - x\sin x\right)$.

6. $f'(x) = -3e^{-3x}\sin 5x + 5e^{-3x}\cos 5x$

The rule for finding the derivative of $y = e^u$ is $\dfrac{dy}{dx} = e^u\dfrac{du}{dx}$, where u is a function of x. Here we will use the Product Rule to find the derivative: $f'(x) = e^{-3x}(-3)\sin 5x + e^{-3x}(5\cos 5x)$, which we can rearrange to $f'(x) = -3e^{-3x}\sin 5x + 5e^{-3x}\cos 5x$.

7. $f'(x) = \pi e^{\pi x} - \pi$

The rule for finding the derivative of $y = e^u$ is $\dfrac{dy}{dx} = e^u\dfrac{du}{dx}$, where u is a function of x and the rule for finding the derivative of $y = \ln u$ is $\dfrac{dy}{dx} = \dfrac{1}{u}\dfrac{du}{dx}$, where u is a function of x.

We get $f'(x) = e^{\pi x}(\pi) - \dfrac{1}{e^{\pi x}}\pi e^{\pi x}$. This can be simplified to $f'(x) = \pi e^{\pi x} - \pi$. You might have noticed that $\ln e^{\pi x} = \pi x$, which would have made the derivative a little easier.

8. $f'(x) = \dfrac{3}{x\ln 12}$

The rule for finding the derivative of $y = \log_a u$ is $\dfrac{dy}{dx} = \dfrac{1}{u\ln a}\dfrac{du}{dx}$, where u is a function of x.

Before we find the derivative, we can use the laws of logarithms to expand the logarithm. We get

$f(x) = \log_{12} x^3 = 3\log_{12} x$. Now we can find the derivative: $f'(x) = \dfrac{3}{x\ln 12}$.

9. $f'(x) = \dfrac{1}{xe^{4x}\ln 4} - \dfrac{4\log_4 x}{e^{4x}}$

The rule for finding the derivative of $y = \log_a u$ is $\dfrac{dy}{dx} = \dfrac{1}{u\ln a}\dfrac{du}{dx}$, and the rule for finding the

derivative of $y = e^u$ is $\dfrac{dy}{dx} = e^u\dfrac{du}{dx}$, where u is a function of x. Here we will use the Quotient

Rule to find the derivative: $f'(x) = \dfrac{\left(e^{4x}\right)\left(\dfrac{1}{x\ln 4}\right) - \left(\log_4 x\right)\left(e^{4x}\right)(4)}{\left(e^{4x}\right)^2}$. This can be simplified to

$f'(x) = \dfrac{1}{xe^{4x}\ln 4} - \dfrac{4\log_4 x}{e^{4x}}$.

10. $f'(x) = \dfrac{3}{2}$

The rule for finding the derivative of $y = \log_a u$ is $\dfrac{dy}{dx} = \dfrac{1}{u\ln a}\dfrac{du}{dx}$, and the rule for finding the

derivative of $y = a^u$ is $\dfrac{dy}{dx} = a^u(\ln a)\dfrac{du}{dx}$, where u is a function of x. Before we find the derivative,

we can use the laws of logarithms to simplify the logarithm. We get $f(x) = \dfrac{1}{2}\log 10^{3x}$. Now if

you are alert, you will remember that $\log 10^{3x} = 3x$, so this simplifies to $f(x) = \dfrac{1}{2}3x = \dfrac{3}{2}x$. The

derivative is simply $f'(x) = \dfrac{3}{2}$.

11. $f'(x) = 3e^{3x} - 3^{ex}(e)(\ln 3)$

The rule for finding the derivative of $y = e^u$ is $\dfrac{dy}{dx} = e^u\dfrac{du}{dx}$, and the rule for finding the derivative

of $y = a^u$ is $\dfrac{dy}{dx} = a^u(\ln a)\dfrac{du}{dx}$, where u is a function of x. We get $f'(x) = 3e^{3x} - 3^{ex}(e)(\ln 3)$.

12. $f'(x) = 10^{\sin x}(\cos x)(\ln 10)$

The rule for finding the derivative of $y = a^u$ is $\dfrac{dy}{dx} = a^u(\ln a)\dfrac{du}{dx}$, where u is a function of x. We get $f'(x) = 10^{\sin x}(\cos x)(\ln 10)$.

13. $f'(x) = \ln 10$

Before we find the derivative, we can use the laws of logarithms to simplify the logarithm. We get $f(x) = \ln(10^x) = x\ln 10$. Now the derivative is simply $f'(x) = \ln 10$.

14. $f'(x) = x^4 5^x(5 + x\ln 5)$

The rule for finding the derivative of $y = a^u$ is $\dfrac{dy}{dx} = a^u(\ln a)\dfrac{du}{dx}$, where u is a function of x. Here we will use the Product Rule to find the derivative: $f'(x) = x^5(5^x\ln 5) + (5^x)(5x^4)$, which simplifies to $f'(x) = x^4 5^x(5 + x\ln 5)$.

SOLUTIONS TO PRACTICE PROBLEM SET 15

1. $\dfrac{16}{15}$

First, we take the derivative of y: $\dfrac{dy}{dx} = 1 - \dfrac{1}{x^2}$. Next, we find the value of x where $y = \dfrac{17}{4}$:

$\dfrac{17}{4} = x + \dfrac{1}{x}$.

With a little algebra, you should get $x = 4$ or $x = \dfrac{1}{4}$. Because $x > 1$, we can ignore the second

answer. Or, if you are permitted, use your calculator.

Now we can use the formula for the derivative of the inverse of $f(x)$ (see page 240):

$\dfrac{d}{dx}f^{-1}(x)\bigg|_{x=c} = \dfrac{1}{\left[\dfrac{d}{dy}f(y)\right]_{y=a}}$, where $f(a) = c$. This formula means that we find the derivative of the inverse of a function at a value a by taking the reciprocal of the derivative and plugging in

the value of x that makes y equal to a.

$$\frac{1}{\frac{dy}{dx}\bigg|_{x=4}} = \frac{1}{1 - \frac{1}{x^2}\bigg|_{x=4}} = \frac{16}{15}$$

2. $-\dfrac{1}{12}$

First, we take the derivative of y: $\dfrac{dy}{dx} = 3 - 15x^2$. Next, we find the value of x where $y = 2$: $2 = 3x - 5x^3$.

With a little algebra, you should get $x = -1$. Or, if you are permitted, use your calculator.

Now we can use the formula for the derivative of the inverse of $f(x)$ (see page 240):

$$\frac{d}{dx} f^{-1}(x)\bigg|_{x=c} = \frac{1}{\left[\dfrac{d}{dy} f(y)\right]_{y=a}} \text{, where } f(a) = c. \text{ This formula means that we find the derivative}$$

of the inverse of a function at a value a by taking the reciprocal of the derivative and plugging in

the value of x that makes y equal to a.

$$\frac{1}{\frac{dy}{dx}\bigg|_{x=-1}} = \frac{1}{3 - 15x^2\big|_{x=-1}} = -\frac{1}{12}$$

3. $\dfrac{1}{e}$

First, we take the derivative of y: $\dfrac{dy}{dx} = e^x$. Next, we find the value of x where $y = e : e = e^x$. It should be obvious that $x = 1$.

Now we can use the formula for the derivative of the inverse of $f(x)$ (see page 240):

$$\frac{d}{dx} f^{-1}(x)\Big|_{x=c} = \frac{1}{\left[\dfrac{d}{dy} f(y)\right]_{y=a}}, \text{ where } f(a)=c.$$ This formula means that we find the derivative

of the inverse of a function at a value a by taking the reciprocal of the derivative and plugging in

the value of x that makes y equal to a.

$$\frac{1}{\dfrac{dy}{dx}\Big|_{x=1}} = \frac{1}{e^x\big|_{x=1}} = \frac{1}{e}$$

4. $\dfrac{1}{4}$

First, we take the derivative of y: $\dfrac{dy}{dx} = 1 + 3x^2$. Next, we find the value of x where $y = -2$:

$-2 = x + x^3$. You should be able to tell by inspection that $x = -1$ is a solution. Or, if you are permit-

ted, use your calculator. Remember that the AP Exam won't give you a problem where it is very

difficult to solve for the inverse value of y. If the algebra looks difficult, look for an obvious solu-

tion, such as $x = 0$ or $x = 1$ or $x = -1$.

Now we can use the formula for the derivative of the inverse of $f(x)$: $\dfrac{d}{dx} f^{-1}(x)\Big|_{x=c} = \dfrac{1}{\left[\dfrac{d}{dy} f(y)\right]_{y=a}}$,

where $f(a)=c$. This formula means that we find the derivative of the inverse of a function at a

value a by taking the reciprocal of the derivative and plugging in the value of x that makes y equal

to a: $\dfrac{1}{\dfrac{dy}{dx}\Big|_{x=-1}} = \dfrac{1}{1 + 3x^2\big|_{x=-1}} = \dfrac{1}{4}$.

5. 1

First, we take the derivative of y: $\dfrac{dy}{dx} = 4 - 3x^2$. Next, we find the value of x where $y = 3$:

$3 = 4x - x^3$. You should be able to tell by inspection that $x = 1$ is a solution. Or, if you are permit-

ted, use your calculator. Remember that the AP Exam won't give you a problem where it is very difficult to solve for the inverse value of y. If the algebra looks difficult, look for an obvious solution, such as $x = 0$ or $x = 1$ or $x = -1$.

Now we can use the formula for the derivative of the inverse of $f(x)$: $\dfrac{d}{dx} f^{-1}(x) \Big|_{x=c} = \dfrac{1}{\left[\dfrac{d}{dy} f(y)\right]_{y=a}}$,

where $f(a) = c$. This formula means that we find the derivative of the inverse of a function at a

value a by taking the reciprocal of the derivative and plugging in the value of x that makes y equal

to a: $\dfrac{1}{\dfrac{dy}{dx}\Big|_{x=1}} = \dfrac{1}{4 - 3x^2}\Big|_{x=1} = 1.$

6.   1

First, we take the derivative of y: $\dfrac{dy}{dx} = \dfrac{1}{x}$. Next, we find the value of x where $y = 0$: $\ln x = 0$. You

should know that $x = 1$ is the solution. Now we can use the formula for the derivative of the inverse

of $f(x)$: $\dfrac{d}{dx} f^{-1}(x)\Big|_{x=c} = \dfrac{1}{\left[\dfrac{d}{dy} f(y)\right]_{y=a}}$, where $f(a) = c$. This formula means that we find the

derivative of the inverse of a function at a value a by taking the reciprocal of the derivative and

plugging in the value of x that makes y equal to a.

$$\dfrac{1}{\dfrac{dy}{dx}\Big|_{x=1}} = \dfrac{1}{\dfrac{1}{x}\Big|_{x=1}} = 1$$

SOLUTIONS TO PRACTICE PROBLEM SET 16

1.   5.002

Recall the differential formula that we use for approximating the value of a function: $f(x + \Delta x) \approx f(x) + f'(x)\Delta x$. Here we want to approximate the value of $\sqrt{25.02}$, so we'll use $f(x) = \sqrt{x}$ with $x = 25$ and $\Delta x = 0.02$. First, we need to find $f'(x)$: $f'(x) = \dfrac{1}{2\sqrt{x}}$. Now, we plug into the formula: $f(x + \Delta x) \approx \sqrt{x} + \dfrac{1}{2\sqrt{x}}\Delta x$. If we plug in $x = 25$ and $\Delta x = 0.02$, we get $\sqrt{25 + 0.02} \approx \sqrt{25} + \dfrac{1}{2\sqrt{25}}(0.02)$. If we evaluate this, we get $\sqrt{25.02} \approx 5 + \dfrac{1}{10}(0.02) = 5.002$.

2.   3.999375

Recall the differential formula that we use for approximating the value of a function: $f(x + \Delta x) \approx f(x) + f'(x)\Delta x$. Here we want to approximate the value of $\sqrt{63.97}$, so we'll use $f(x) = \sqrt[3]{x}$ with $x = 64$ and $\Delta x = -0.03$. First, we need to find $f'(x)$: $f'(x) = \dfrac{1}{3}x^{-\frac{2}{3}} = \dfrac{1}{3\sqrt[3]{x^2}}$. Now, we plug into the formula: $f(x + \Delta x) \approx \sqrt[3]{x} + \dfrac{1}{3\sqrt[3]{x^2}}\Delta x$. If we plug in $x = 64$ and $\Delta x = -0.03$, we get $\sqrt[3]{64 - 0.03} \approx \sqrt[3]{64} + \dfrac{1}{3\sqrt[3]{64^2}}(-0.03)$. If we evaluate this, we get $\sqrt[3]{63.97} \approx 4 + \dfrac{1}{48}(-0.03) = 3.999375$.

3.   1.802

Recall the differential formula that we use for approximating the value of a function: $f(x + \Delta x) \approx f(x) + f'(x)\Delta x$. Here we want to approximate the value of $\tan 61°$. Be careful! Whenever we work with trigonometric functions, it is *very* important to work with radians, *not* degrees! Remember that $60° = \dfrac{\pi}{3}$ radians and $1° = \dfrac{\pi}{180}$ radians, so we'll use $f(x) = \tan x$ with $x = \dfrac{\pi}{3}$ and $\Delta x = \dfrac{\pi}{180}$. First, we need to find $f'(x)$: $f'(x) = \sec^2 x$. Now, we plug into the formula: $f(x + \Delta x) \approx \tan x + \sec^2 x \, \Delta x$. If we plug in $x = \dfrac{\pi}{3}$ and $\Delta x = \dfrac{\pi}{180}$, we get $\tan\left(\dfrac{\pi}{3} + \dfrac{\pi}{180}\right) \approx \tan\left(\dfrac{\pi}{3}\right) + \sec^2\left(\dfrac{\pi}{3}\right)\left(\dfrac{\pi}{180}\right)$. If we evaluate this, we get $\tan\left(\dfrac{61\pi}{180}\right) \approx \sqrt{3} + (2)^2\left(\dfrac{\pi}{180}\right) \approx 1.802$.

4. ± 2.16 in.^{3}

Recall the formula that we use when we want to approximate the error in a measurement: $dy = f'(x)dx$. Here we want to approximate the error in the volume of a cube when we know that it has a side of length 6 in. with an error of ± 0.02 in., where $V(x) = x^3$ (the volume of a cube of side x) with $dx = \pm 0.02$. We find the derivative of the volume: $V'(x) = 3x^2$. Now we can plug into the formula: $dV = 3x^2dx$. If we plug in $x = 6$ and $dx = \pm 0.02$, we get $dV = 3(6)^2(\pm 0.02) = \pm 2.16$.

5. $\pi \approx 3.142$ mm^3

Recall the formula that we use when we want to approximate the error in a measurement:

$dy = f'(x)dx$. Here we want to approximate the increase in the volume of a sphere when we know that

it has a radius of length 5 mm with an increase of 0.01 mm, where $V(r) = \dfrac{4}{3}\pi r^3$ (the volume of a

sphere of radius r) with $dr = 0.01$. We find the derivative of the volume: $V'(r) = \dfrac{4}{3}\pi(3r^2) = 4\pi r^2$.

Now we can plug into the formula: $dV = 4\pi r^2 dr$. If we plug in $r = 5$ and $dr = 0.01$, we get

$dV = 4\pi(5)^2(0.01) = \pi \approx 3.142$.

6.   (a)   1.963 m^3;   (b)   15.708 m^3

(a) Recall the formula that we use when we want to approximate the error in a measurement: $dy = f'(x)dx$. Here we want to approximate the error in the volume of a cylinder when we know that it has a diameter of length 5 meters (which means that its radius is 2.5 meters) and its height is 20 meters, with an error in the height of 0.1 meters, where $V = \pi r^2 h$ with $dh = 0.1$. Note that the radius is exact, so when we take the derivative, we will treat only the height as a variable. We find the derivative of the volume: $V' = \pi r^2$. Now we can plug into the formula: $dV = \pi r^2\, dh$. If we plug in $r = 2.5$, $h = 20$, and $dh = 0.1$, we get $dV = \pi(2.5)^2(0.1) = 0.625\pi \approx 1.963$.

(b) Here we want to approximate the error in the volume of a cylinder when we know that it has a diameter of length 5 meters (which means that its radius is 2.5 meters) and its height is 20 meters, with an error in the diameter of 0.1 meters (which means that the error in the radius is 0.05 meters), where $V = \pi r^2 h$ with $dr = 0.05$. Note that the height is exact, so when we take the derivative, we will treat only the radius as a variable. We find the derivative of the volume: $V' = 2\pi rh$. Now we can plug into the formula: $dV = 2\pi rh\, dr$. If we plug in $r = 2.5$, $h = 20$, and $dr = 0.05$, we get $dV = 2\pi(2.5)(20)(0.05) = 5\pi \approx 15.708$.

SOLUTIONS TO PRACTICE PROBLEM SET 17

1. $\dfrac{3}{4}$

Recall L'Hôpital's Rule: If $f(c) = g(c) = 0$, or if $f(c) = g(c) = \infty$, and if $f'(c)$ and $g'(c)$ exist, and if $g'(c) \ne 0$, then $\lim\limits_{x \to c} \dfrac{f(x)}{g(x)} = \lim\limits_{x \to c} \dfrac{f'(x)}{g'(x)}$. Here $f(x) = \sin 3x$, and $g(x) = \sin 4x$, and $\sin 0 = 0$. This means that we can use L'Hôpital's Rule to find the limit. We take the derivative of the numerator and the denominator: $\lim\limits_{x \to 0} \dfrac{\sin 3x}{\sin 4x} = \lim\limits_{x \to 0} \dfrac{3 \cos 3x}{4 \cos 4x}$. If we take the new limit, we get $\lim\limits_{x \to 0} \dfrac{3 \cos 3x}{4 \cos 4x} = \dfrac{3 \cos 0}{4 \cos 0} = \dfrac{(3)(1)}{(4)(1)} = \dfrac{3}{4}$.

2. -1

Recall L'Hôpital's Rule: If $f(c) = g(c) = 0$, or if $f(c) = g(c) = \infty$, and if $f'(c)$ and $g'(c)$ exist, and if $g'(c) \ne 0$, then $\lim\limits_{x \to c} \dfrac{f(x)}{g(x)} = \lim\limits_{x \to c} \dfrac{f'(x)}{g'(x)}$. Here $f(x) = x - \pi$ and $g(x) = \sin x$. We can see that $x - \pi = 0$ when $x = \pi$, and that $\sin \pi = 0$. This means that we can use L'Hôpital's Rule to find the limit. We take the derivative of the numerator and the denominator: $\lim\limits_{x \to \pi} \dfrac{x - \pi}{\sin x} = \lim\limits_{x \to \pi} \dfrac{1}{\cos x}$. If we take the new limit, we get $\lim\limits_{x \to \pi} \dfrac{1}{\cos x} = \dfrac{1}{\cos \pi} = \dfrac{1}{-1} = -1$.

3. $\dfrac{1}{6}$

Recall L'Hôpital's Rule: If $f(c) = g(c) = 0$, or if $f(c) = g(c) = \infty$, and if $f'(c)$ and $g'(c)$ exist, and if $g'(c) \ne 0$, then $\lim\limits_{x \to c} \dfrac{f(x)}{g(x)} = \lim\limits_{x \to c} \dfrac{f'(x)}{g'(x)}$. Here $f(x) = x - \sin x$ and $g(x) = x^3$. We can see that $x - \sin x = 0$ when $x = 0$, and that $0^3 = 0$. This means that we can use L'Hôpital's Rule to find the limit. We take the derivative of the numerator and the denominator: $\lim\limits_{x \to 0} \dfrac{x - \sin x}{x^3} = \lim\limits_{x \to 0} \dfrac{1 - \cos x}{3x^2}$.

If we take the new limit, we get $\lim\limits_{x \to 0} \dfrac{1 - \cos x}{3x^2} = \dfrac{1 - \cos 0}{3(0)^2} = \dfrac{1 - 1}{0} = \dfrac{0}{0}$. But this is still indeterminate, so what do we do? Use L'Hôpital's Rule again! We take the derivative of the numerator and

the denominator: $\lim\limits_{x \to 0} \dfrac{1 - \cos x}{3x^2} = \lim\limits_{x \to 0} \dfrac{\sin x}{6x}$. If we take the limit, we get $\lim\limits_{x \to 0} \dfrac{\sin x}{6x} = \dfrac{\sin 0}{6(0)} = \dfrac{0}{0}$.

We need to use L'Hôpital's Rule one more time. We get $\lim\limits_{x \to 0} \dfrac{\sin x}{6x} = \lim\limits_{x \to 0} \dfrac{\cos x}{6}$. Now, if we take

the limit, we get $\lim\limits_{x \to 0} \dfrac{\cos x}{6} = \dfrac{\cos 0}{6} = \dfrac{1}{6}$. Notice that if L'Hôpital's Rule results in an indeterminate

form, we can use the rule again and again (but not infinitely often).

4. -2

Recall L'Hôpital's Rule: If $f(c) = g(c) = 0$, or if $f(c) = g(c) = \infty$, and if $f'(c)$ and $g'(c)$

exist, and if $g'(c) \neq 0$, then $\lim\limits_{x \to c} \dfrac{f(x)}{g(x)} = \lim\limits_{x \to c} \dfrac{f'(x)}{g'(x)}$. Here $f(x) = e^{3x} - e^{5x}$ and $g(x) = x$. We

can see that $e^{3x} - e^{5x} = 0$ when $x = 0$, and that the denominator is obviously zero at $x = 0$.

This means that we can use L'Hôpital's Rule to find the limit. We take the derivative of the

numerator and the denominator: $\lim\limits_{x \to 0} \dfrac{e^{3x} - e^{5x}}{x} = \lim\limits_{x \to 0} \dfrac{3e^{3x} - 5e^{5x}}{1}$. If we take the new limit, we get

$\lim\limits_{x \to 0} \dfrac{3e^{3x} - 5e^{5x}}{1} = \dfrac{3e^0 - 5e^0}{1} = \dfrac{3(1) - 5(1)}{1} = -2$.

5. -2

Recall L'Hôpital's Rule: If $f(c) = g(c) = 0$, or if $f(c) = g(c) = \infty$, and if $f'(c)$ and $g'(c)$

exist, and if $g'(c) \neq 0$, then $\lim\limits_{x \to c} \dfrac{f(x)}{g(x)} = \lim\limits_{x \to c} \dfrac{f'(x)}{g'(x)}$. Here $f(x) = \tan x - x$ and $g(x) = \sin x - x$.

We can see that $\tan x - x = 0$ and $\sin x - x = 0$ when $x = 0$. This means that we can use

L'Hôpital's Rule to find the limit. We take the derivative of the numerator and the denominator:

$\lim\limits_{x \to 0} \dfrac{\tan x - x}{\sin x - x} = \lim\limits_{x \to 0} \dfrac{\sec^2 x - 1}{\cos x - 1}$. If we take the new limit, we get $\lim\limits_{x \to 0} \dfrac{\sec^2 x - 1}{\cos x - 1} = \dfrac{\sec^2(0) - 1}{\cos(0) - 1} = \dfrac{1 - 1}{1 - 1} = \dfrac{0}{0}$.

But this is still indeterminate, so what do we do? Use L'Hôpital's Rule again! We take the derivative

of the numerator and the denominator: $\lim\limits_{x \to 0} \dfrac{\sec^2 x - 1}{\cos x - 1} = \lim\limits_{x \to 0} \dfrac{2\sec x (\sec x \tan x)}{-\sin x} = \lim\limits_{x \to 0} \dfrac{2\sec^2 x \tan x}{-\sin x}$.

If we take the limit, we get $\lim\limits_{x \to 0} \dfrac{2\sec^2 (0) \tan (0)}{-\sin (0)} = \dfrac{0}{0}$. At this point you might be getting nervous

as the derivatives start to get messier. Let's try using trigonometric identities to simplify the limit.

If we rewrite the numerator in terms of sin x and cos x, we get $\dfrac{2\sec^2 x \tan x}{-\sin x} = \dfrac{2\left(\dfrac{1}{\cos^2 x}\right)\left(\dfrac{\sin x}{\cos x}\right)}{-\sin x}$.

We can simplify this to $\dfrac{2\left(\dfrac{1}{\cos^2 x}\right)\left(\dfrac{\sin x}{\cos x}\right)}{-\sin x} = -2\dfrac{\dfrac{\sin x}{\cos^3 x}}{\sin x}$. This simplifies to $-2\dfrac{\sin x}{\cos^3 x}\dfrac{1}{\sin x} = \dfrac{-2}{\cos^3 x}$.

Now, if we take the limit, we get $\lim\limits_{x \to 0} \dfrac{-2}{\cos^3 x} = \dfrac{-2}{\cos^3 (0)} = \dfrac{-2}{1^3} = -2$.

Note that we could have used trigonometric identities on either of the first two limits as well. Remember that when you have a limit that is an indeterminate form, you can sometimes use algebra or trigonometric identities (or both) to simplify the limit. Sometimes this will get rid of the problem.

6.   0

Recall L'Hôpital's Rule: If $f(c) = g(c) = 0$, or if $f(c) = g(c) = \infty$, and if $f'(c)$ and $g'(c)$

exist, and if $g'(c) \neq 0$, then $\lim\limits_{x \to c} \dfrac{f(x)}{g(x)} = \lim\limits_{x \to c} \dfrac{f'(x)}{g'(x)}$. Here $f(x) = x^5$ and $g(x) = e^{5x}$. We

can see that $x^5 = \infty$ when $x = \infty$, and that $e^\infty = \infty$. This means that we can use L'Hôpital's

Rule to find the limit. We take the derivative of the numerator and the denominator:

$\lim\limits_{x \to \infty} \dfrac{x^5}{e^{5x}} = \lim\limits_{x \to \infty} \dfrac{5x^4}{5e^{5x}} = \lim\limits_{x \to \infty} \dfrac{x^4}{e^{5x}}$. If we take the new limit, we get $\lim\limits_{x \to 0} \dfrac{x^4}{e^{5x}} = \dfrac{\infty}{\infty}$. But this is still

indeterminate, so what do we do? Use L'Hôpital's Rule again! We take the derivative of the numerator and the denominator: $\lim_{x \to \infty} \dfrac{x^4}{e^{5x}} = \lim_{x \to \infty} \dfrac{4x^3}{5e^{5x}}$. We are going to need to use L'Hôpital's

Rule again. In fact, we can see that each time we use the rule we are reducing the power of the

x in the numerator and that we are going to need to keep doing so until the x term is gone. Let's

take the derivative: $\lim_{x \to \infty} \dfrac{4x^3}{5e^{5x}} = \lim_{x \to \infty} \dfrac{12x^2}{25e^{5x}}$. And again: $\lim_{x \to \infty} \dfrac{12x^2}{25e^{5x}} = \lim_{x \to \infty} \dfrac{24x}{125e^{5x}}$. And one more

time: $\lim_{x \to \infty} \dfrac{24x}{125e^{5x}} = \lim_{x \to \infty} \dfrac{24}{625e^{5x}}$. Now, if we take the limit, we get $\lim_{x \to \infty} \dfrac{24}{625e^{5x}} = 0$. Notice that

if L'Hôpital's Rule results in an indeterminate form, we can use the rule again and again (but not

infinitely often).

7. $\dfrac{1}{7}$

Recall L'Hôpital's Rule: If $f(c) = g(c) = 0$, or if $f(c) = g(c) = \infty$, and if $f'(c)$ and $g'(c)$

exist, and if $g'(c) \neq 0$, then $\lim_{x \to c} \dfrac{f(x)}{g(x)} = \lim_{x \to c} \dfrac{f'(x)}{g'(x)}$. Here $f(x) = x^5 + 4x^3 - 8$ and

$g(x) = 7x^5 - 3x^2 - 1$. We can see that $x^5 + 4x^3 - 8 = \infty$ and $7x^5 - 3x^2 - 1 = \infty$ when $x = \infty$. This

means that we can use L'Hôpital's Rule to find the limit. We take the derivative of the numera-

tor and the denominator: $\lim_{x \to \infty} \dfrac{x^5 + 4x^3 - 8}{7x^5 - 3x^2 - 1} = \lim_{x \to \infty} \dfrac{5x^4 + 12x^2}{35x^4 - 6x}$. If we take the new limit, we get

$\lim_{x \to \infty} \dfrac{5x^4 + 12x^2}{35x^4 - 6x} = \dfrac{\infty}{\infty}$. But this is still indeterminate, so what do we do? Use L'Hôpital's Rule again!

We take the derivative of the numerator and the denominator: $\lim_{x \to \infty} \dfrac{5x^4 + 12x^2}{35x^4 - 6x} = \lim_{x \to \infty} \dfrac{20x^3 + 24x}{140x^3 - 6}$.

We are going to need to use L'Hôpital's Rule again. In fact, we can see that each time we use

the rule we are reducing the power of the x terms in the numerator and denominator and that

we are going to need to keep doing so until the x terms are gone. Let's take the derivative:

$$\lim_{x\to\infty}\frac{20x^3+24x}{140x^3-6}=\lim_{x\to\infty}\frac{60x^2+24}{420x^2}$$. And again, $\lim_{x\to\infty}\dfrac{60x^2+24}{420x^2}=\lim_{x\to\infty}\dfrac{120x}{840x}$. Now we can take the

limit: $\lim_{x\to\infty}\dfrac{120x}{840x}=\lim_{x\to\infty}\dfrac{1}{7}=\dfrac{1}{7}$. Notice that if L'Hôpital's Rule results in an indeterminate form, we

can use the rule again and again (but not infinitely often).

8.　　　1

Recall L'Hôpital's Rule: If $f(c)=g(c)=0$, or if $f(c)=g(c)=\infty$, and if $f'(c)$ and $g'(c)$

exist, and if $g'(c)\ne0$, then $\lim_{x\to c}\dfrac{f(x)}{g(x)}=\lim_{x\to c}\dfrac{f'(x)}{g'(x)}$. Here $f(x)=\ln(\sin x)$ and $g(x)=\ln(\tan x)$,

and both of these approach infinity as x approaches 0 from the right. This means that we

can use L'Hôpital's Rule to find the limit. We take the derivative of the numerator and the

denominator: $\lim_{x\to0^+}\dfrac{\ln(\sin x)}{\ln(\tan x)}=\lim_{x\to0^+}\dfrac{\dfrac{\cos x}{\sin x}}{\dfrac{\sec^2 x}{\tan x}}$. Let's use trigonometric identities to simplify the

limit: $\dfrac{\dfrac{\cos x}{\sin x}}{\dfrac{\sec^2 x}{\tan x}}=\dfrac{\cos x}{\sin x}\dfrac{\tan x}{\sec^2 x}=\dfrac{\cos x}{\sin x}\dfrac{\dfrac{\sin x}{\cos x}}{\sec^2 x}=\dfrac{1}{\sec^2 x}$. Now, if we take the new limit, we get

$$\lim_{x\to0^+}\frac{1}{\sec^2 x}=1.$$

9. $\dfrac{1}{2}$

Recall L'Hôpital's Rule: If $f(c)=g(c)=0$, or if $f(c)=g(c)=\infty$, and if $f'(c)$ and

$g'(c)$ exist, and if $g'(c)\neq 0$, then $\lim\limits_{x\to c}\dfrac{f(x)}{g(x)}=\lim\limits_{x\to c}\dfrac{f'(x)}{g'(x)}$. Here $f(x)$ = cot $2x$ and

$g(x)$ = cot x, and both of these approach infinity as x approaches 0 from the right. This

means that we can use L'Hôpital's Rule to find the limit. We take the derivative of the

numerator and the denominator: $\lim\limits_{x\to 0^+}\dfrac{\cot 2x}{\cot x}=\lim\limits_{x\to 0^+}\dfrac{-2\cot 2x\csc 2x}{-\cot x\csc x}$. This seems to be

worse than what we started with. Instead, let's use trigonometric identities to simplify the

limit: $\dfrac{\cot 2x}{\cot x}=\dfrac{\dfrac{\cos 2x}{\sin 2x}}{\dfrac{\cos x}{\sin x}}=\dfrac{\cos 2x}{\sin 2x}\dfrac{\sin x}{\cos x}=\dfrac{\cos^2 x-\sin^2 x}{2\sin x\cos x}\dfrac{\sin x}{\cos x}$. Notice that our problem is

with sin x as x approaches zero (because it becomes zero), not with cos x. As long as we are

multiplying the numerator and denominator by sin x, we are going to get an indetermi-

nate form. So, thanks to trigonometric identities, we can eliminate the problem term:

$\dfrac{\cos^2 x-\sin^2 x}{2\sin x\cos x}\dfrac{\sin x}{\cos x}=\dfrac{\cos^2 x-\sin^2 x}{2\cos^2 x}$. If we take the limit of this expression, it is not indetermi-

nate. We get $\lim\limits_{x\to 0^+}\dfrac{\cos^2 x-\sin^2 x}{2\cos^2 x}=\dfrac{\cos^2 0-\sin^2 0}{2\cos^2 0}=\dfrac{1^2-0}{2(1^2)}=\dfrac{1}{2}$. Notice that we didn't need to use

L'Hôpital's Rule here. You should bear in mind that just because a limit is indeterminate does not

mean that the best way to evaluate it is with L'Hôpital's Rule.

10.   1

Recall L'Hôpital's Rule: if $f(c) = g(c) = 0$, or if $f(c) = g(c) = \infty$, and if $f'(c)$ and $g'(c)$ exist, and if $g'(c) \neq 0$, then $\lim\limits_{x \to c} \dfrac{f(x)}{g(x)} = \lim\limits_{x \to c} \dfrac{f'(x)}{g'(x)}$. Here $f(x) = x$ and $g(x) = \ln(x+1)$,

and both of these approach zero as x approaches 0 from the right. This means that we can use

L'Hôpital's Rule to find the limit. We take the derivative of the numerator and the denominator:

$\lim\limits_{x \to 0^+} \dfrac{1}{\dfrac{1}{x+1}} = \lim\limits_{x \to 0^+} (x+1)$. Now, if we take the new limit, we get $\lim\limits_{x \to 0^+} (x+1) = 1$.

SOLUTIONS TO PRACTICE PROBLEM SET 18

1. $-\dfrac{1}{3x^3} + C$

Here we will use the Power Rule, which says that $\int x^n dx = \dfrac{x^{n+1}}{n+1} + C$. The integral is
$\int \dfrac{1}{x^4} dx = \int x^{-4} dx = \dfrac{x^{-3}}{-3} + C = -\dfrac{1}{3x^3} + C$.

2. $10\sqrt{x} + C$

Here we will use the Power Rule, which says that $\int x^n dx = \dfrac{x^{n+1}}{n+1} + C$. The integral is
$\int \dfrac{5}{\sqrt{x}} dx = 5\int x^{-\frac{1}{2}} dx = 5\dfrac{x^{\frac{1}{2}}}{\frac{1}{2}} + C = 10\sqrt{x} + C$.

3. $x^5 - x^3 + x^2 + 6x + C$

Here we will use the Power Rule, which says that $\int x^n dx = \dfrac{x^{n+1}}{n+1} + C$. The integral is
$\int \left(5x^4 - 3x^2 + 2x + 6\right) dx = 5\dfrac{x^5}{5} - 3\dfrac{x^3}{3} + 2\dfrac{x^2}{2} + 6x + C = x^5 - x^3 + x^2 + 6x + C$.

4.
$$-\frac{3}{2x^2}+\frac{2}{x}+\frac{x^5}{5}+2x^8+C$$

Here we will use the Power Rule, which says that $\int x^n dx = \frac{x^{n+1}}{n+1}+C.$ The integral is

$$\int\left(3x^{-3}-2x^{-2}+x^4+16x^7\right)dx=3\frac{x^{-2}}{-2}-2\frac{x^{-1}}{-1}+\frac{x^5}{5}+16\frac{x^8}{8}+C=-\frac{3}{2x^2}+\frac{2}{x}+\frac{x^5}{5}+2x^8+C.$$

5.
$$\frac{3x^{\frac{4}{3}}}{2}+\frac{3x^{\frac{7}{3}}}{7}+C$$

Here we will use the Power Rule, which says that $\int x^n dx = \frac{x^{n+1}}{n+1}+C.$ First, let's simplify the integrand: $\int x^{\frac{1}{3}}(2+x)\,dx=\int\left(2x^{\frac{1}{3}}+x^{\frac{4}{3}}\right)dx.$ Now, we can evaluate the integral:

$$\int\left(2x^{\frac{1}{3}}+x^{\frac{4}{3}}\right)dx=2\frac{x^{\frac{4}{3}}}{\frac{4}{3}}+\frac{x^{\frac{7}{3}}}{\frac{7}{3}}+C=\frac{3x^{\frac{4}{3}}}{2}+\frac{3x^{\frac{7}{3}}}{7}+C.$$

6.
$$\frac{x^7}{7}+\frac{2x^5}{5}+\frac{x^3}{3}+C$$

Here we will use the Power Rule, which says that $\int x^n dx = \frac{x^{n+1}}{n+1}+C.$ First, let's simplify the integrand: $\int(x^3+x)^2\,dx=\int(x^6+2x^4+x^2)dx.$ Now, we can evaluate the integral:

$$\int\left(x^6+2x^4+x^2\right)dx=\frac{x^7}{7}+\frac{2x^5}{5}+\frac{x^3}{3}+C.$$

7.
$$\frac{x^5}{5}-\frac{2x^3}{3}-\frac{1}{x}+C$$

Here we will use the Power Rule, which says that $\int x^n dx = \frac{x^{n+1}}{n+1}+C.$

First, let's simplify the integrand: $\int\frac{x^6-2x^4+1}{x^2}dx=\int\left(\frac{x^6}{x^2}-2\frac{x^4}{x^2}+\frac{1}{x^2}\right)dx=\int\left(x^4-2x^2+x^{-2}\right)dx.$

Now, we can evaluate the integral: $\int\left(x^4-2x^2+x^{-2}\right)dx=\frac{x^5}{5}-\frac{2x^3}{3}+\frac{x^{-1}}{-1}+C=\frac{x^5}{5}-\frac{2x^3}{3}-\frac{1}{x}+C.$

8. $\dfrac{x^5}{5} - \dfrac{3x^4}{4} + x^3 - \dfrac{x^2}{2} + C$

Here we will use the Power Rule, which says that $\displaystyle\int x^n\,dx = \dfrac{x^{n+1}}{n+1} + C$.

First, let's simplify the integrand: $\displaystyle\int x(x-1)^3\,dx = \int x(x^3 - 3x^2 + 3x - 1)\,dx = \int (x^4 - 3x^3 + 3x^2 - x)\,dx$.

Now, we can evaluate the integral:

$$\int\left(x^4 - 3x^3 + 3x^2 - x\right)dx = \dfrac{x^5}{5} - \dfrac{3x^4}{4} + \dfrac{3x^3}{3} - \dfrac{x^2}{2} + C = \dfrac{x^5}{5} - \dfrac{3x^4}{4} + x^3 - \dfrac{x^2}{2} + C.$$

9. $\tan x + \sec x + C$

Here we will use the Rules for the Integrals of Trig Functions, namely: $\displaystyle\int \sec^2 x\,dx = \tan x + C$ and $\displaystyle\int (\sec x \tan x)\,dx = \sec x + C$. First, let's expand the integrand: $\displaystyle\int \sec x\,(\sec x + \tan x)\,dx = \int (\sec^2 x + \sec x \tan x)\,dx$. We get $\displaystyle\int (\sec^2 x + \sec x \tan x)\,dx = \tan x + \sec x + C$.

10. $\tan x + \dfrac{x^2}{2} + C$

Here we will use the Rules for the Integrals of Trig Functions, namely: $\displaystyle\int \sec^2 x\,dx = \tan x + C$ and the Power Rule, which says that $\displaystyle\int x^n\,dx = \dfrac{x^{n+1}}{n+1} + C$. We get $\displaystyle\int\left(\sec^2 x + x\right)dx = \tan x + \dfrac{x^2}{2} + C$.

11. $\sin x + 4\tan x + C$

Here we will use the Rules for the Integrals of Trig Functions, namely: $\displaystyle\int \sec^2 x\,dx = \tan x + C$ and $\displaystyle\int \cos x\,dx = \sin x + C$.

First, we need to rewrite the integrand, using trig identities:

$\displaystyle\int \dfrac{\cos^3 x + 4}{\cos^2 x}\,dx = \int\left(\dfrac{\cos^3 x}{\cos^2 x} + \dfrac{4}{\cos^2 x}\right)dx = \int\left(\cos x + 4\sec^2 x\right)dx$. Now, we can evaluate the integral:

$\displaystyle\int (\cos x + 4\sec^2 x)\,dx = \sin x + 4\tan x + C$.

12. $-2\cos x + C$

Here we will use the Rules for the Integrals of Trig Functions, namely: $\displaystyle\int \sin x\,dx = -\cos x + C$. First, we need to rewrite the integrand, using trig identities: $\displaystyle\int \dfrac{\sin 2x}{\cos x}\,dx = \int \dfrac{2\sin x \cos x}{\cos x}\,dx = \int 2\sin x\,dx$.

Now, we can evaluate the integral: $\displaystyle\int 2\sin x\,dx = -2\cos x + C$.

13. $x + \sin x + C$

Here we will use the Rules for the Integrals of Trig Functions, namely: $\int \cos x\ dx = \sin x + C$. First, we need to rewrite the integrand, using trig identities: $\int \left(1 + \cos^2 x \sec x\right)dx = \int \left(1 + \dfrac{\cos^2 x}{\cos x}\right)dx = \int (1 + \cos x)dx$. Now, we can evaluate the integral: $\int (1 + \cos x)dx = x + \sin x + C$.

14. $-\cos x + C$

Here we will use the Rules for the Integrals of Trig Functions, namely: $\int \sin x\ dx = -\cos x + C$. First, we need to rewrite the integrand, using trig identities: $\int \dfrac{1}{\csc x}dx = \int \sin x\ dx$. Now we can evaluate the integral: $\int \sin x\ dx = -\cos x + C$.

15. $\dfrac{x^2}{2} - 2\tan x + C$

Here we will use the Rules for the Integrals of Trig Functions, namely: $\int \sec^2 x\ dx = \tan x + C$. First, we need to rewrite the integrand, using trig identities: $\int \left(x - \dfrac{2}{\cos^2 x}\right)dx = \int \left(x - 2\sec^2 x\right)dx$. Now, we can evaluate the integral: $\int \left(x - 2\sec^2 x\right)dx = \dfrac{x^2}{2} - 2\tan x + C$.

SOLUTIONS TO PRACTICE PROBLEM SET 19

1. $\dfrac{\sin^2 2x}{4} + C$

If we let $u = \sin 2x$, then $du = 2\cos 2x\ dx$. First, we need to substitute for $\cos 2x\ dx$, so we can divide the du term by 2: $\dfrac{du}{2} = \cos 2x\ dx$. Next, we can substitute into the integral: $\int \sin 2x \cos 2x\ dx = \dfrac{1}{2}\int u\ du$. Now we can integrate: $\dfrac{1}{2}\int u\ du = \dfrac{1}{2}\left(\dfrac{u^2}{2}\right) + C = \dfrac{u^2}{4} + C$.

Last, we substitute back and get $\dfrac{\sin^2 2x}{4} + C$.

2. $-\dfrac{9}{4}\left(10-x^2\right)^{\frac{2}{3}}+C$

First, pull the constant out of the integrand: $\displaystyle\int\dfrac{3x\,dx}{\sqrt[3]{10-x^2}}\,dx=3\int\dfrac{x\,dx}{\sqrt[3]{10-x^2}}\,dx$. If we let $u=10-x^2$,

then $du=-2x\,dx$. We need to substitute for $x\,dx$, so we can divide the du term by -2: $-\dfrac{du}{2}=x\,dx$.

Next, we can substitute into the integral: $3\displaystyle\int\dfrac{x\,dx}{\sqrt[3]{10-x^2}}\,dx=-\dfrac{3}{2}\int u^{-\frac{1}{3}}\,du$. Now we can integrate:

$-\dfrac{3}{2}\displaystyle\int u^{-\frac{1}{3}}\,du=-\dfrac{3}{2}\dfrac{u^{\frac{2}{3}}}{\frac{2}{3}}+C=-\dfrac{9}{4}u^{\frac{2}{3}}+C$. Last, we substitute back and get $-\dfrac{9}{4}\left(10-x^2\right)^{\frac{2}{3}}+C$.

3. $\dfrac{1}{30}\left(5x^4+20\right)^{\frac{3}{2}}+C$

If we let $u=5x^4+20$, then $du=20x^3\,dx$. We need to substitute for $x^3\,dx$, so we can divide the du

term by 20: $\dfrac{du}{20}=x^3\,dx$. Next, we can substitute into the integral: $\displaystyle\int x^3\sqrt{5x^4+20}\,dx=\dfrac{1}{20}\int u^{\frac{1}{2}}\,du$.

Now we can integrate: $\dfrac{1}{20}\displaystyle\int u^{\frac{1}{2}}\,du=\dfrac{1}{20}\dfrac{u^{\frac{3}{2}}}{\frac{3}{2}}+C=\dfrac{1}{30}u^{\frac{3}{2}}+C$. Last, we substitute back and get

$\dfrac{1}{30}\left(5x^4+20\right)^{\frac{3}{2}}+C$.

4. $-\dfrac{1}{12\left(x^3+3x\right)^4}+C$

If we let $u=x^3+3x$, then $du=(3x^2+3)\,dx$. We need to substitute for $(x^2+1)\,dx$, so we can

divide the du term by 3: $\dfrac{du}{3}=\left(x^2+1\right)dx$. Next, we can substitute into the integral:

$\displaystyle\int\left(x^2+1\right)\left(x^3+3x\right)^{-5}dx=\dfrac{1}{3}\int u^{-5}\,du$. Now we can integrate: $\dfrac{1}{3}\displaystyle\int u^{-5}\,du=\dfrac{1}{3}\dfrac{u^{-4}}{-4}+C=-\dfrac{1}{12}\dfrac{1}{u^4}+C$.

Last, we substitute back and get $-\dfrac{1}{12\left(x^3+3x\right)^4}+C$.

5. $-2\cos\sqrt{x}+C$

If we let $u=\sqrt{x}$, then $du=\dfrac{1}{2\sqrt{x}}dx$. We need to substitute for $\dfrac{1}{\sqrt{x}}dx$, so we can multiply the

du term by 2: $2du=\dfrac{1}{\sqrt{x}}dx$. Next, we can substitute into the integral: $\displaystyle\int\dfrac{1}{\sqrt{x}}\sin\sqrt{x}\,dx=2\int\sin u\,du$.

Now we can integrate: $2\displaystyle\int\sin u\,du=-2\cos u+C$. Last, we substitute back and get $-2\cos\sqrt{x}+C$.

6. $\dfrac{1}{3}\tan\left(x^3\right)+C$

If we let $u=x^3$, then $du=3x^2dx$. We need to substitute for x^2dx, so we can divide the du term by 3:

$\dfrac{du}{3}=x^2dx$. Next, we can substitute into the integral: $\displaystyle\int x^2\sec^2\left(x^3\right)dx=\dfrac{1}{3}\int\sec^2 u\,du$. Now we can

integrate: $\dfrac{1}{3}\displaystyle\int\sec^2 u\,du=\dfrac{1}{3}\tan u+C$. Last, we substitute back and get $\dfrac{1}{3}\tan\left(x^3\right)+C$.

7. $-\dfrac{1}{3}\sin\left(\dfrac{3}{x}\right)+C$

If we let $u=\dfrac{3}{x}$, then $du=-\dfrac{3}{x^2}dx$. We need to substitute for $\dfrac{1}{x^2}dx$, so we can divide the du term

by -3: $-\dfrac{du}{3}=\dfrac{1}{x^2}dx$. Next, we can substitute into the integral: $\displaystyle\int\dfrac{\cos\left(\dfrac{3}{x}\right)}{x^2}dx=-\dfrac{1}{3}\int\cos u\,du$.

Now we can integrate: $-\dfrac{1}{3}\displaystyle\int\cos u\,du=-\dfrac{1}{3}\sin u+C$. Last, we substitute back and get

$-\dfrac{1}{3}\sin\left(\dfrac{3}{x}\right)+C$.

8. $-\cos(\sin x)+C$

If we let $u=\sin x$, then $du=\cos x\,dx$. Next, we can substitute into the integral:
$\displaystyle\int\sin(\sin x)\cos x\,dx=\int\sin u\,du$. Now we can integrate: $\displaystyle\int\sin u\,du=-\cos u+C$. Last, we substitute
back and get $-\cos(\sin x)+C$.

SOLUTIONS TO PRACTICE PROBLEM SET 20

1. $\dfrac{25}{32}$

First, let's draw a picture.

$f(x) = 2x - x^2$

The width of each rectangle is found by taking the difference between the endpoints and dividing by n. Here, the width of each rectangle is $\dfrac{2-1}{4} = \dfrac{1}{4}$.

We find the heights of the rectangles by evaluating $y = 2x - x^2$ at the appropriate endpoints.

$$y(1) = 2(1) - (1)^2 = 1 \; ; \quad y\left(\frac{5}{4}\right) = 2\left(\frac{5}{4}\right) - \left(\frac{5}{4}\right)^2 = \frac{15}{16}; \quad y\left(\frac{3}{2}\right) = 2\left(\frac{3}{2}\right) - \left(\frac{3}{2}\right)^2 = \frac{3}{4}$$

and $y\left(\dfrac{7}{4}\right) = 2\left(\dfrac{7}{4}\right) - \left(\dfrac{7}{4}\right)^2 = \dfrac{7}{16}$.

Therefore, the area is $\left(\dfrac{1}{4}\right)\left(1 + \dfrac{15}{16} + \dfrac{3}{4} + \dfrac{7}{16}\right) = \dfrac{25}{32}$.

2. $\dfrac{17}{32}$

First, let's draw a picture.

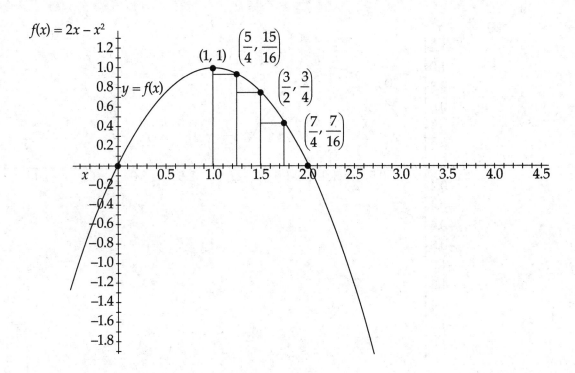

The width of each rectangle is found by taking the difference between the endpoints and dividing by n. Here, the width of each rectangle is $\dfrac{2-1}{4} = \dfrac{1}{4}$.

We find the heights of the rectangles by evaluating $y = 2x - x^2$ at the appropriate endpoints.

$$y\left(\frac{5}{4}\right) = 2\left(\frac{5}{4}\right) - \left(\frac{5}{4}\right)^2 = \frac{15}{16}; \quad y\left(\frac{3}{2}\right) = 2\left(\frac{3}{2}\right) - \left(\frac{3}{2}\right)^2 = \frac{3}{4}; \quad y\left(\frac{7}{4}\right) = 2\left(\frac{7}{4}\right) - \left(\frac{7}{4}\right)^2 = \frac{7}{16};$$

and $y(2) = 2(2) - (2)^2 = 0$.

Therefore, the area is $\left(\dfrac{1}{4}\right)\left(\dfrac{15}{16} + \dfrac{3}{4} + \dfrac{7}{16} + 0\right) = \dfrac{17}{32}$.

3. $\dfrac{21}{32}$

First, let's draw a picture.

The height of each trapezoid is found by taking the difference between the endpoints and dividing

by n. Here, the width of each trapezoid is $\dfrac{2-1}{4} = \dfrac{1}{4}$.

We find the bases of the trapezoids by evaluating $y = 2x - x^2$ at the appropriate endpoints.

$y(1) = 2(1) - (1)^2 = 1;$ $y\left(\dfrac{5}{4}\right) = 2\left(\dfrac{5}{4}\right) - \left(\dfrac{5}{4}\right)^2 = \dfrac{15}{16};$ $y\left(\dfrac{3}{2}\right) = 2\left(\dfrac{3}{2}\right) - \left(\dfrac{3}{2}\right)^2 = \dfrac{3}{4};$

$y\left(\dfrac{7}{4}\right) = 2\left(\dfrac{7}{4}\right) - \left(\dfrac{7}{4}\right)^2 = \dfrac{7}{16};$ and $y(2) = 2(2) - (2)^2 = 0.$

Therefore, the area is $\left(\dfrac{1}{2}\right)\left(\dfrac{1}{4}\right)\left[1 + (2)\left(\dfrac{15}{16}\right) + (2)\left(\dfrac{3}{4}\right) + (2)\left(\dfrac{7}{16}\right) + 0\right] = \dfrac{21}{32}.$

4. $\dfrac{43}{64}$

First, let's draw a picture.

The height of each rectangle is found by taking the difference between the endpoints and dividing by n. Here, the width of each rectangle is $\dfrac{2-1}{4} = \dfrac{1}{4}$.

We find the bases of the rectangles by evaluating $y = 2x - x^2$ at the appropriate endpoints.

$y\left(\dfrac{9}{8}\right) = 2\left(\dfrac{9}{8}\right) - \left(\dfrac{9}{8}\right)^2 = \dfrac{63}{64}$; $y\left(\dfrac{11}{8}\right) = 2\left(\dfrac{11}{8}\right) - \left(\dfrac{11}{8}\right)^2 = \dfrac{55}{64}$; $y\left(\dfrac{13}{8}\right) = 2\left(\dfrac{13}{8}\right) - \left(\dfrac{13}{8}\right)^2 = \dfrac{39}{64}$;

and $y\left(\dfrac{15}{8}\right) = 2\left(\dfrac{15}{8}\right) - \left(\dfrac{15}{8}\right)^2 = \dfrac{15}{64}$.

Therefore, the area is $\left(\dfrac{1}{4}\right)\left(\dfrac{63}{64} + \dfrac{55}{64} + \dfrac{39}{64} + \dfrac{15}{64}\right) = \dfrac{43}{64}$.

5. $\dfrac{2}{3}$

We will find the exact area by evaluating the integral $\int_1^2 \left(2x - x^2\right) dx$. According to the Fundamental Theorem of Calculus, $\int_1^2 \left(2x - x^2\right) dx = \left(\dfrac{2x^2}{2} - \dfrac{x^3}{3}\right)\Big|_1^2 = \left(x^2 - \dfrac{x^3}{3}\right)\Big|_1^2$. If we evaluate the integral at the limits, we get $\left[(2)^2 - \left(\dfrac{2^3}{3}\right)\right] - \left[(1)^2 - \left(\dfrac{1^3}{3}\right)\right] = \dfrac{2}{3}$.

6.   2

According to the Fundamental Theorem of Calculus, $\int_{-\frac{\pi}{2}}^{\frac{\pi}{2}} \cos x\, dx = \sin x\Big|_{-\frac{\pi}{2}}^{\frac{\pi}{2}} = \sin\dfrac{\pi}{2} - \sin\left(-\dfrac{\pi}{2}\right) = 2$.

7. $\dfrac{968}{5}$

First, let's rewrite the integrand: $\int_1^9 \left(2x\sqrt{x}\right) dx = \int_1^9 \left(2x^{\frac{3}{2}}\right) dx$. Now, according to the Fundamental

Theorem of Calculus, $\int_1^9 \left(2x^{\frac{3}{2}}\right) dx = \left(\dfrac{2x^{\frac{5}{2}}}{\frac{5}{2}}\right)\Big|_1^9 = \left(\dfrac{4x^{\frac{5}{2}}}{5}\right)\Big|_1^9 = \left(\dfrac{4(9)^{\frac{5}{2}}}{5} - \dfrac{4(1)^{\frac{5}{2}}}{5}\right) = \dfrac{968}{5}$.

8. $-\dfrac{161}{20}$

According to the Fundamental Theorem of Calculus,

$\int_0^1 \left(x^4 - 5x^3 + 3x^2 - 4x - 6\right) dx = \left(\dfrac{x^5}{5} - 5\dfrac{x^4}{4} + x^3 - 2x^2 - 6x\right)\Big|_0^1 =$

$\left(\dfrac{(1)^5}{5} - 5\dfrac{(1)^4}{4} + (1)^3 - 2(1)^2 - 6(1)\right) - 0 = -\dfrac{161}{20}$.

9.   0

According to the Fundamental Theorem of Calculus,

$$\int_{-\frac{\pi}{2}}^{\frac{\pi}{2}} \sin x \, dx = -\cos x \Big|_{-\frac{\pi}{2}}^{\frac{\pi}{2}} = -\cos\frac{\pi}{2} + \cos\left(-\frac{\pi}{2}\right) = 0.$$

10.   216

Recall that the formula for finding the area under the curve using the left end-points is $\left(\dfrac{b-a}{n}\right)[y_0 + y_1 + y_2 + \ldots + y_{n-1}]$. This formula assumes that the x-values are evenly spaced, but they aren't here, so we will replace the values of $\left(\dfrac{b-a}{n}\right)$ with the appropriate widths of each rectangle. The width of the first rectangle is $1 - 0 = 1$; the second width is $3 - 1 = 2$; the third is $5 - 3 = 2$; the fourth is $9 - 5 = 4$; and the fifth is $14 - 9 = 5$. We find the height of each rectangle by evaluating $g(x)$ at the appropriate value of x, the left endpoint of each interval on the x-axis. Here $y_0 = 10$, $y_1 = 8$, $y_2 = 11$, $y_3 = 17$, and $y_4 = 20$. Therefore, we can approximate the integral with:

$$\int_0^{14} g(x)\,dx = (1)(10) + (2)(8) + (2)(11) + (4)(17) + (5)(20) = 216.$$

SOLUTIONS TO PRACTICE PROBLEM SET 21

1. $2\sqrt{\dfrac{2}{\pi}}$

We find the average value of the function, $f(x)$, on the interval $[a, b]$ using the formula $f(c) = \dfrac{1}{b-a}\int_a^b f(x)\,dx$. Here we are looking for the average value of $f(x) = 4x\cos x^2$ on the interval $\left[0, \sqrt{\dfrac{\pi}{2}}\right]$. Using the formula, we need to find $\dfrac{1}{\sqrt{\dfrac{\pi}{2}} - 0}\int_0^{\sqrt{\frac{\pi}{2}}} \left(4x\cos x^2\right)\,dx = \sqrt{\dfrac{2}{\pi}}\int_0^{\sqrt{\frac{\pi}{2}}} \left(4x\cos x^2\right)\,dx$.

We will need to use *u*-substitution to evaluate the integral. Let $u = x^2$ and $du = 2x\, dx$. We need to

substitute for $4x\, dx$, so we multiply by 2 to get $2\, du = 4x\, dx$. Now we can substitute into the integral:

$\int (4x \cos x^2)dx = 2\int \cos u\, du = 2 \sin u$. If we substitute back, we get $2 \sin x^2$. Now we can evaluate

the integral: $\sqrt{\dfrac{2}{\pi}} \int_0^{\sqrt{\frac{\pi}{2}}} (4x \cos x^2)dx = \sqrt{\dfrac{2}{\pi}} \left(2 \sin x^2\right)\Big|_0^{\sqrt{\frac{\pi}{2}}} = \sqrt{\dfrac{2}{\pi}} \left(2 \sin \dfrac{\pi}{2} - 2 \sin 0\right) = 2\sqrt{\dfrac{2}{\pi}}$.

2. $\dfrac{8}{3}$

We find the average value of the function, $f(x)$, on the interval $[a, b]$ using the formula

$f(c) = \dfrac{1}{b-a} \int_a^b f(x)dx$. Here we are looking for the average value of $f(x) = \sqrt{x}$ on

the interval $[0, 16]$. Using the formula, we need to find $\dfrac{1}{16-0} \int_0^{16} \left(\sqrt{x}\right)dx$. We get

$\dfrac{1}{16} \int_0^{16} \left(\sqrt{x}\right)dx = \dfrac{1}{16}\left(\dfrac{x^{\frac{3}{2}}}{\frac{3}{2}}\right)\Bigg|_0^{16} = \dfrac{1}{24}\left(x^{\frac{3}{2}}\right)\Big|_0^{16} = \dfrac{1}{24}\left((16)^{\frac{3}{2}} - 0\right) = \dfrac{8}{3}$.

3.   1

We find the average value of the function, $f(x)$, on the interval $[a, b]$ using the formula

$f(c) = \dfrac{1}{b-a} \int_a^b f(x)dx$. Here we are looking for the average value of $f(x) = 2|x|$ on the interval

$[-1, 1]$. Using the formula, we need to find $\dfrac{1}{1-(-1)} \int_{-1}^1 (2|x|)dx$. Recall that the absolute value

function must be rewritten as a piecewise function: $|x| = \begin{cases} x; x \geq 0 \\ -x; x < 0 \end{cases}$. Thus, we need to split the

integral into two separate integrals in order to evaluate it: $\dfrac{1}{1-(-1)} \int_{-1}^1 (2|x|)dx = \int_{-1}^0 (-x)dx + \int_0^1 x\, dx$.

We get $\int_{-1}^0 (-x)dx + \int_0^1 x\, dx = \left(-\dfrac{x^2}{2}\right)\Big|_{-1}^0 + \left(\dfrac{x^2}{2}\right)\Big|_{-1}^0 = \left(0 - \left(-\dfrac{1}{2}\right)\right) + \left(\dfrac{1}{2} - 0\right) = 1$.

4. $\sin^2 x$

We find the derivative of an integral using the Second Fundamental Theorem of Calculus: $\dfrac{d}{dx}\displaystyle\int_a^x f(t)\,dt = f(x)$. We get $\dfrac{d}{dx}\displaystyle\int_1^x \sin^2 t\,dt = \sin^2 x$.

5. $27x^2 - 9x$

We find the derivative of an integral using the Second Fundamental Theorem of Calculus: $\dfrac{d}{dx}\displaystyle\int_a^x f(t)\,dt = f(x)$. We get $\dfrac{d}{dx}\displaystyle\int_1^{3x}\left(t^2 - t\right)dt = 3\left[(3x)^2 - (3x)\right] = 27x^2 - 9x$. Don't forget that because the upper limit is a function, we need to multiply the answer by the derivative of that function.

6. $-2\cos x$

We find the derivative of an integral using the Second Fundamental Theorem of Calculus: $\dfrac{d}{dx}\displaystyle\int_a^x f(t)\,dt = f(x)$. We get $\dfrac{d}{dx}\displaystyle\int_1^x -2\cos t\,dt = -2\cos x$.

SOLUTIONS TO PRACTICE PROBLEM SET 22

1. $\dfrac{1}{16+x^2}$

We find the derivative of the inverse tangent using the formula $\dfrac{d}{dx}\left(\tan^{-1} u\right) = \dfrac{1}{1+u^2}\dfrac{du}{dx}$. Here we have $u = \dfrac{x}{4}$, so $\dfrac{du}{dx} = \dfrac{1}{4}$. Therefore, $\dfrac{d}{dx}\left(\dfrac{1}{4}\tan^{-1}\dfrac{x}{4}\right) = \dfrac{1}{4}\dfrac{1}{1+\left(\dfrac{x}{4}\right)^2}\left(\dfrac{1}{4}\right) = \dfrac{1}{16}\dfrac{1}{1+\dfrac{x^2}{16}} = \dfrac{1}{16+x^2}$.

2. $$\frac{-1}{|x|\sqrt{x^2-1}}$$

We find the derivative of the inverse sine using the formula $\frac{d}{dx}\left(\sin^{-1}u\right)=\frac{1}{\sqrt{1-u^2}}\frac{du}{dx}$. Here we have $u=\frac{1}{x}$, so $\frac{du}{dx}=-\frac{1}{x^2}$. Therefore,

$$\frac{d}{dx}\sin^{-1}\left(\frac{1}{x}\right)=\frac{1}{\sqrt{1-\left(\frac{1}{x}\right)^2}}\left(-\frac{1}{x^2}\right)=\frac{-1}{\sqrt{1-\frac{1}{x^2}}}\left(\frac{1}{x^2}\right)=\frac{-1}{|x|\sqrt{x^2-1}}.$$

3. $$\frac{e^x}{1+e^{2x}}$$

We find the derivative of the inverse tangent using the formula $\frac{d}{dx}\left(\tan^{-1}u\right)=\frac{1}{1+u^2}\frac{du}{dx}$. Here we

have $u=e^x$, so $\frac{du}{dx}=e^x$. Therefore, $\frac{d}{dx}\left(\tan^{-1}e^x\right)=\frac{1}{1+\left(e^x\right)^2}\left(e^x\right)=\frac{e^x}{1+e^{2x}}$.

4. $$\frac{1}{\sqrt{\pi}}\sec^{-1}\frac{x}{\sqrt{\pi}}+C$$

Recall that $\int\frac{du}{u\sqrt{u^2-1}}=\sec^{-1}u+C$. Here we have $\int\frac{dx}{x\sqrt{x^2-\pi}}$, and we just need to rearrange

the integrand so that it is in the proper form to use the integral formula. If we factor π out of

the radicand, we get $\int\frac{dx}{x\sqrt{\pi}\sqrt{\frac{x^2}{\pi}-1}}$. Next, we do u-substitution. Let $u=\frac{x}{\sqrt{\pi}}$ and $du=\frac{1}{\sqrt{\pi}}dx$.

Multiply both by $\sqrt{\pi}$ so that $u\sqrt{\pi}=x$ and $du\sqrt{\pi}=dx$. Substituting into the integrand, we get

$\int\frac{du\sqrt{\pi}}{\pi u\sqrt{u^2-1}}=\frac{1}{\sqrt{\pi}}\int\frac{du}{u\sqrt{u^2-1}}$. Now we get $\frac{1}{\sqrt{\pi}}\int\frac{du}{u\sqrt{u^2-1}}=\frac{1}{\sqrt{\pi}}\sec^{-1}u+C$. Substituting back,

we get $\frac{1}{\sqrt{\pi}}\sec^{-1}\frac{x}{\sqrt{\pi}}+C$.

5. $\dfrac{1}{\sqrt{7}}\tan^{-1}\dfrac{x}{\sqrt{7}}+C$

Recall that $\displaystyle\int\dfrac{du}{1+u^2}=\tan^{-1}u+C$. Here we have $\displaystyle\int\dfrac{dx}{7+x^2}$, and we just need to rearrange the

integrand so that it is in the proper form to use the integral formula. If we factor 7 out of the

denominator, we get $\displaystyle\int\dfrac{dx}{7\left(1+\dfrac{x^2}{7}\right)}=\dfrac{1}{7}\int\dfrac{dx}{\left(1+\left(\dfrac{x}{\sqrt{7}}\right)^2\right)}$. Next, we do u-substitution. Let $u=\dfrac{x}{\sqrt{7}}$

and $du=\dfrac{1}{\sqrt{7}}dx$. Multiply du by $\sqrt{7}$ so that $du\sqrt{7}=dx$. Substituting into the integrand, we get

$\dfrac{1}{7}\displaystyle\int\dfrac{du\sqrt{7}}{1+u^2}=\dfrac{1}{\sqrt{7}}\int\dfrac{du}{1+u^2}$. Now we get $\dfrac{1}{\sqrt{7}}\displaystyle\int\dfrac{du}{1+u^2}=\dfrac{1}{\sqrt{7}}\tan^{-1}u+C$. Substituting back, we get

$\dfrac{1}{\sqrt{7}}\tan^{-1}\dfrac{x}{\sqrt{7}}+C$.

6. $\tan^{-1}(\ln x)+C$

Recall that $\displaystyle\int\dfrac{du}{1+u^2}=\tan^{-1}u+C$. Here we have $\displaystyle\int\dfrac{dx}{x\left(1+\ln^2 x\right)}$, and we will need to do u-substi-

tution. Let $u=\ln x$ and $du=\dfrac{1}{x}dx$. Substituting into the integrand, we get $\displaystyle\int\dfrac{du}{1+u^2}=\tan^{-1}u+C$.

Substituting back, we get $\tan^{-1}\left(\ln x\right)+C$.

7. $\sin^{-1}(\tan x)+C$

Recall that $\displaystyle\int\dfrac{du}{\sqrt{1-u^2}}=\sin^{-1}u+C$. Here we have $\displaystyle\int\dfrac{\sec^2 x\, dx}{\sqrt{1-\tan^2 x}}$, and we will need to do

u-substitution. Let $u=\tan x$ and $du=\sec^2 x\, dx$. Substituting into the integrand, we get

$\displaystyle\int\dfrac{du}{\sqrt{1-u^2}}=\sin^{-1}u+C$. Substituting back, we get $\sin^{-1}\left(\tan x\right)+C$.

8. $\dfrac{1}{2}\sin^{-1}\dfrac{2x}{3}+C$

Recall   that $\displaystyle\int\dfrac{du}{\sqrt{1-u^2}}=\sin^{-1}u+C$.   Here   we   have $\displaystyle\int\dfrac{dx}{\sqrt{9-4x^2}}$,   and   we

just need to rearrange the integrand so that it is in the proper form to use the

integral formula. If we factor 9 out of the radicand, we get $\displaystyle\int\dfrac{dx}{3\sqrt{1-\dfrac{4x^2}{9}}}=\dfrac{1}{3}\int\dfrac{dx}{\sqrt{1-\left(\dfrac{2x}{3}\right)^2}}$. Next,

we do u-substitution. Let $u=\dfrac{2x}{3}$ and $du=\dfrac{2}{3}dx$. Multiply du by $\dfrac{3}{2}$ so that $\dfrac{3}{2}\,du=dx$. Sub-

stituting into the integrand, we get $\dfrac{1}{3}\left(\dfrac{3}{2}\right)\displaystyle\int\dfrac{du}{\sqrt{1-u^2}}=\dfrac{1}{2}\sin^{-1}u+C$. Substituting back, we get

$\dfrac{1}{2}\sin^{-1}\dfrac{2x}{3}+C$.

SOLUTIONS TO PRACTICE PROBLEM SET 23

1. $\ln|\tan x|+C$

Whenever we have an integral in the form of a quotient, we check to see if the solution

is a logarithm. A clue is whether the numerator is the derivative of the denominator, as

it is here. Let's use u-substitution. If we let $u=\tan x$, then $du=\sec^2x\,dx$. If we substitute

into the integrand, we get $\displaystyle\int\dfrac{\sec^2x}{\tan x}dx=\int\dfrac{du}{u}$. Recall that $\displaystyle\int\dfrac{du}{u}=\ln|u|+C$. Substituting back, we get

$\displaystyle\int\dfrac{\sec^2x}{\tan x}dx=\ln|\tan x|+C$.

2. $-\ln|1 - \sin x| + C$

Whenever we have an integral in the form of a quotient, we check to see if the solution is a logarithm. A clue is whether the numerator is the derivative of the denominator, as it is here. Let's use u-substitution. If we let $u = 1 - \sin x$, then $du = -\cos x \, dx$. If we substitute into the integrand, we get $\int \dfrac{\cos x}{1 - \sin x} dx = -\int \dfrac{du}{u}$. Recall that $\int \dfrac{du}{u} = \ln|u| + C$. Substituting back, we get $\int \dfrac{\cos x}{1 - \sin x} dx = -\ln|1 - \sin x| + C$.

3. $\ln|\ln x| + C$

Whenever we have an integral in the form of a quotient, we check to see if the solution is a logarithm. A clue is whether the numerator is the derivative of the denominator, as it is here. Let's use u-substitution. If we let $u = \ln x$, then $du = \dfrac{1}{x} dx$. If we substitute into the integrand, we get $\int \dfrac{1}{x \ln x} dx = \int \dfrac{du}{u}$. Recall that $\int \dfrac{du}{u} = \ln|u| + C$. Substituting back, we get $\int \dfrac{1}{x \ln x} dx = \ln|\ln x| + C$.

4. $-\ln|\cos x| - x + C$

First, let's rewrite the integrand.

$\int \dfrac{\sin x - \cos x}{\cos x} dx = \int \left(\dfrac{\sin x}{\cos x} - \dfrac{\cos x}{\cos x} \right) dx = \int \left(\dfrac{\sin x}{\cos x} - 1 \right) dx = \int \left(\dfrac{\sin x}{\cos x} \right) dx - \int dx$. Whenever we have

an integral in the form of a quotient, we check to see if the solution is a logarithm. A clue is whether the numerator is the derivative of the denominator, as it is in the first integral.

Let's use u-substitution. If we let $u = \cos x$, then $du = -\sin x \, dx$. If we substitute into the integrand, we get $\int \left(\dfrac{\sin x}{\cos x} \right) dx = -\int \dfrac{du}{u}$. Recall that $\int \dfrac{du}{u} = \ln|u| + C$. Substituting back, we get

$\int \left(\dfrac{\sin x}{\cos x} \right) dx = -\ln|\cos x| + C$. The second integral is simply $\int dx = x + C$. Therefore, the integral is

$\int \left(\dfrac{\sin x}{\cos x} \right) dx - \int dx = -\ln|\cos x| - x + C$.

5. $\ln\left(1+2\sqrt{x}\right)+C$

Whenever we have an integral in the form of a quotient, we check to see if the solution is a logarithm. A clue is whether the numerator is the derivative of the denominator, as it is here. Let's use u-substitution. If we let $u=1+2\sqrt{x}$, then $du=\dfrac{1}{\sqrt{x}}dx$. If we substitute into the integrand, we get $\displaystyle\int\frac{1}{\sqrt{x}\left(1+2\sqrt{x}\right)}dx=\int\frac{du}{u}$. Recall that $\displaystyle\int\frac{du}{u}=\ln|u|+C$. Substituting back, we get $\displaystyle\int\frac{1}{\sqrt{x}\left(1+2\sqrt{x}\right)}dx=\ln\left(1+2\sqrt{x}\right)+C$. Notice that we don't need the absolute value bars because $1+2\sqrt{x}$ is never negative.

6      $\ln(1+e^x)+C$

Whenever we have an integral in the form of a quotient, we check to see if the solution is a logarithm. A clue is whether the numerator is the derivative of the denominator, as it is here. Let's use u-substitution. If we let $u=1+e^x$, then $du=e^x dx$. If we substitute into the integrand, we get $\displaystyle\int\frac{e^x}{1+e^x}dx=\int\frac{du}{u}$. Recall that $\displaystyle\int\frac{du}{u}=\ln|u|+C$. Substituting back, we get $\displaystyle\int\frac{e^x}{1+e^x}dx=\ln\left(1+e^x\right)+C$. Notice that we don't need the absolute value bars because $1+e^x$ is never negative.

7. $\dfrac{1}{10}e^{5x^2-1}+C$

Recall that $\int e^u du = e^u + C$. Let's use u-substitution. If we let $u = 5x^2 - 1$, then $du = 10x\ dx$. But we need to substitute for $x\ dx$, so if we divide du by 10, we get $\dfrac{1}{10}du = x\ dx$. Now, if we substitute into the integrand, we get $\displaystyle\int xe^{5x^2-1}dx=\frac{1}{10}\int e^u du=\frac{1}{10}e^u+C$. Substituting back, we get $\displaystyle\int xe^{5x^2-1}dx=\frac{1}{10}e^{5x^2-1}+C$.

8. $\ln|e^x - e^{-x}| + C$

Recall that $\int e^u du = e^u + C$. Let's use u-substitution. If we let $u = e^x - e^{-x}$, then $du = e^x + e^{-x}dx$. If we

substitute into the integrand, we get $\int \dfrac{e^x + e^{-x}}{e^x - e^{-x}} dx = \int \dfrac{du}{u}$. Recall that $\int \dfrac{du}{u} = \ln|u| + C$. Substituting

back, we get $\int \dfrac{e^x + e^{-x}}{e^x - e^{-x}} dx = \ln|e^x - e^{-x}| + C$.

9. $-\dfrac{4^{-x^2}}{\ln 16} + C$

Recall that $\int a^u du = \dfrac{1}{\ln a} a^u + C$. Let's use u-substitution. If we let $u = -x^2$, then

$du = -2x\, dx$. But we need to substitute for $x\, dx$, so if we divide du by -2, we get $-\dfrac{1}{2} du = x\, dx$.

Now, if we substitute into the integrand, we get $\int x 4^{-x^2} dx = -\dfrac{1}{2}\int 4^u du = -\dfrac{1}{2}\left(\dfrac{1}{\ln 4}\right) 4^u + C$. Substi-

tuting back, we get $\int x 4^{-x^2} dx = -\dfrac{1}{2}\left(\dfrac{1}{\ln 4}\right) 4^{-x^2} + C = -\dfrac{1}{\ln 16} 4^{-x^2} + C$.

10. $\dfrac{7^{\sin x}}{\ln 7} + C$

Recall that $\int a^u du = \dfrac{1}{\ln a} a^u + C$. Let's use u-substitution. If we let $u = \sin x$, then

$du = \cos x\, dx$. If we substitute into the integrand, we get $\int 7^{\sin x} \cos x\, dx = \int 7^u du = \dfrac{1}{\ln 7} 7^u + C$.

Substituting back, we get $\int 7^{\sin x} \cos x\, dx = \dfrac{7^{\sin x}}{\ln 7} + C$.

SOLUTIONS TO PRACTICE PROBLEM SET 24

1. $\dfrac{32}{3}$

We find the area of a region bounded by $f(x)$ above and $g(x)$ below at all points of the interval $[a, b]$ using the formula $\int_a^b \left[f(x) - g(x) \right] dx$. Here $f(x) = 2$ and $g(x) = x^2 - 2$. First, let's make a sketch of the region.

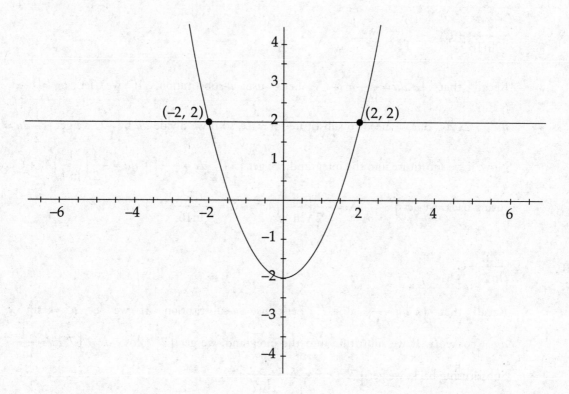

Next, we need to find where the two curves intersect, which will be the endpoints of the region.

We do this by setting the two curves equal to each other. We get $x^2 - 2 = 2$. The solutions are

$(-2, 0)$ and $(2, 0)$. Therefore, in order to find the area of the region, we need to evaluate the integral

$\int_{-2}^2 \left(2 - \left(x^2 - 2 \right) \right) dx = \int_{-2}^2 \left(4 - x^2 \right) dx$. We get

$$\int_{-2}^2 \left(4 - x^2 \right) dx = \left(4x - \frac{x^3}{3} \right) \Bigg|_{-2}^2 = \left(4(2) - \frac{(2)^3}{3} \right) - \left(4(-2) - \frac{(-2)^3}{3} \right) = \frac{32}{3}.$$

2. $\dfrac{8}{3}$

We find the area of a region bounded by $f(x)$ above and $g(x)$ below at all points of the interval $[a, b]$ using the formula $\displaystyle\int_{a}^{b}\left[f(x)-g(x)\right]dx$. Here $f(x) = 4x - x^2$ and $g(x) = x^2$.

First, let's make a sketch of the region.

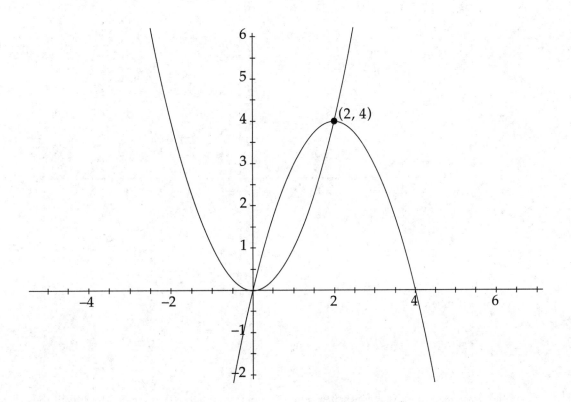

Next, we need to find where the two curves intersect, which will be the end-points of the region. We do this by setting the two curves equal to each other. We get

$4x - x^2 = x^2$. The solutions are $(0, 0)$ and $(2, 4)$. Therefore, in order to find the area of the region, we need to evaluate the integral $\displaystyle\int_{0}^{2}\left(\left(4x-x^2\right)-x^2\right)dx = \int_{0}^{2}\left(4x-2x^2\right)dx$. We get

$$\int_{0}^{2}\left(4x-2x^2\right)dx = \left(2x^2 - \frac{2x^3}{3}\right)\Bigg|_{0}^{2} = \left(2\left(2\right)^2 - \frac{2\left(2\right)^3}{3}\right) - 0 = \frac{8}{3}.$$

3.   36

We find the area of a region bounded by $f(x)$ above and $g(x)$ below at all points of the interval $[a, b]$

using the formula $\int_a^b \left[f(x) - g(x) \right] dx$. Here $f(x) = 2x - 5$ and $g(x) = x^2 - 4x - 5$.

First, let's make a sketch of the region.

Next, we need to find where the two curves intersect, which will be the endpoints of the region. We

do this by setting the two curves equal to each other. We get $x^2 - 4x - 5 = 2x - 5$. The solutions are

$(0, -5)$ and $(6, 7)$. Therefore, in order to find the area of the region, we need to evaluate the integral

$\int_0^6 \left[(2x - 5) - \left(x^2 - 4x - 5 \right) \right] dx = \int_0^6 \left(-x^2 + 6x \right) dx$. We get

$\int_0^6 \left(-x^2 + 6x \right) dx = \left(-\dfrac{x^3}{3} + 3x^2 \right) \Bigg|_0^6 = \left(-\dfrac{(6)^3}{3} + 3(6)^2 \right) - 0 = 36.$

4. $\dfrac{17}{4}$

We find the area of a region bounded by $f(x)$ above and $g(x)$ below at all points of the interval $[a, b]$

using the formula $\displaystyle\int_a^b \left[f(x) - g(x) \right] dx$.

First, let's make a sketch of the region.

Notice that in the region from $x = -1$ to $x = 0$ the top curve is $f(x) = 0$ (the x-axis), and the

bottom curve is $g(x) = x^3$, but from $x = 0$ to $x = 2$ the situation is reversed, so the top curve is

$f(x) = x^3$, and the bottom curve is $g(x) = 0$. Thus, we split the region into two pieces and find the

area by evaluating two integrals and adding the answers: $\displaystyle\int_{-1}^0 \left(0 - x^3 \right) dx$ and $\displaystyle\int_0^2 \left(x^3 - 0 \right) dx$. We get

$\displaystyle\int_{-1}^0 \left(0 - x^3 \right) dx = \left(-\dfrac{x^4}{4} \right)\Bigg|_{-1}^0 = 0 - \left(-\dfrac{(-1)^4}{4} \right) = \dfrac{1}{4}$ and $\displaystyle\int_0^2 \left(x^3 - 0 \right) dx = \left(\dfrac{x^4}{4} \right)\Bigg|_0^2 = \dfrac{(2)^4}{4} - 0 = 4$. There-

fore, the area of the region is $\dfrac{17}{4}$.

5. $\dfrac{9}{2}$

We find the area of a region bounded by $f(y)$ on the right and $g(y)$ on the left at all points of the interval $[c, d]$ using the formula $\int_{c}^{d} \left[f(y) - g(y) \right] dy$. Here, $f(y) = y + 2$ and $g(y) = y^2$.

First, let's make a sketch of the region.

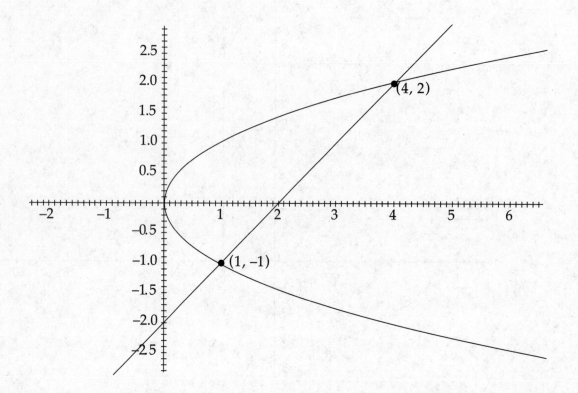

Next, we need to find where the two curves intersect, which will be the endpoints of the region.

We do this by setting the two curves equal to each other. We get $y^2 = y + 2$. The solutions are $(4, 2)$

and $(1, -1)$. Therefore, in order to find the area of the region, we need to evaluate the integral

$\int_{-1}^{2} \left(y + 2 - y^2 \right) dy$. We get

$$\int_{-1}^{2} \left(y + 2 - y^2 \right) dy = \left(\frac{y^2}{2} + 2y - \frac{y^3}{3} \right)\Bigg|_{-1}^{2} = \left(\frac{(2)^2}{2} + 2(2) - \frac{(2)^3}{3} \right) - \left(\frac{(-1)^2}{2} + 2(-1) - \frac{(-1)^3}{3} \right) = \frac{9}{2}.$$

6.   4

We find the area of a region bounded by $f(y)$ on the right and $g(y)$ on the left at all points of the interval $[c, d]$ using the formula $\int_c^d \left[f(y) - g(y) \right] dy$. Here, $f(y) = 3 - 2y^2$ and $g(y) = y^2$.

First, let's make a sketch of the region.

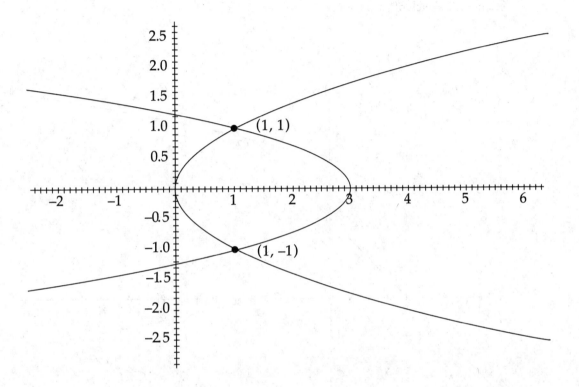

Next, we need to find where the two curves intersect, which will be the endpoints of the region. We do this by setting the two curves equal to each other. We get $y^2 = 3 - 2y^2$. The solutions are $(1, 1)$ and $(1, -1)$. Therefore, in order to find the area of the region, we need to evaluate the integral $\int_{-1}^{1} \left(3 - 2y^2 - y^2 \right) dy = \int_{-1}^{1} \left(3 - 3y^2 \right) dy$. We get

$$\int_{-1}^{1} \left(3 - 3y^2 \right) dy = \left(3y - y^3 \right) \Big|_{-1}^{1} = \left(3(1) - (1)^3 \right) - \left(3(-1) - (-1)^3 \right) = 4.$$

7. $\dfrac{9}{2}$

We find the area of a region bounded by $f(y)$ on the right and $g(y)$ on the left at all points of the interval $[c, d]$ using the formula $\int_c^d \left[f(y) - g(y) \right] dy$. Here $f(y) = y - 2$ and $g(y) = y^2 - 4y + 2$. First, let's make a sketch of the region.

Next, we need to find where the two curves intersect, which will be the endpoints of the region.

We do this by setting the two curves equal to each other. We get $y^2 - 4y + 2 = y - 2$. The solutions are $(2, 4)$ and $(-1, 1)$. Therefore, in order to find the area of the region, we need to evaluate the

integral $\int_1^4 \left[(y - 2) - (y^2 - 4y + 2) \right] dy = \int_1^4 \left[(-y^2 + 5y - 4) \right] dy$. We get $\int_1^4 \left[(-y^2 + 5y - 4) \right] dy =$

$\left(-\dfrac{y^3}{3} + \dfrac{5y^2}{2} - 4y \right) \Big|_1^4 = \left(-\dfrac{(4)^3}{3} + \dfrac{5(4)^2}{2} - 4(4) \right) - \left(-\dfrac{(1)^3}{3} + \dfrac{5(1)^2}{2} - 4(1) \right) = \dfrac{9}{2}$.

8. $\dfrac{12}{5}$

We find the area of a region bounded by $f(y)$ on the right and $g(y)$ on the left at all points of the interval $[c, d]$ using the formula $\int_c^d \left[f(y) - g(y) \right] dy$. Here $f(y) = 2 - y^4$ and $g(y) = y^{\frac{2}{3}}$. First, let's make a sketch of the region.

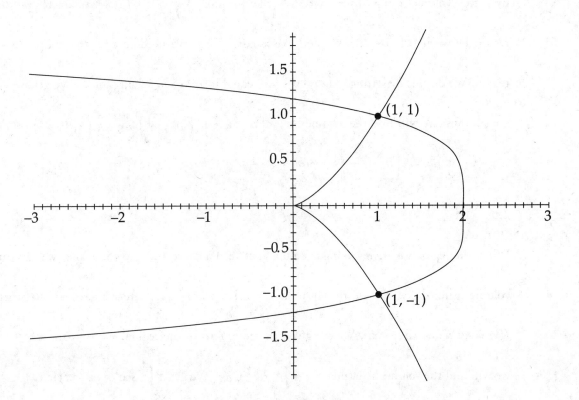

Next, we need to find where the two curves intersect, which will be the end-points of the region. We do this by setting the two curves equal to each other. We get

$2 - y^4 = y^{\frac{2}{3}}$. The solutions are $(1,\ 1)$ and $(1,\ -1)$. Therefore, in order to find the area of the region, we need to evaluate the integral $\int_{-1}^{1} \left(2 - y^4 - y^{\frac{2}{3}} \right) dy$. We get

$\int_{-1}^{1} \left(2 - y^4 - y^{\frac{2}{3}} \right) dy = \left(2y - \dfrac{y^5}{5} - \dfrac{3}{5} y^{\frac{5}{3}} \right) \Bigg|_{-1}^{1} = \left(2(1) - \dfrac{(1)^5}{5} - \dfrac{3}{5}(1)^{\frac{5}{3}} \right) - \left(2(-1) - \dfrac{(-1)^5}{5} - \dfrac{3}{5}(-1)^{\frac{5}{3}} \right) = \dfrac{12}{5}$.

SOLUTIONS TO PRACTICE PROBLEM SET 25

1. 36π

    When the region we are revolving is defined between a curve $f(x)$ and the x-axis, we can find the volume using disks. We use the formula $V = \pi \int_a^b \left[f(x) \right]^2 dx$. Here we have a region between $f(x) = \sqrt{9 - x^2}$ and the x-axis. First, we need to find the endpoints of the region. We do this by setting $f(x) = \sqrt{9 - x^2}$ equal to zero and solving for x. We get $x = -3$ and $x = 3$. Thus, we will find the volume by evaluating $\pi \int_{-3}^3 \left(\sqrt{9 - x^2} \right)^2 dx = \pi \int_{-3}^3 \left(9 - x^2 \right) dx$. We get $\pi \int_{-3}^3 \left(9 - x^2 \right) dx = \pi \left(9x - \frac{x^3}{3} \right) \Big|_{-3}^3 = 36\pi$.

2. 2π

    When the region we are revolving is defined between a curve $f(x)$ and the x-axis, we can find the volume using disks. We use the formula $V = \pi \int_a^b \left[f(x) \right]^2 dx$. Here we have a region between $f(x) = \sec x$ and the x-axis. We are given the endpoints of the region: $x = -\frac{\pi}{4}$ and $x = \frac{\pi}{4}$. Thus, we will find the volume by evaluating $\pi \int_{-\frac{\pi}{4}}^{\frac{\pi}{4}} \sec^2 x \, dx$. We get $\pi \int_{-\frac{\pi}{4}}^{\frac{\pi}{4}} \sec^2 x \, dx = \pi \left(\tan x \right) \Big|_{-\frac{\pi}{4}}^{\frac{\pi}{4}} = 2\pi$.

3. $\dfrac{16\pi}{15}$

    When the region we are revolving is defined between a curve $f(y)$ and the y-axis, we can find the volume using disks. We use the formula $V = \pi \int_a^b \left[f(y) \right]^2 dy$ (see page 339 and note that when $g(y) = 0$ we get disks instead of washers). Here we have a region between $f(y) = 1 - y^2$ and the y-axis. First, we need to find the endpoints of the region. We do this by setting $f(y) = 1 - y^2$ equal to zero and solving for y. We get $y = -1$ and $y = 1$.

Thus, we will find the volume by evaluating $\pi \int_{-1}^{1} \left(1-y^2\right)^2 dy = \pi \int_{-1}^{1} \left(1-2y^2+y^4\right) dy$. We get

$$\pi \int_{-1}^{1} \left(1-2y^2+y^4\right) dy = \pi \left(y - \frac{2y^3}{3} + \frac{y^5}{5} \right)\Bigg|_{-1}^{1} = \frac{16\pi}{15}.$$

4. 2π

When the region we are revolving is defined between a curve $f(y)$ and the y-axis, we can find the volume using disks. We use the formula $V = \pi \int_{a}^{b} \left[f(y) \right]^2 dy$ (see page 339 and note that when $g(y) = 0$ we get disks instead of washers). Here we have a region between $f(y) = \sqrt{5}y^2$ and the y-axis. We are given the endpoints of the region: $y = -1$ and $y = 1$. Thus, we will find the volume by evaluating $\pi \int_{-1}^{1} \left(\sqrt{5}y^2\right)^2 dy = \pi \int_{-1}^{1} \left(5y^4\right) dy$. We get $\pi \int_{-1}^{1} \left(5y^4\right) dy = \pi \left(y^5\right)\Big|_{-1}^{1} = 2\pi$.

5. 8π

When the region we are revolving is defined between curves $f(x)$ and $g(x)$, we can find the volume using cylindrical shells. We use the formula $V = 2\pi \int_{a}^{b} x \left[f(x) - g(x) \right] dx$. Here we have $f(x) = x$ and $g(x) = -\dfrac{x}{2}$. Thus, the height of each shell is $f(x) - g(x) = x - \left(-\dfrac{x}{2}\right) = \dfrac{3x}{2}$, and the radius is simply x. The endpoints of our region are $x = 0$ and $x = 2$. Therefore, we will find the volume by evaluating $2\pi \int_{0}^{2} \left(x \left(\dfrac{3x}{2}\right) \right) dx = 2\pi \int_{0}^{2} \left(\dfrac{3x^2}{2}\right) dx = 3\pi \int_{0}^{2} x^2 dx$. We get

$$3\pi \int_{0}^{2} x^2 dx = 3\pi \left(\frac{x^3}{3}\right)\Bigg|_{0}^{2} = 8\pi.$$

6. $\dfrac{7\pi}{15}$

When the region we are revolving is defined between curves $f(x)$ and $g(x)$, we can find the volume using cylindrical shells. We use the formula $V = 2\pi \int_{a}^{b} x \left[f(x) - g(x) \right] dx$. Here we have $f(x) = \sqrt{x}$ and $g(x) = 2x - 1$. Thus, the height of each shell is $f(x) - g(x) = \sqrt{x} - (2x-1) = \sqrt{x} - 2x + 1$, and the radius is simply x. The left endpoint of our region is $x = 0$, and we find the right endpoint by finding where $f(x) = \sqrt{x}$ intersects $g(x) = 2x - 1$. We get $x = 1$. Therefore, we will find the

volume by evaluating $2\pi \int_0^1 \left(x \left(\sqrt{x} - 2x + 1 \right) \right) dx = 2\pi \int_0^1 \left(x^{\frac{3}{2}} - 2x^2 + x \right) dx$. We get

$2\pi \int_0^1 \left(x^{\frac{3}{2}} - 2x^2 + x \right) dx = 2\pi \left(\frac{2x^{\frac{5}{2}}}{5} - \frac{2x^3}{3} + \frac{x^2}{2} \right) \Bigg|_0^1 = \frac{7\pi}{15}$.

7. $\dfrac{128\pi}{5}$

When the region we are revolving is defined between curves $f(y)$ and $g(y)$, we can find the

volume using cylindrical shells. We use the formula $V = 2\pi \int_a^b y \left[f(y) - g(y) \right] dy$. Here we have

$f(y) = \sqrt{y}$ (which we get by solving $y = x^2$ for x) and $g(y) = 0$ (the y-axis). We can easily see

that the top endpoint is $y = 4$ and the bottom one is $y = 0$. Therefore, we will find the volume by

evaluating $2\pi \int_0^4 \left(y\sqrt{y} \right) dy = 2\pi \int_0^4 \left(y^{\frac{3}{2}} \right) dy$. We get $2\pi \int_0^4 \left(y^{\frac{3}{2}} \right) dy = 2\pi \left(\frac{2y^{\frac{5}{2}}}{5} \right) \Bigg|_0^4 = \frac{128\pi}{5}$.

8. $\dfrac{256\pi}{5}$

When the region we are revolving is defined between curves $f(x)$ and $g(x)$, we can find the vol-

ume using cylindrical shells. We use the formula $V = 2\pi \int_a^b x \left[f(x) - g(x) \right] dx$. Here we have

$f(x) = 2\sqrt{x}$ and $g(x) = 0$. Thus, the height of each shell is $f(x) - g(x) = 2\sqrt{x}$, and the

radius is simply x. We can easily see that the left endpoint is $x = 0$ and that the right endpoint is

$x = 4$. Therefore, we will find the volume by evaluating $2\pi \int_0^4 \left(x \left(2\sqrt{x} \right) \right) dx = 2\pi \int_0^4 \left(2x^{\frac{3}{2}} \right) dx$. We

get $2\pi \int_0^4 \left(2x^{\frac{3}{2}} \right) dx = 4\pi \left(\frac{2x^{\frac{5}{2}}}{5} \right) \Bigg|_0^4 = \frac{256\pi}{5}$.

9. $\dfrac{256}{3}$

To find the volume of a solid with a cross-section of an isosceles right triangle, we integrate the area of the square (*side²*) over the endpoints of the interval. Here the sides of the squares are found by $f(x) - g(x) = \sqrt{16 - x^2} - 0$, and the intervals are found by setting $y = \sqrt{16 - x^2}$ equal to zero. We get $x = -4$ and $x = 4$. Thus, we find the volume by evaluating the integral $\int_{-4}^{4} \left(\sqrt{16 - x^2} \right)^2 dx = \int_{-4}^{4} \left(16 - x^2 \right) dx$. We get $\int_{-4}^{4} \left(16 - x^2 \right) dx = \left(16x - \dfrac{x^3}{3} \right) \Big|_{-4}^{4} = \dfrac{256}{3}$.

10. $\dfrac{128}{15}$

To find the volume of a solid with a cross-section of a square, we integrate the area of the isosceles right triangle $\dfrac{hypotenuse^2}{4}$ over the endpoints of the interval. Here, the hypotenuses of the triangles are found by $f(x) - g(x) = 4 - x^2$, and the intervals are found by setting $y = x^2$ equal to $y = 4$. We get $x = -2$ and $x = 2$. Thus, we find the volume by evaluating the integral $\int_{-2}^{2} \dfrac{\left(4 - x^2 \right)^2}{4} dx = \int_{-2}^{2} \dfrac{\left(16 - 8x^2 + x^4 \right)}{4} dx = \int_{-2}^{2} \left(4 - 2x^2 + \dfrac{x^4}{4} \right) dx$.

We get $\int_{-2}^{2} \left(4 - 2x^2 + \dfrac{x^4}{4} \right) dx = \left(4x - \dfrac{2x^3}{3} + \dfrac{x^5}{20} \right) \Big|_{-2}^{2} = \dfrac{128}{15}$.

SOLUTIONS TO PRACTICE PROBLEM SET 26

1. $y = \sqrt[4]{\dfrac{28x^3}{3} - 236}$

    We solve this differential equation by separation of variables. We want to get all of the y variables on one side of the equals sign and all of the x variables on the other side. We can do this easily by cross-multiplying. We get $y^3\,dy = 7x^2\,dx$. Next, we integrate both sides.

    $$\int y^3\,dy = \int 7x^2\,dx$$

    $$\frac{y^4}{4} = 7\frac{x^3}{3} + C$$

    $$y^4 = \frac{28x^3}{3} + C$$

    Now we solve for C. We plug in $x = 3$ and $y = 2$.

    $$16 = 252 + C$$

    $$C = -236$$

    Therefore, $y^4 = \dfrac{28x^3}{3} - 236$.

    Now we isolate y. We get the equation $y = \sqrt[4]{\dfrac{28x^3}{3} - 236}$.

2. $y = 6e^{\frac{5x^3}{3}}$

    We solve this differential equation by separation of variables. We want to get all of the y variables on one side of the equals sign and all of the x variables on the other side. We can do this easily by dividing both sides by y and multiplying both sides by dx. We get $\dfrac{dy}{y} = 5x^2\,dx$. Next, we integrate both sides.

    $$\int \frac{dy}{y} = \int 5x^2\,dx$$

    $$\ln y = \frac{5x^3}{3} + C_0$$

    Now we isolate y: $y = e^{\frac{5x^3}{3} + C}$. We can rewrite this as $y = e^{\frac{5x^3}{3}}\left(e^{C_0}\right) = Ce^{\frac{5x^3}{3}}$.

Note that we are using the letter C in the last equation. This is to distinguish it from the C_0 in the first equation. Now we solve for C. We plug in $x = 0$ and $y = 6$: $6 = Ce^0 = C$. Therefore, the equation is $y = 6e^{\frac{5x^3}{3}}$.

3. $y = \sqrt[3]{3e^x - 2}$

We solve this differential equation by separation of variables. We want to get all of the y variables on one side of the equals sign and all of the x variables on the other side. We can do this easily by cross-multiplying. We get $y^2 dy = e^x dx$. Next, we integrate both sides.

$$\int y^2 dy = \int e^x dx$$

$$\frac{y^3}{3} = e^x + C$$

Now we isolate y.

$$y^3 = 3e^x + C$$

$$y = \sqrt[3]{3e^x + C}$$

Now we solve for C. We plug in $x = 0$ and $y = 1$.

$$1 = \sqrt[3]{3e^0 + C}$$

$$C = -2$$

Therefore, the equation is $y = \sqrt[3]{3e^x - 2}$.

4. $y = 2x^2$

We solve this differential equation by separation of variables. We want to get all of the y variables on one side of the equals sign and all of the x variables on the other side. We can do this easily by dividing both sides by y^2 and multiplying both sides by dx. We get $\frac{dy}{y^2} = \frac{dx}{x^3}$. Next, we integrate both sides.

$$\int \frac{dy}{y^2} = \int \frac{dx}{x^3}$$

$$-\frac{1}{y} = -\frac{1}{2x^2} + C$$

Now we isolate y.

$$\frac{1}{y} = \frac{1}{2x^2} + C$$

Now we solve for C. We plug in $x = 1$ and $y = 2$.

$$\frac{1}{2} = \frac{1}{2(1)^2} + C$$

$$C = 0$$

Therefore, the equation is $\frac{1}{y} = \frac{1}{2x^2} + C$, which can be rewritten as $y = 2x^2$.

5.　　　$y = \sin^{-1}(-\cos x)$

We solve this differential equation by separation of variables. We want to get all of the y variables on one side of the equals sign and all of the x variables on the other side. We can do this easily by cross-multiplying. We get $\cos y\, dy = \sin x\, dx$. Next, we integrate both sides.

$$\int \cos y\, dy = \int \sin x\, dx$$

$$\sin y = -\cos x + C$$

Now we solve for C. We plug in $x = 0$ and $y = \frac{3\pi}{2}$.

$$\sin\frac{3\pi}{2} = -\cos 0 + C$$

$$-1 = -1 + C$$

$$C = 0$$

Now we isolate y to get the equation $y = \sin^{-1}(-\cos x)$.

6.　　　20,000 (approximately)

The phrase "grows exponentially" means that we can represent the situation with the differential equation $\frac{dy}{dt} = ky$, where k is a constant and y is the population at a time t. Here we are also told that $y = 4,000$ at $t = 0$ and $y = 6,500$ at $t = 3$. We solve this differential equation by separation of variables. We want to get all of the y variables on one side of the equals sign and all of the t variables on the other side. We can do this easily by dividing both sides by y and multiplying both sides by dt. We get $\frac{dy}{y} = k\, dt$. Next, we integrate both sides.

$$\int \frac{dy}{y} = \int k\, dt$$

$$\ln y = kt + C_0$$

$$y = Ce^{kt}$$

Next, we plug in $y = 4{,}000$ and $t = 0$ to solve for C.

$$4{,}000 = Ce^0$$

$$C = 4{,}000$$

Now, we have $y = 4{,}000e^{kt}$. Next, we plug in $y = 6{,}500$ and $t = 3$ to solve for k.

$$6{,}500 = 4{,}000e^{3k}$$

$$k = \frac{1}{3}\ln\frac{13}{8} \approx 0.162$$

Therefore, our equation for the population of bacteria, y, at time, t, is $y \approx 4{,}000e^{0.162t}$, or if we want an exact solution, it is $y = 4{,}000\left(\frac{13}{8}\right)^{\frac{t}{3}}$. Finally, we can solve for the population at time $t = 10$: $y \approx 4{,}000e^{0.162(10)} \approx 20{,}212$ or $y = 4{,}000\left(\frac{13}{8}\right)^{\frac{10}{3}} \approx 20{,}179$. (Notice how even with an "exact" solution, we still have to round the answer. And, if we are concerned with significant figures, the population can be written as $20{,}000$.)

7.   45 m

Because acceleration is the derivative of velocity, we can write $\dfrac{dv}{dt} = -9$. We are also told that $v = 18$ and $h = 45$ when $t = 0$. We solve this differential equation by separation of variables. We want to get all of the v variables on one side of the equals sign and all of the t variables on the other side. We can do this easily by multiplying both sides by dt: $dv = -9\,dt$. Next, we integrate both sides:

$$\int dv = \int -9\,dt$$

$$v = -9 + C_0$$

Now we can solve for C_0 by plugging in $v = 18$ and $t = 0$: $18 = -9\,(0) + C_0$ so $C_0 = 18$. Thus, our equation for the velocity of the rock is $v = -9t + 18$. Next, because the height of the rock is the derivative of the velocity, we can write $\frac{dh}{dt} = -9t + 18$. We again separate the variables by multiplying both sides by dt: $dh = (-9t + 18)\,dt$. We integrate both sides:

$$\int dh = \int (-9t + 18)\,dt$$

$$h = -\frac{9t^2}{2} + 18t + C_1$$

Next, we plug in $h = 45$ and $t = 0$ to solve for C_1: $45 = -\frac{9(0)^2}{2} + 18(0) + C_1$, so $C_1 = 45$ m.

Therefore, the equation for the height of the rock, h, at time t is $h = -\frac{9t^2}{2} + 18t + 45$.

Finally, we can solve for the height of the rock at time $t = 4$: $h = -\frac{9t^2}{2} + 18t + 45$, so $h = -\frac{9(4)^2}{2} + 18(4) + 45 = 45$ m.

8.   8,900 grams (approximately)

We can express this situation with the differential equation $\frac{dm}{dt} = -km$, where m is the mass at time t. We are also given that $m = 10{,}000$ when $t = 0$ and $m = 5{,}000$ when $t = 5{,}750$. We solve this differential equation by separation of variables. We want to get all of the m variables on one side of the equals sign and all of the t variables on the other side. We can do this easily by dividing both sides by m and multiplying both sides by dt. We get $\frac{dm}{m} = -k\,dt$. Next, we integrate both sides.

$$\int \frac{dm}{m} = \int -k\,dt$$

$$\ln m = -kt + C_0$$

$$m = Ce^{-kt}$$

Now we can solve for C by plugging in $m = 10,000$ and $t = 0$: $10,000 = Ce^0$, so $C = 10,000$. This gives us the equation $m = 10,000e^{-kt}$. Next, we can solve for k by plugging in $m = 5,000$ and $t = 5,750$.

$$5,000 = 10,000e^{-5,750k}$$

$$\frac{1}{2} = e^{-5,750k}$$

$$-\frac{1}{5,750}\ln\frac{1}{2} = k \approx 0.000121$$

Therefore, the equation for the mass m of the element at time t is $m \approx 10,000e^{-0.000121t}$, or if we want an exact solution, it is $m = 10,000\left(\frac{1}{2}\right)^{\frac{t}{5,750}}$.

Finally, we can solve for the mass of the element at time $t = 1,000$: $m \approx 10,000e^{-0.000121(1,000)} \approx 8,860$ grams or $m = 10,000\left(\frac{1}{2}\right)^{\frac{1,000}{5,750}} \approx 8,864$ grams. (And, if we are concerned with significant figures, the mass can be written as 8,900.)

9.

10.

Chapter 20
Solutions to End of Chapter Drills

CHAPTER 3

1. **C**   Simply plug in $x = 2$: $\lim\limits_{x \to 2}\left(x^4 - 3x^2 + 5\right) = (2)^4 - 3(2)^2 + 5 = 16 - 12 + 5 = 9$

   The answer is (C).

2. **A**   Plug in $x = 2$: $\lim\limits_{x \to 2} \dfrac{x^2 - x - 2}{x^2 - 8x + 15} = \dfrac{(2)^2 - (2) - 2}{(2)^2 - 8(2) + 15} = \dfrac{0}{3} = 0$

   The answer is (A)

3. **B**   Plug in $x = 7$: $\lim\limits_{x \to 7} \dfrac{(7)^2 - 3(7) - 28}{(7)^2 - 5(7) - 14} = \dfrac{0}{0}$. When we get a limit of $\dfrac{0}{0}$, this does not necessarily

   mean that the limit does not exist. First, let's factor the numerator and denominator of the limit:

   $\lim\limits_{x \to 7} \dfrac{x^2 - 3x - 28}{x^2 - 5x - 14} = \lim\limits_{x \to 7} \dfrac{(x-7)(x+4)}{(x-7)(x+2)}$. Note that when we plug in $x = 7$, the factor $(x - 7)$ is what

   makes the function zero in the numerator and denominator. Cancel the factors to reduce the limit

   to: $\lim\limits_{x \to 7} \dfrac{(x+4)}{(x+2)}$. Now plug in $x = 7$: $\lim\limits_{x \to 7} \dfrac{(x+4)}{(x+2)} = \dfrac{7+4}{7+2} = \dfrac{11}{9}$.

   The answer is (B).

4. **C**   Remember that when we are finding the limit at infinity of a rational function, we compare the

   degree of the highest term in the numerator to the degree of the highest term in the denominator.

   The degrees are both 3, so the limit will be the ratio of the coefficients of the terms, namely $\dfrac{4}{5}$.

   The answer is (C).

5. **A**   When we have a limit of the form $\lim\limits_{x \to 0} \dfrac{a \sin bx}{c \tan dx}$, the limit is $\dfrac{ab}{cd}$. Here we get

   $\lim\limits_{x \to 0} \dfrac{4 \sin 3x}{2 \tan 5x} = \dfrac{(4)(3)}{(2)(5)} = \dfrac{6}{5}$.

   The answer is (A).

6. **D**   Plug in $x = 5$: $\lim\limits_{x \to 5} \dfrac{11}{x^2 - 25} = \dfrac{11}{(5)^2 - 25} = \dfrac{11}{0}$. This is going to give us an infinite limit or the limit

   doesn't exist. We need to take the limit from both sides of 5. First, find $\lim\limits_{x \to 5^-} \dfrac{11}{x^2 - 25}$. The

   denominator will be just less than zero, so $\lim\limits_{x \to 5^-} \dfrac{11}{x^2 - 25} = -\infty$. Next, find $\lim\limits_{x \to 5^+} \dfrac{11}{x^2 - 25}$. Now the

   denominator will be just greater than zero, so $\lim\limits_{x \to 5^+} \dfrac{11}{x^2 - 25} = \infty$. The two limits do not match, so

   the limit does not exist.

   The answer is (D).

CHAPTER 4

1. **C** First, let's factor the numerator and denominator. We get $f(x) = \dfrac{x^2 - 8x + 15}{x^2 - 9} = \dfrac{(x-3)(x-5)}{(x-3)(x+3)}$.

   At a value of x where the denominator is zero and the numerator is not also zero, we get a vertical asymptote, which is an essential discontinuity. At a value of x where the denominator is zero and the numerator is also zero, we get a hole, which is a removable discontinuity. Thus we have a removable discontinuity at $x = 3$ and an essential discontinuity at $x = -3$.

   The answer is (C).

2. **D** The only value where we might have a discontinuity is at $x = 4$. Plug $x = 4$ into both pieces of the function: $f(x) = \begin{cases} 5(4) - 11 = 9 \\ 2x + 1 = 9 \end{cases}$. The function is 9 for both pieces, so the function is continuous for all real numbers.

   The answer is (D).

3. **C** The only value where we might have a discontinuity is at $x = 2$. Plug $x = 2$ into both pieces of the function: $f(x) = \begin{cases} 6(2) + 1 = 13 \\ 5(2) + 2 = 12 \end{cases}$. The function "jumps" from 12 to 13 at $x = 2$, so it has a jump discontinuity at $x = 2$.

   The answer is (C).

4. **A** We need to find a value of k where the value of f in the top piece and the bottom piece is the same at $x = 1$. Plug $x = 1$ into the top piece and we get $(1)^2 + k(1) - 4 = k - 3$. Now plug $x = 1$ into the bottom piece and we get $6k(1) + 7 = 6k + 7$. Set the two pieces equal to each other and solve for k.

   $$k - 3 = 6k + 7$$
   $$5k = -10$$
   $$k = -2$$

   The answer is (A).

5. **B** Here, in order for f to be continuous for all real numbers, the value of f in the top piece at $x = 2$ has to equal the value of f in the middle piece at $x = 2$, and the value of f in the middle piece at $x = -2$ has to equal the value of f in the bottom piece at $x = -2$. So, first let's plug $x = 2$ into the top piece: $a(2)^2 + 2b(2) - 3 = 4a + 4b - 3$. Next plug $x = 2$ into the middle piece: $4b(2)^2 - 3a(2) + 3 = 16b - 6a + 3$.

   Now set the two pieces equal to each other:

   $$4a + 4b - 3 = 16b - 6a + 3$$

   $10a - 12b = 6$, which simplifies to $5a - 6b = 3$.

Now, let's plug $x = -2$ into the middle piece: $4b(-2)^2 - 3a(-2) + 3 = 16b + 6a + 3$. Next plug $x = -2$ into the bottom piece: $4a(-2)^2 + 5b(-2) + 25 = 16a - 10b + 25$. Now set the two pieces equal to each other:

$$16a - 10b + 25 = 16b + 6a + 3$$

$$10a - 26b = -22$$

Finally, we solve the pair of equations for a and b.

$$5a - 6b = 3$$

$$10a - 26b = -22$$

Multiply the top equation by 2:

$$10a - 12b = 6$$

$$10a - 26b = -22$$

Subtract the bottom equation from the top equation: $14b = 28$, so $b = 2$. Plug in for b to get $5a - 6(2) = 3$, so $a = 3$.

The answer is (B).

CHAPTER 5

1. **C** We can find the derivative using the formula $f'(x) = \lim_{h \to 0} \dfrac{f(x+h) - f(x)}{h}$. We get $f'(x) = \lim_{h \to 0} \dfrac{8(x+h) - 11 - (8x - 11)}{h}$.

Simplify the numerator: $\lim_{h \to 0} \dfrac{8(x+h) - 11 - (8x - 11)}{h} = \lim_{h \to 0} \dfrac{8x + 8h - 11 - 8x + 11}{h}$.

$$\lim_{h \to 0} \frac{8x + 8h - 11 - 8x + 11}{h} = \lim_{h \to 0} \frac{8h}{h}$$

Cancel h out of the numerator and denominator: $\lim_{h \to 0} \dfrac{8h}{h} = \lim_{h \to 0} 8$.

And take the limit: $\lim_{h \to 0} 8 = 8$.

Therefore, the derivative is 8.

The answer is (C).

2. We can find the derivative using the formula $f'(x) = \lim\limits_{h \to 0} \dfrac{f(x+h) - f(x)}{h}$. We get

$$f'(x) = \lim\limits_{h \to 0} \frac{3(x+h)^2 + 7(x+h) - 2 - \left(3x^2 + 7x - 2\right)}{h}.$$

Simplify the numerator: $\lim\limits_{h \to 0} \dfrac{3\left(x^2 + 2xh + h^2\right) + 7x + 7h - 2 - 3x^2 - 7x + 2}{h}$

$\lim\limits_{h \to 0} \dfrac{3x^2 + 6xh + 3h^2 + 7x + 7h - 2 - 3x^2 - 7x + 2}{h}$

$\lim\limits_{h \to 0} \dfrac{6xh + 3h^2 + 7h}{h}$

Factor h out of the numerator: $\lim\limits_{h \to 0} \dfrac{h(6x + 3h + 7)}{h}$

Cancel the h in the numerator and the denominator: $\lim\limits_{h \to 0} 6x + 3h + 7$

And take the limit: $\lim\limits_{h \to 0} 6x + 3h + 7 = 6x + 7$.

Therefore, the derivative is $6x + 7$.

The answer is (C).

3. **D** We can find the derivative using the formula $f'(x) = \lim\limits_{h \to 0} \dfrac{f(x+h) - f(x)}{h}$. We get

$$f'(x) = \lim\limits_{h \to 0} \frac{(x+h)^3 - 4(x+h)^2 + 11 - \left(x^3 - 4x^2 + 11\right)}{h}.$$

Simplify the numerator: $\lim\limits_{h \to 0} \dfrac{x^3 + 3x^2h + 3xh^2 + h^3 - 4\left(x^2 + 2xh + h^2\right) + 11 - x^3 + 4x^2 - 11}{h}$

$\lim\limits_{h \to 0} \dfrac{x^3 + 3x^2h + 3xh^2 + h^3 - 4x^2 - 8xh - 4h^2 + 11 - x^3 + 4x^2 - 11}{h}$

$\lim\limits_{h \to 0} \dfrac{3x^2h + 3xh^2 + h^3 - 8xh - 4h^2}{h}$

Factor h out of the numerator: $\lim\limits_{h \to 0} \dfrac{h\left(3x^2 + 3xh + h^2 - 8x - 4h\right)}{h}$

Cancel the h in the numerator and the denominator: $\lim\limits_{h \to 0} 3x^2 + 3xh + h^2 - 8x - 4h$

And take the limit: $\lim\limits_{h \to 0} 3x^2 + 3xh + h^2 - 8x - 4h = 3x^2 - 8x$.

Therefore, the derivative is $3x^2 - 8x$.

The answer is (D).

4. **A** We can find the derivative using the formula $f'(x) = \lim\limits_{h \to 0} \dfrac{f(x+h) - f(x)}{h}$, and we can plug in $x = 25$ either after we find the derivative or before. Let's do it afterward.

We get $f'(x) = \lim\limits_{h \to 0} \dfrac{6\sqrt{(x+h)} - 6\sqrt{x}}{h}$

Multiply the numerator and the denominator by the conjugate of the numerator:

$$\lim\limits_{h \to 0} \frac{\left(6\sqrt{(x+h)} - 6\sqrt{x}\right)\left(6\sqrt{(x+h)} + 6\sqrt{x}\right)}{h\left(6\sqrt{(x+h)} + 6\sqrt{x}\right)}$$

Multiply out the numerator: $\lim\limits_{h \to 0} \dfrac{36(x+h) - 36x}{h\left(6\sqrt{(x+h)} + 6\sqrt{x}\right)}$

Simplify the numerator and denominator: $\lim\limits_{h \to 0} \dfrac{36x + 36h - 36x}{h\left(6\sqrt{(x+h)} + 6\sqrt{x}\right)} = \lim\limits_{h \to 0} \dfrac{36h}{6h\left(\sqrt{(x+h)} + \sqrt{x}\right)}$

Cancel the $6h$ in the numerator and the denominator: $\lim\limits_{h \to 0} \dfrac{6}{\sqrt{(x+h)} + \sqrt{x}}$

And take the limit: $\dfrac{6}{\sqrt{x+h} + \sqrt{x}} = \dfrac{6}{2\sqrt{x}} = \dfrac{3}{\sqrt{x}}$.

Finally, plug in $x = 25$: $\dfrac{3}{\sqrt{25}} = \dfrac{3}{5}$.

The answer is (A).

5. **A** We can find the derivative using the formula $f'(x) = \lim\limits_{h \to 0} \dfrac{f(x+h) - f(x)}{h}$, and we can plug in $x = 2$ either after we find the derivative or before. Let's do it at the beginning.

We get $f'(x) = \lim\limits_{h \to 0} \dfrac{\dfrac{1}{4(x+h)^2} - \dfrac{1}{4x^2}}{h}$. Plug in $x = 2$: $f'(x) = \lim\limits_{h \to 0} \dfrac{\dfrac{1}{4(2+h)^2} - \dfrac{1}{16}}{h}$

Combine the two rational expressions in the numerator:

$$\lim\limits_{h \to 0} \frac{\dfrac{1}{4(2+h)^2} - \dfrac{1}{16}}{h} = \lim\limits_{h \to 0} \frac{\dfrac{16}{4(2+h)^2 (16)} - \dfrac{4(2+h)^2}{4(2+h)^2 16}}{h} = \lim\limits_{h \to 0} \frac{\dfrac{16 - 4(2+h)^2}{4(2+h)^2 (16)}}{h}$$

Multiply out the numerator: $\lim\limits_{h \to 0} \dfrac{\dfrac{16 - 4\left(4 + 4h + h^2\right)}{64(2+h)^2}}{h} = \lim\limits_{h \to 0} \dfrac{\dfrac{16 - 16 - 16h - 4h^2}{64(2+h)^2}}{h}$

Simplify the numerator and denominator: $\lim\limits_{h \to 0} \dfrac{\dfrac{16 - 16 - 16h - 4h^2}{64(2+h)^2}}{h} = \lim\limits_{h \to 0} \dfrac{-16h - 4h^2}{64h(2+h)^2}$

Factor $-4h$ out of the numerator $\lim\limits_{h \to 0} \dfrac{-4h(4+h)}{64h(2+h)^2}$

Cancel the $4h$ in the numerator and the denominator: $\displaystyle\lim_{h \to 0} \frac{-(4+h)}{16(2+h)^2}$

And take the limit: $\dfrac{-4}{64} = -\dfrac{1}{16}$.

The answer is (A).

CHAPTER 6

1. **D**   Use the Power Rule:

   $$y = 5x^3 - 7x^2 + 11x + 2$$

   $$\frac{dy}{dx} = 15x^2 - 14x + 11$$

   The answer is (D).

2. **B**   Use the Chain Rule:

   $$\frac{dy}{dx} = 4\sec^2(4x)$$

   The answer is (B).

3. **C**   Use the Chain Rule:

   $\dfrac{dy}{dx} = 2\cos(x^2)(-\sin(x^2))(2x)$, which simplifies to $\dfrac{dy}{dx} = -4x\cos(x^2)\sin(x^2)$

   The answer is (C).

4. **A**   It might help to rewrite the function as $y = 4(x-1)^{-\frac{1}{2}}$. Now we can use the Chain Rule:

   $\dfrac{dy}{dx} = 4\left(-\dfrac{1}{2}\right)(x-1)^{-\frac{3}{2}}(1)$, which simplifies to $\dfrac{dy}{dx} = -\dfrac{2}{(x-1)^{\frac{3}{2}}}$

   The answer is (A).

5. **D**   It might help to rewrite the function as $y = \left(\dfrac{x^2}{\sec x}\right)^{\frac{1}{2}}$. Now we can use the Chain Rule and the

   Quotient Rule: $\dfrac{dy}{dx} = \dfrac{1}{2}\left(\dfrac{x^2}{\sec x}\right)^{-\frac{1}{2}}\left(\dfrac{\sec x(2x) - (x^2)\sec x \tan x}{\sec^2 x}\right)$.

   The answer is (D).

CHAPTER 7

1. **B** Take the derivative of each term with respect to x. Remember that when you are taking the derivative of a function of y, we have to multiply that derivative by $\dfrac{dy}{dx}$. We get

$$2x + 6y^2\frac{dy}{dx} = 3x^2 - 8y\frac{dy}{dx}.$$

Now we have to isolate $\dfrac{dy}{dx}$. Group the terms that contain $\dfrac{dy}{dx}$ on one side of the equals sign and the terms that do not contain $\dfrac{dy}{dx}$ on the other side: $6y^2\dfrac{dy}{dx} + 8y\dfrac{dy}{dx} = 3x^2 - 2x$.

Factor out $\dfrac{dy}{dx}$: $\left(6y^2 + 8y\right)\dfrac{dy}{dx} = 3x^2 - 2x$

And divide: $\dfrac{dy}{dx} = \dfrac{3x^2 - 2x}{6y^2 + 8y}$

The answer is (B).

2. **A** Take the derivative of each term with respect to x. Remember that when you are taking the derivative of a function of y, we have to multiply that derivative by $\dfrac{dy}{dx}$. We get

$$2x + 3x\left(2y\frac{dy}{dx}\right) + 3y^2 - 3y^2\frac{dy}{dx} = 0.$$

Now we have to isolate $\dfrac{dy}{dx}$. Group the terms that contain $\dfrac{dy}{dx}$ on one side of the equals sign and the terms that do not contain $\dfrac{dy}{dx}$ on the other side: $6xy\dfrac{dy}{dx} - 3y^2\dfrac{dy}{dx} = -2x - 3y^2$.

Factor out $\dfrac{dy}{dx}$: $\left(6xy - 3y^2\right)\dfrac{dy}{dx} = -2x - 3y^2$

And divide: $\dfrac{dy}{dx} = \dfrac{-2x - 3y^2}{6xy - 3y^2}$

The answer is (A).

3. **C** Take the derivative of each term with respect to x. Remember that when you are taking the derivative of a function of y, we have to multiply that derivative by $\dfrac{dy}{dx}$. We get $4\cos\left(x^2\right)(2x) = 3\dfrac{dy}{dx} - 2y\dfrac{dy}{dx}$.

Now, simply plug in $\left(\sqrt{\dfrac{\pi}{6}}, 2\right)$: $4\cos\left(\dfrac{\pi}{6}\right)\left(2\sqrt{\dfrac{\pi}{6}}\right) = 3\dfrac{dy}{dx} - 4\dfrac{dy}{dx}$.

This simplifies to $4\dfrac{\sqrt{3}}{2}\left(2\sqrt{\dfrac{\pi}{6}}\right) = -\dfrac{dy}{dx}$

$4\sqrt{3}\left(\sqrt{\dfrac{\pi}{6}}\right) = -\dfrac{dy}{dx}$

$-4\sqrt{\dfrac{\pi}{2}} = \dfrac{dy}{dx}$

The answer is (C).

4. **A**   Take the derivative of each term with respect to x. Remember that when you are taking the derivative of a function of y, we have to multiply that derivative by $\dfrac{dy}{dx}$. We get

$x\left(2y\dfrac{dy}{dx}\right)+y^2-4x^3\left(2y\dfrac{dy}{dx}\right)-12x^2y^2=0$. Now, simply plug in (1,1): $2\dfrac{dy}{dx}+1-8\dfrac{dy}{dx}-12=0$.

This simplifies to $-6\dfrac{dy}{dx}=11$

$\dfrac{dy}{dx}=-\dfrac{11}{6}$

The answer is (A).

5. **A**   Take the derivative of each term with respect to x. Remember that when you are taking the derivative of a function of y, we have to multiply that derivative by $\dfrac{dy}{dx}$. We get $\dfrac{dy}{dx}=3y^2\dfrac{dy}{dx}-2x$.

Now we have to isolate $\dfrac{dy}{dx}$. Group the terms that contain $\dfrac{dy}{dx}$ on one side of the equals sign and the terms that do not contain $\dfrac{dy}{dx}$ on the other side: $2x=3y^2\dfrac{dy}{dx}-\dfrac{dy}{dx}$.

Factor out $\dfrac{dy}{dx}$: $2x=\left(3y^2-1\right)\dfrac{dy}{dx}$

And divide: $\dfrac{dy}{dx}=\dfrac{2x}{3y^2-1}$.

Now we need to find the second derivative so do implicit differentiation again:

$\dfrac{d^2y}{dx^2}=\dfrac{\left(3y^2-1\right)(2)-(2x)\left(6y\dfrac{dy}{dx}\right)}{\left(3y^2-1\right)^2}$

Next, we substitute $\dfrac{dy}{dx}=\dfrac{2x}{3y^2-1}$: $\dfrac{d^2y}{dx^2}=\dfrac{\left(3y^2-1\right)(2)-(12xy)\left(\dfrac{2x}{3y^2-1}\right)}{\left(3y^2-1\right)^2}$. Which simplifies to

$\dfrac{d^2y}{dx^2}=\dfrac{\left(3y^2-1\right)^2(2)-\left(24x^2y\right)}{\left(3y^2-1\right)^3}$

The answer is (A).

CHAPTER 8

1. **C** First, find the y-coordinate: $y(2) = 5(2)^3 - 20(2) + 10 = 10$.

    Second, take the derivative: $\frac{dy}{dx} = 15x^2 - 20$. Plug in $x = 2$ to get the slope of the tangent line: $\frac{dy}{dx} = 15(2)^2 - 20 = 40$.

    Now we have a slope and a point, so the equation is $y - 10 = 40(x - 2)$.

    The answer is (C).

2. **D** First, find the y-coordinate: $y\left(\frac{\pi}{8}\right) = 4\sec\left(\frac{\pi}{4}\right) = 4\sqrt{2}$.

    Second, take the derivative: $\frac{dy}{dx} = 8\sec(2x)\tan(2x)$. Plug in $x = \frac{\pi}{8}$ to get the slope of the tangent line: $\frac{dy}{dx} = 8\sec\left(\frac{\pi}{4}\right)\tan\left(\frac{\pi}{4}\right) = 8\sqrt{2}$. Thus, the slope of the normal line is the negative reciprocal of the slope of the tangent line: $-\frac{1}{8\sqrt{2}}$.

    Now we have a slope and a point, so the equation is $y - 4\sqrt{2} = -\frac{1}{8\sqrt{2}}\left(x - \frac{\pi}{8}\right)$.

    The answer is (D).

3. **B** First, find $f(3)$ and $f(8)$: $f(3) = 3^2 - 6(3) + 5 = -4$ and $f(8) = 8^2 - 6(8) + 5 = 21$. Then $\frac{21 - (-4)}{8 - 3} = 5 = f'(c)$. Next, find $f'(x)$: $f'(x) = 2x - 6$. Then, according to the MVTD: $2c - 6 = 5$, so $c = \frac{11}{2}$.

    The answer is (B).

4. **D** First, because the function is not continuous on the interval, there may not be a solution. Let's show that this is true. First, find $f(2)$ and $f(6)$: $f(2) = \frac{2}{2 - 4} = -1$ and $f(6) = \frac{2}{6 - 4} = 1$.

    Then $\frac{1 - (-1)}{6 - 2} = \frac{1}{2} = f'(c)$. Next, find $f'(x)$: $f'(x) = \frac{-2}{(x - 4)^2}$. Then, according to the MVTD: $\frac{-2}{(c - 4)^2} = \frac{1}{2}$. This gives us $(c - 4)^2 = -4$, which has no solution.

    The answer is (D).

5. **D** First, find $f(-3)$ and $f(3)$: $f(-3)=(-3)^4-5(-3)^2+4=40$ and $f(3)=(3)^4-5(3)^2+4=40$.

Next, find $f'(x)$: $f'(x)=4x^3-10x$. By setting $f'(c)=4c^3-10c=0$, we get

$$c\left(4c^2-10\right)=0$$

$c=0$ or $c=\pm\sqrt{\dfrac{5}{2}}$, all of which are in the interval.

The answer is (D).

CHAPTER 9

1. **C** First, let's draw a picture:

Note that the base of the rectangle is $2x$ and the height of the rectangle is y. Thus the area of the

rectangle is $A=2xy$. Because $y=\sqrt{200-x^2}$, we can substitute and the area is $A=2x\sqrt{200-x^2}$.

Take the derivative: $\dfrac{dA}{dx}=2\sqrt{200-x^2}+2x\left(\dfrac{1}{2}\left(200-x^2\right)^{-\frac{1}{2}}(-2x)\right)=2\sqrt{200-x^2}+\dfrac{-2x^2}{\sqrt{200-x^2}}$

Set the derivative equal to zero and solve for x: $2\sqrt{200-x^2}+\dfrac{-2x^2}{\sqrt{200-x^2}}=0$

$2\sqrt{200-x^2}=\dfrac{2x^2}{\sqrt{200-x^2}}$

$2\left(200-x^2\right)=2x^2$

$200-x^2=x^2$

$200=2x^2$

$x=\pm10$. We can throw out the negative answer so the maximum is at $x=10$.

The answer is (C).

2.   **D**   First, let's draw a picture:

Note that the area of the base is x^2 and the height is y, so the volume of the box is $V = x^2 y = 108$

and the surface area is $A = x^2 + 4xy$. If we solve the volume for y, we get $y = \dfrac{108}{x^2}$. We can then

plug this into the area formula to get the area in terms of x only: $A = x^2 + 4x\left(\dfrac{108}{x^2}\right) = x^2 + \dfrac{432}{x}$.

Take the derivative: $\dfrac{dA}{dx} = 2x - \dfrac{432}{x^2}$.

Set it equal to zero and solve for x: $2x - \dfrac{432}{x^2} = 0$.

$2x^3 = 432$

$x^3 = 216$

$x = 6$ in. and thus $y = \dfrac{108}{6^2} = 3$ in.

The answer is (D).

3.   **B**   First, take the derivative: $\dfrac{dy}{dx} = 3x^2 - 6x - 24$. Set the derivative equal to zero and solve for x: $3x^2 - 6x - 24 = 0$.

$x^2 - 2x - 8 = 0$
$(x - 4)(x + 2) = 0$
$x = 4, -2$

Now that we have the critical values, we can find the y-coordinates by plugging the critical values back into the original equation. We get $y = 4^3 - 3(4)^2 - 24(4) + 6 = -74$.

$y = (-2)^3 - 3(-2)^2 - 24(-2) + 6 = 34$

Finally, we can determine which gives us the maximum and which gives us the minimum by using

the second derivative test. The second derivative is $\dfrac{d^2 y}{dx^2} = 6x - 6$.

At $x = 4$, the second derivative is $6(4) - 6 > 0$, so $(4, -74)$ is at a minimum.

At $x = -2$, the second derivative is $6(-2) - 6 < 0$, so $(-2, 34)$ is at a maximum.

The answer is (B).

4. **A** First, find the second derivative.

$$\frac{dy}{dx} = 3x^2 - 24x$$

$$\frac{d^2y}{dx^2} = 6x - 24$$

Now, set the second derivative equal to zero and solve for x.

$$6x - 24 = 0$$

$$x = 4$$

Now that we have the x-coordinate of the point of inflection, we can find the y-coordinate by plugging x into the original equation. We get $y = (4)^3 - 12(4)^2 + 120 = -8$.

Therefore, the point of inflection is at $(4, -8)$.

The answer is (A).

5. **B** First, take the derivative: $\frac{dy}{dx} = \cos x - \sqrt{3}\sin x$. Next, set the derivative equal to zero and solve for x: $\cos x - \sqrt{3}\sin x = 0$.

$$\cos x = \sqrt{3}\sin x$$

$$\frac{1}{\sqrt{3}} = \frac{\sin x}{\cos x} = \tan x$$

$$x = \frac{\pi}{6}$$

Now we need to figure out the interval where the function is increasing or decreasing. Put the critical value on a number line:

We can test values in each interval to find the sign of the derivative. Let's try a value between $x = 0$

and $x = \frac{\pi}{6}$: $\frac{dy}{dx} = \cos\frac{\pi}{12} - \sqrt{3}\sin\frac{\pi}{12} > 0$

And a value between $x = \frac{\pi}{6}$ and $x = \pi$: $\frac{dy}{dx} = \cos\frac{\pi}{2} - \sqrt{3}\sin\frac{\pi}{2} < 0$.

Thus the function is increasing on the interval $\left(0, \frac{\pi}{6}\right)$ and decreasing on $\left(\frac{\pi}{6}, \pi\right)$.

The answer is (B).

CHAPTER 10

1.   **C**   First, let's draw a picture:

Let's call the distance from the top of the ladder to the ground y and the distance from the wall to the bottom of the ladder x. The relationship between x and y is $x^2 + y^2 = 25^2$ (Pythagorean Theorem). We are given $\dfrac{dy}{dt} = -6$ and we are looking for $\dfrac{dx}{dt}$ when $y = 20$.

Let's take the derivative with respect to t: $2x\dfrac{dx}{dt} + 2y\dfrac{dy}{dt} = 0$, which simplifies to $x\dfrac{dx}{dt} + y\dfrac{dy}{dt} = 0$.

We have $y = 20$ and $\dfrac{dy}{dt}$, so in order to find $\dfrac{dx}{dt}$, we will need to find x. Plugging into the original equation, we get $x^2 + (20)^2 = 25^2$, so $x = 15$. Now we can plug this into the derivative and solve for $\dfrac{dx}{dt}$: $15\dfrac{dx}{dt} + (20)(-6) = 0$.

Therefore, $\dfrac{dx}{dt} = \dfrac{120}{15} = 8$ ft/s.

The answer is (C).

2. **A** First, let's draw a picture:

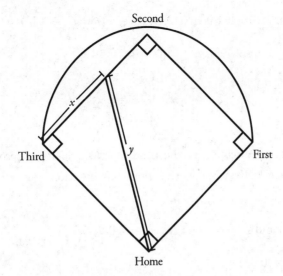

Call the distance from the runner to third base x and the distance from the runner to home plate y.

Then $\dfrac{dx}{dt} = -30$ (it's negative because the distance is shrinking), and we are looking for $\dfrac{dy}{dt}$ when

$x = 45$ (halfway from second base to third base). The relationship between x and y is $x^2 + 90^2 = y^2$

(Pythagorean Theorem). Take the derivative with respect to t: $2x\dfrac{dx}{dt} = 2y\dfrac{dy}{dt}$, which simplifies to

$x\dfrac{dx}{dt} = y\dfrac{dy}{dt}$. We will need to find y next. We have $45^2 + 90^2 = y^2$, so $y = 45\sqrt{5}$. Plugging every-

thing in, we get $(45)(-30) = \left(45\sqrt{5}\right)\dfrac{dy}{dt}$. Thus $\dfrac{dy}{dt} = -\dfrac{30}{\sqrt{5}}$ ft/s.

The answer is (A).

3. **B** We are given $\dfrac{dV}{dt} = -24\pi$ and we are looking for $\dfrac{dS}{dt}$ when $r = 6$. We know that the volume

of a sphere is $V = \dfrac{4}{3}\pi r^3$ and the surface area is $S = 4\pi r^2$. We could solve the volume for r

and plug it into the surface area equation to get S in terms of V, but this is messy. It's easier

to work with the equations separately. First, let's take the derivative of V with respect to t:

$\dfrac{dV}{dt} = \dfrac{4}{3}\pi\left(3r^2\dfrac{dr}{dt}\right) = 4\pi r^2\dfrac{dr}{dt}$. Next we plug in to solve for $\dfrac{dr}{dt}$: $-24\pi = 4\pi(6)^2\dfrac{dr}{dt}$.

$\dfrac{dr}{dt} = -\dfrac{1}{6}$

Next, let's take the derivative of S with respect to t: $\dfrac{dS}{dt} = 8\pi r\dfrac{dr}{dt}$. Now we can plug in $r = 6$ and

$\dfrac{dr}{dt} = -\dfrac{1}{6}$: $\dfrac{dS}{dt} = 8\pi(6)\dfrac{-1}{6} = -8\pi$ in.^{2}/s.

The answer is (B).

4. **C** The velocity function is simply the first derivative: $v(t) = t\left(\dfrac{1}{t}\right) + \ln t = 1 + \ln t$.

Set this equal to zero and solve for t: $1 + \ln t = 0$.

$\ln t = -1$, so $t = e^{-1} = \dfrac{1}{e}$.

The answer is (C).

5. **D** The particle is changing direction at the time that its velocity is zero and the velocity changes sign there. The first derivative of the position function gives us the velocity function: $x'(t) = v(t) = 12t^2 - 24t$.

Set this equal to zero and solve for t: $12t^2 - 24t = 0$

$t(12t - 24) = 0$, so $t = 0$ or $t = 2$.

We can throw out $t = 0$, so the particle is changing direction at $t = 2$. We can verify this by making sure that the velocity changes sign at $t = 2$. At $t = 1$, $v(1) = 12(1)^2 - 24(1) < 0$ and at $t = 3$, $v(3) = 12(3)^2 - 24(3) > 0$, so the position is indeed changing direction at $t = 2$.

The answer is (D).

CHAPTER 11

1. **C** Remember that if $y = \ln(f(x))$, then $\dfrac{dy}{dx} = \dfrac{f'(x)}{f(x)}$.

Here we get $\dfrac{dy}{dx} = \dfrac{3x^2 - 7}{x^3 - 7x + 5}$.

The answer is (C).

2. **A** First, use the log laws to rewrite the function:

$$y = \ln\left(\frac{\sqrt{x^2 - 11}}{x^2 + 9}\right) = \ln\sqrt{x^2 - 11} - \ln(x^2 + 9) = \frac{1}{2}\ln(x^2 - 11) - \ln(x^2 + 9)$$

Now take the derivative: $\dfrac{dy}{dx} = \dfrac{1}{2}\dfrac{2x}{x^2 - 11} - \dfrac{2x}{x^2 + 9} = \dfrac{x}{x^2 - 11} - \dfrac{2x}{x^2 + 9}$

The answer is (A).

3. **D** Remember that if $y = e^{f(x)}$, then $\dfrac{dy}{dx} = e^{f(x)}f'(x)$.

Here we get $\dfrac{dy}{dx} = 2\cos x e^{2\sin x}$

The answer is (D).

4. **B** Remember that if $y = \log_B(f(x))$, then $\dfrac{dy}{dx} = \dfrac{f'(x)}{f(x)}\dfrac{1}{\ln B}$.

Here we get $\dfrac{dy}{dx} = \dfrac{4x^3 + \sec x}{x^4 + \tan x}\dfrac{1}{\ln 8}$

The answer is (B).

5. **C** Remember that if $y = B^{f(x)}$, then $\dfrac{dy}{dx} = B^{f(x)}f'(x)\ln B$.

Here we get $\dfrac{dy}{dx} = 7^{x^2+1}(2x\ln 7)$

The answer is (C).

CHAPTER 12

1. **C** First, by inspection we can see that when $y = -7$, $x = 0$. Now let's use the formula for the derivative

of the inverse. The derivative is $\dfrac{dy}{dx} = 15x^2 + 1$. Now we use the formula to find the derivative of

the inverse: $\dfrac{1}{\left.\dfrac{dy}{dx}\right|_{x=0}} = \dfrac{1}{15x^2 + 1\big|_{x=0}} = 1$.

The answer is (C).

2. **D** First, let's find the slope of the tangent line using $\dfrac{dy}{dx} = \dfrac{\frac{dy}{dt}}{\frac{dx}{dt}}$: $\dfrac{dy}{dt} = -2\sin t$ and $\dfrac{dx}{dt} = 8\sin t\cos t$.

Therefore, $\dfrac{dy}{dx} = \dfrac{-2\sin t}{8\sin t\cos t} = \dfrac{-1}{4\cos t}$. Next, plug in $t = \dfrac{\pi}{3}$ to get the slope of the tan-

gent line: $\dfrac{-1}{4\cos\left(\dfrac{\pi}{3}\right)} = -\dfrac{1}{2}$. We will also need to find a pair of coordinates. At $t = \dfrac{\pi}{3}$,

$x = 4\sin^2\dfrac{\pi}{3} = 4\left(\dfrac{\sqrt{3}}{2}\right)^2 = 4\left(\dfrac{3}{4}\right) = 3$ and $y = 2\cos\dfrac{\pi}{3} = 1$. Therefore, the equation of the tangent

line is $y - 1 = -\dfrac{1}{2}(x - 3)$.

The answer is (D).

3. **B** Because $\lim_{x\to\infty} x^{-\frac{4}{3}} = 0$ and $\lim_{x\to\infty} \sin\left(\frac{1}{x}\right) = 0$, the limit is an indeterminate form $\frac{0}{0}$. We can try to

find the limit using L'Hôpital's Rule. Take the derivative of the numerator and the denominator:

$$\lim_{x\to\infty} \frac{x^{-\frac{4}{3}}}{\sin\left(\frac{1}{x}\right)} = \lim_{x\to\infty} \frac{-\frac{4}{3}x^{-\frac{7}{3}}}{\left(-\frac{1}{x^2}\right)\cos\left(\frac{1}{x}\right)}.$$

This simplifies to: $\lim_{x\to\infty} \dfrac{-\frac{4}{3}x^{-\frac{7}{3}}}{\left(-\frac{1}{x^2}\right)\cos\left(\frac{1}{x}\right)} = \lim_{x\to\infty} \dfrac{\frac{4}{3}x^{-\frac{1}{3}}}{\cos\left(\frac{1}{x}\right)}$. Now when we take the limit, we get

$\lim_{x\to\infty} \dfrac{\frac{4}{3}x^{-\frac{1}{3}}}{\cos\left(\frac{1}{x}\right)} = \dfrac{0}{1} = 0$.

The answer is (B).

4. **C** Let $x = 2$, $\Delta x = +0.03$, and $f(x) = x^5$. Using differentials, we get $f(2+.03) \approx f(2) + f'(2)(0.03)$. Take the derivative: $f'(x) = 5x^4$. At $x = 2$, we get $f(2) = 2^5 = 32$ and $f'(2) = 5(2)^4 = 80$, so $(2.03)^5 \approx 32 + (80)(0.03) \approx 34.4$.

The answer is (C).

5. **A** Take the log of both sides: $\ln y = \ln \sqrt[4]{\dfrac{(8x+1)^3}{(2x-11)^5}}$. Simplify using the log rules:

$$\ln y = \frac{1}{4}\ln\frac{(8x+1)^3}{(2x-11)^5} = \frac{1}{4}\ln(8x+1)^3 - \frac{1}{4}\ln(2x-11)^5 = \frac{3}{4}\ln(8x+1) - \frac{5}{4}\ln(2x-11).$$

Now take the derivative of both sides: $\dfrac{1}{y}\dfrac{dy}{dx} = \dfrac{3}{4}\dfrac{8}{8x+1} - \dfrac{5}{4}\dfrac{2}{2x-11}$.

Multiply by y: $\dfrac{dy}{dx} = y\left(\dfrac{3}{4}\dfrac{8}{8x+1} - \dfrac{5}{4}\dfrac{2}{2x-11}\right)$.

Note that we could now replace y with $\sqrt[4]{\dfrac{(8x+1)^3}{(2x-11)^5}}$, but shouldn't do so unless asked to.

The answer is (A).

CHAPTER 13

1. **C** Using the Power Rule, we get $\int \left(3x^4 - 6x^3 + 2x + 7\right) dx = \dfrac{3x^5}{5} - \dfrac{6x^4}{4} + x^2 + 7x + C$

   The answer is (C).

2. **B** $\int 3\cos x - \dfrac{4}{\sqrt{x}} dx = 3\sin x - 8\sqrt{x} + C$

   The answer is (B).

3. **A** Use u-substitution. Let $u = 1 - x^3$ and $du = -3x^2 dx$, so $-3 du = 9x^2 dx$.

   Substitute: $\int \dfrac{9x^2}{\left(1-x^3\right)^2} dx = -3\int \dfrac{du}{u^2} = -3\int u^{-2} du = 3u^{-1} + C = \dfrac{3}{1-x^3} + C$

   The answer is (A).

4. **D** Use u-substitution. Let $u = \dfrac{1}{x}$ and $du = -\dfrac{1}{x^2} dx$.

   Substitute: $\int \dfrac{\cos\left(\dfrac{1}{x}\right)}{x^2} dx = -\int \cos u \, du = -\sin u = -\sin\left(\dfrac{1}{x}\right) + C$

   The answer is (D).

5. **B** Use u-substitution. Let $u = \cos(3x)$ and $du = -3\sin(3x) dx$, so $-\dfrac{1}{3} du = \sin(3x) dx$. Substitute:

   $\int \cos^3(3x)\sin(3x) dx = -\dfrac{1}{3}\int u^3 du = -\dfrac{u^4}{12} = -\dfrac{\cos^4(3x)}{12} + C$.

   The answer is (B).

CHAPTER 14

1. **C** The width of each rectangle is $\dfrac{5-1}{4} = 1$. Using the formula, the area is approximately:

   $(1)\left[y(1) + y(2) + y(3) + y(4)\right] =$

   $\left[\left(3+(1)^2\right) + \left(3+(2)^2\right) + \left(3+(3)^2\right) + \left(3+(4)^2\right)\right] =$

   $[4+7+12+19] = 42$

   The answer is (C).

2. **B** The width of each trapezoid is $\dfrac{8-0}{4}=2$. Using the formula, the area is approximately:

$$\frac{1}{2}(2)\big[y(0)+2y(2)+2y(4)+2y(6)+y(8)\big]=$$

$$\Big[\big(0^3+4\big)+2\big(2^3+4\big)+2\big(4\big(4^3+4\big)\big)+2\big(6^3+4\big)+\big(8^3+4\big)\Big]=$$

$$\big[4+2(12)+2(68)+2(220)+(516)\big]=1{,}120$$

The answer is (B).

3. **D** The width of each rectangle is $\dfrac{16-0}{4}=4$. Using the formula, the area is approximately:

$$(4)\big[y(2)+y(6)+y(10)+y(14)\big]=$$

$$(4)\left[\frac{3}{2}+\frac{3}{6}+\frac{3}{10}+\frac{3}{14}\right]=(4)\left(\frac{88}{35}\right)=\frac{352}{35}$$

The answer is (D).

4. **B** We need to evaluate $\dfrac{1}{4-2}\displaystyle\int_2^4 \dfrac{x^2}{\left(1-x^3\right)^2}\,dx=\dfrac{1}{2}\displaystyle\int_2^4 \dfrac{x^2}{\left(1-x^3\right)^2}\,dx$. We can evaluate $\displaystyle\int \dfrac{x^2}{\left(1-x^3\right)^2}\,dx$ using

u-substitution. Let $u=1-x^3$ and $du=-3x^2\,dx$, so $-\dfrac{1}{3}\,du=x^2\,dx$.

Substitute: $\displaystyle\int \dfrac{x^2}{\left(1-x^3\right)^2}\,dx=-\frac{1}{3}\int \frac{du}{u^2}=-\frac{1}{3}\int u^{-2}\,du=\frac{1}{3}u^{-1}+C=\frac{1}{3\left(1-x^3\right)}+C$. This is a definite in-

tegral, so we evaluate: $\dfrac{1}{2}\left(\dfrac{1}{3\left(1-x^3\right)}\right)\Bigg|_2^4=\dfrac{1}{2}\left(\dfrac{1}{3(-63)}-\dfrac{1}{3(-7)}\right)=\dfrac{4}{189}$.

The answer is (B).

5. **D** $\dfrac{d}{dx}\displaystyle\int_2^{x^2} \sin^3 t\,dt=\left(\sin^3 x\right)(2x)=2x\sin^3 x$

The answer is (D).

CHAPTER 15

1. **A** Use u-substitution. Let $u=4x^3+1$ and $du=12x^2\,dx$, so $\dfrac{1}{12}\,du=x^2\,dx$.

Substitute: $\displaystyle\int \dfrac{x^2}{\left(4x^3+1\right)}\,dx=\dfrac{1}{12}\int \dfrac{du}{u}$. And integrate: $\dfrac{1}{12}\displaystyle\int \dfrac{du}{u}=\dfrac{1}{12}\ln|u|+C$. Finally, we substitute

back and get $\dfrac{1}{12}\ln\left|4x^3+1\right|+C$.

The answer is (A).

2. **C**   Use u-substitution. Let $u = x^2$ and $du = 2x\,dx$, so $\dfrac{1}{2}du = x\,dx$.

Substitute: $\displaystyle\int x\tan x^2\,dx = \dfrac{1}{2}\int \tan u\,du$. And integrate: $\dfrac{1}{2}\displaystyle\int \tan u\,du = -\dfrac{1}{2}\ln|\cos u| + C$. Finally, we substitute back and get $-\dfrac{1}{2}\ln|\cos x^2| + C$.

The answer is (C).

3. **D**   Use u-substitution. Let $u = 1 + x^4$ and $du = 4x^3\,dx$, so $\dfrac{1}{4}du = x^3\,dx$.

Substitute: $\displaystyle\int x^3 e^{1+x^4}\,dx = \dfrac{1}{4}\int e^u\,du$. And integrate: $\dfrac{1}{4}\displaystyle\int e^u\,du = \dfrac{1}{4}e^u + C$. Finally, we substitute back and get $\dfrac{1}{4}e^{1+x^4} + C$.

The answer is (D).

4. **A**   Use u-substitution. Let $u = 2 + x^3$ and $du = 3x^2\,dx$, so $\dfrac{1}{3}du = x^2\,dx$.

Substitute: $\displaystyle\int x^2 5^{2+x^3}\,dx = \dfrac{1}{3}\int 5^u\,du$. And integrate: $\dfrac{1}{3}\displaystyle\int 5^u\,du = \dfrac{1}{3\ln 5}5^u + C$. Finally, we substitute back and get $\dfrac{1}{3\ln 5}5^{2+x^3} + C$.

The answer is (A).

5. **B**   Use u-substitution. Let $u = x^2$ and $du = 2x\,dx$, so $2\,du = 4x\,dx$.

Substitute: $\displaystyle\int 4x\sec(x^2)\,dx = 2\int \sec(u)\,du$. And integrate: $2\displaystyle\int \sec(u)\,du = 2\ln|\sec(u) + \tan(u)| + C$. Finally, we substitute back and get $2\ln|\sec(x^2) + \tan(x^2)| + C$.

The answer is (B).

CHAPTER 16

1. **B**   First, let's draw a picture.

In order to find the area, we will need to evaluate the integral $\int_0^{\frac{\pi}{2}} \cos x \, dx$.

We get $\int_0^{\frac{\pi}{2}} \cos x \, dx = \sin x \Big|_0^{\frac{\pi}{2}} = \sin\frac{\pi}{2} - \sin 0 = 1$.

The answer is (B).

2. **B**   First, let's draw a picture.

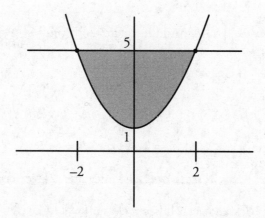

Note that the curve $y = 5$ is on top and the curve $y = x^2 + 1$ is on the bottom. Next, we need to find where the curves intersect. Set them equal to each other and solve for x:

$$x^2 + 1 = 5$$

$$x^2 = 4$$

$$x = \pm 2$$

Thus, in order to find the area, we will need to evaluate the integral $\int_{-2}^{2} 5 - (x^2 + 1) \, dx = \int_{-2}^{2} 4 - x^2 \, dx$.

We get $\int_{-2}^{2} 4 - x^2 \, dx = \left(4x - \frac{x^3}{3} \right) \Big|_{-2}^{2} = \left(8 - \frac{8}{3} \right) - \left(-8 + \frac{8}{3} \right) = \frac{32}{3}$.

The answer is (B).

3. **B**   First, let's draw a picture.

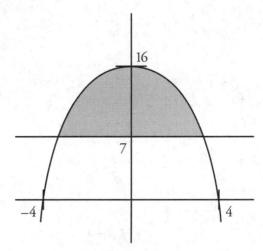

Note that the curve $y = 16 - x^2$ is on top and the curve $y = 7$ is on the bottom. Next, we need to find where the curves intersect. Set them equal to each other and solve for x:

$16 - x^2 = 7$
$x^2 = 9$
$x = \pm 3$

Thus, in order to find the area, we will need to evaluate the integral $\int_{-3}^{3} (16 - x^2) - 7 \, dx = \int_{-3}^{3} 9 - x^2 \, dx$.

We get $\int_{-3}^{3} 9 - x^2 \, dx = \left(9x - \frac{x^3}{3} \right)\Big|_{-3}^{3} = (27 - 9) - (-27 + 9) = 36$.

The answer is (B).

4. **D**   First, let's draw a picture.

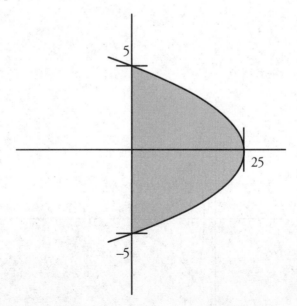

Next, we need to find where the curve $x = 25 - y^2$ intersects the y-axis. Set $25 - y^2 = 0$. We get

$y = \pm 5$. Thus, in order to find the area, we will need to evaluate the integral $\int_{-5}^{5} 25 - y^2 \, dy$.

We get $\int_{-5}^{5} 25 - y^2 \, dy = \left(25y - \dfrac{y^3}{3}\right)\Bigg|_{-5}^{5} = \left(125 - \dfrac{125}{3}\right) - \left(-125 + \dfrac{125}{3}\right) = \dfrac{500}{3}$.

The answer is (D).

5. **C**   First, let's draw a picture.

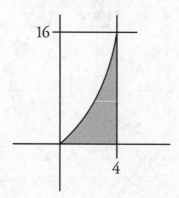

Note that the curve $x = 4$ is farther away from the y-axis than the curve $x = \sqrt{y}$. Next, we need

to find where the two curves intersect. Set them equal to each other and solve for y: $\sqrt{y} = 4$, so

$y = 16$. Thus, in order to find the area, we will need to evaluate the integral $\int_{0}^{16} 4 - \sqrt{y} \, dy$. We get

$\int_{0}^{16} 4 - \sqrt{y} \, dy = \left(4y - \dfrac{2}{3} y^{\frac{3}{2}}\right)\Bigg|_{0}^{16} = \left(4y - \dfrac{2}{3}\sqrt{y^3}\right)\Bigg|_{0}^{16} = \left(64 - \dfrac{128}{3}\right) - 0 = \dfrac{64}{3}$.

The answer is (C).

CHAPTER 17

1. **B**   First, let's draw a picture.

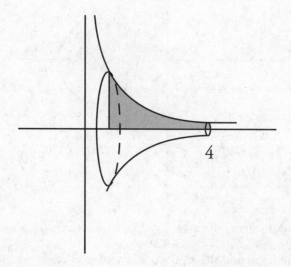

Next, we need to find the boundaries of the region. We are given that they are $x = 1$ and $x = 4$.

Thus, in order to find the volume, we will need to evaluate the integral $\pi \int_1^4 \left(\dfrac{1}{x}\right)^2 dx$. We get

$$\pi \int_1^4 \left(\frac{1}{x}\right)^2 dx = \pi \int_1^4 x^{-2} \, dx = \pi \left(\frac{x^{-1}}{-1}\right)\Bigg|_1^4 = \pi\left(-\frac{1}{x}\right)\Bigg|_1^4 = \frac{3\pi}{4}.$$

The answer is (B).

2. **C** First, let's draw a picture.

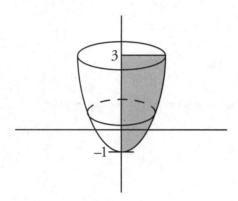

Next, we need to find the boundaries of the region. One of them is $y = 3$ and the other is where the curve intersects the y-axis, which is at $y = -1$. Thus, in order to find the volume, we will need to evaluate the integral $\pi \int_{-1}^3 \left(\sqrt{1+y}\right)^2 dy$. We get

$$\pi \int_{-1}^3 \left(\sqrt{1+y}\right)^2 dy = \pi \int_{-1}^3 (1+y)\, dy = \pi\left(y + \frac{y^2}{2}\right)\Bigg|_{-1}^3 = \pi\left(3 + \frac{9}{2}\right) - \pi\left(-1 + \frac{1}{2}\right) = 8\pi.$$

The answer is (C).

3.   **D**   First, let's draw a picture.

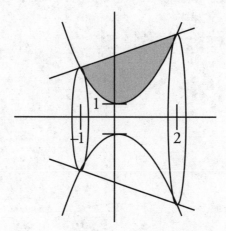

Next, we need to find the boundaries of the region. Set the curves equal to each other and solve for x: $x^2 + 1 = x + 3$

$x^2 - x - 2 = 0$

$(x - 2)(x + 1) = 0$

$x = -1, 2$

Note that the curve $y = x + 3$ is always on top and the curve $y = x^2 + 1$ is always on the bottom.

Thus, in order to find the volume, we will need to evaluate the integral $\pi \int\limits_{-1}^{2} (x+3)^2 - (x^2+1)^2 \, dx$.

We get $\pi \int\limits_{-1}^{2} (x+3)^2 - (x^2+1)^2 \, dx = \pi \int\limits_{-1}^{2} -x^4 - x^2 + 6x + 8 \, dx = \pi \left(-\dfrac{x^5}{5} - \dfrac{x^3}{3} + 3x^2 + 8x \right)\Bigg|_{-1}^{2} = \dfrac{117\pi}{5}$.

The answer is (D).

4. **C**   First, let's draw a picture.

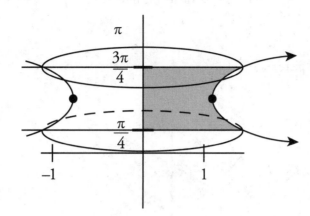

Next, we need to find the boundaries of the region. We are given that they are $y = \dfrac{\pi}{4}$ and $y = \dfrac{3\pi}{4}$.

Thus, in order to find the volume, we will need to evaluate the integral $\pi\displaystyle\int_{\frac{\pi}{4}}^{\frac{3\pi}{4}} \left(\csc y\right)^2 dy$. We get

$$\pi\int_{\frac{\pi}{4}}^{\frac{3\pi}{4}} \left(\csc y\right)^2 dy = -\pi \cot y \Big|_{\frac{\pi}{4}}^{\frac{3\pi}{4}} = 2\pi.$$

The answer is (C).

5. **A**   First, let's draw a picture.

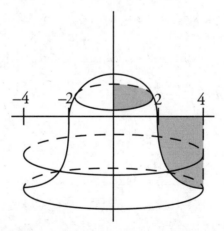

We need to rewrite the curve in terms of x. We get $y = \left(4 - x^2\right)^{\frac{1}{3}}$. Next, we need to find the boundaries of the region. We are given that they are $x = 0$ and $x = 4$, but note that from $x = 0$ to $x = 2$ the curve is above the *x*-axis and that from *x* = 2 to *x* = 4 the curve is below the *x*-axis. Thus, in order

to find the volume, we will need to evaluate the integrals $2\pi\int_{0}^{2} x\left(4-x^2\right)^{\frac{1}{3}}dx - 2\pi\int_{2}^{4} x\left(4-x^2\right)^{\frac{1}{3}}dx$.

The answer is (A).

6.　**C**　First, let's draw a picture.

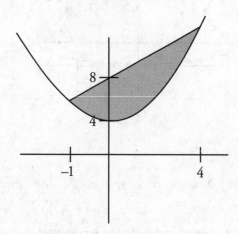

Note that the curve $y = 3x + 8$ is always on top and $y = x^2 + 4$ is always on the bottom of the
region. Next, we need to find the boundaries of the region. Set the curves equal to each other and
solve for *x*: $x^2 + 4 = 3x + 8$

$x^2 - 3x - 4 = 0$

$(x-4)(x+1) = 0$

$x = -1, 4$

The "slice" that will be the side of the square is the top curve minus the bottom curve, so in

order to find the volume, we will need to evaluate the integral $\int_{-1}^{4}\left[(3x+8)-\left(x^2+4\right)\right]^2 dx$. We get

$\int_{-1}^{4}\left[(3x+8)-\left(x^2+4\right)\right]^2 dx = \int_{-1}^{4}\left[-x^2+3x+4\right]^2 dx = \int_{-1}^{4} x^4 - 6x^3 + x^2 + 24x + 16\, dx = \dfrac{625}{6}$.

The answer is (C).

CHAPTER 18

1.　**C**　First, separate the variables: $\dfrac{dy}{y^2} = -4x\,dx$.

Integrate both sides: $\int \dfrac{dy}{y^2} = \int -4x\,dx$, which can be rewritten as $\int y^{-2}\,dy = -4\int x\,dx$.

$\dfrac{y^{-1}}{-1} = -4\dfrac{x^2}{2} + C$, which can be rewritten as $-\dfrac{1}{y} = -2x^2 + C$.

Isolate y: $\dfrac{1}{2x^2+C}=y$. Substitute using the initial condition: $\dfrac{1}{2(0)^2+C}=1$, so $C=1$. Therefore, the equation is $y=\dfrac{1}{2x^2+1}$.

The answer is (C).

2. **D** First, separate the variables: $\dfrac{1}{\tan y}\,dy=\dfrac{4}{\sec x}\,dx$. This can be rewritten as $\cot y\,dy=4\cos x\,dx$. Integrate both sides: $\displaystyle\int\cot y\,dy=4\int\cos x\,dx$.

$\ln|\sin y|=4\sin x+C$. Exponentiate both sides: $\sin y=e^{4\sin x+C}=Ae^{4\sin x}$.

Substitute using the initial condition: $\sin\left(\dfrac{\pi}{6}\right)=Ae^0=A$, so $A=\dfrac{1}{2}$. Therefore, the equation is $\sin y=\dfrac{1}{2}e^{4\sin x}$.

The answer is (D).

3. **B** The exponential decay means that $\dfrac{dy}{dt}=-ky$, where k is a constant. Separate the variables: $\dfrac{dy}{y}=-k\,dt$. Integrate both sides: $\displaystyle\int\dfrac{dy}{y}=-k\int dt$.

$\ln|y|=-kt+C$. Exponentiate both sides: $y=e^{-kt+C}=Ae^{-kt}$. Next, we are given that the half-life is 140 days, so $50=100e^{-k(140)}$. Now we can solve for k: $50=100e^{-k(140)}$, and $\dfrac{1}{2}=e^{-k(140)}$.

$\ln\dfrac{1}{2}=-140k$, so $k=-\dfrac{1}{140}\ln\dfrac{1}{2}$. If we substitute this for k into the equation, we get $y=100e^{\left(\frac{1}{140}\ln\frac{1}{2}\right)t}$. Use the rules of logs to rewrite the exponent: $y=100e^{\left(\ln\frac{1}{2}^{\frac{1}{140}}\right)t}$, and $y=100\dfrac{1}{2}^{\frac{t}{140}}$.

Finally, plug in $t=70$ (10 weeks): $y=100\dfrac{1}{2}^{\frac{1}{2}}\approx71\,\mathrm{g}$.

The answer is (B).

4. **C** We start with $x_0=0$ and $y_0=3$. The slope is $\dfrac{dy}{dx}=0+3=3$, $y_0'=3$.

Increase x_0 by $h=0.25$: $x_1=0.25$.

Multiply $h=0.25$ by $y_0'=3$ and add to y_0: $y_1=3+(3)(0.25)=3.75$.

Plug into the differential equation to get $y_1'=0.25+3.75=4$.

Repeat:

$$x_2 = 0.25 + 0.25 = 0.5$$

$$y_2 = 4 + (4)(0.25) = 5$$

$$y'_2 = 0.5 + 5 = 5.5$$

Repeat:

$$x_3 = 0.5 + 0.25 = 0.75$$

$$y_2 = 5 + (5)(0.25) = 6.25$$

$$y'_2 = 0.75 + 6.25 = 7$$

One more time! $x_4 = 0.75 + 0.25 = 1$

$$y_3 = 6.25 + (6.25)(0.25) = 7.8125$$

$$y'_3 = 1 + 7.8125 = 8.8125$$

The answer is (C).

5. **B**   The exponential growth means that $\dfrac{dy}{dt} = ky$, where k is a constant. Separate the variables $\dfrac{dy}{y} = kdt$. Integrate both sides: $\displaystyle\int \dfrac{dy}{y} = k \int dt$.

$\ln|y| = kt + C$. Exponentiate both sides: $y = e^{kt+C} = Ae^{kt}$. Next, we are given that initially there are 16 cm^2, so $16 = Ae^{k(0)}$ and thus $A = 16$. Now the equation is $y = 16e^{kt}$. Next, solve for k:

$$60 = 16e^{k(30)}$$

$$\frac{60}{16} = \frac{15}{4} = e^{k(30)}$$

$\ln \dfrac{15}{4} = 30k$, so $k = \dfrac{1}{30} \ln \dfrac{15}{4}$. Substitute this for k into the equation, to get $y = 16e^{\left(\frac{1}{30}\ln\frac{15}{4}\right)t}$. Use the rules of logs to rewrite the exponent: $y = 16e^{\left(\frac{1}{30}\ln\frac{15}{4}\right)t} = 16e^{\left(\ln\frac{15}{4}\right)^{\frac{t}{30}}} = 16\left(\dfrac{15}{4}\right)^{\frac{t}{30}}$.

Finally, plug in $t = 92$: $y = 16\left(\dfrac{15}{4}\right)^{\frac{92}{30}} \approx 921 \, \text{cm}^2$.

The answer is (B).

Part VI
Practice Tests 2 and 3

Practice Test 2

AP® Calculus AB Exam

SECTION I: Multiple-Choice Questions

DO NOT OPEN THIS BOOKLET UNTIL YOU ARE TOLD TO DO SO.

At a Glance

Total Time
1 hour and 45 minutes
Number of Questions
45
Percent of Total Grade
50%
Writing Instrument
Pencil required

Instructions

Section I of this examination contains 45 multiple-choice questions. Fill in only the ovals for numbers 1 through 45 on your answer sheet.

CALCULATORS MAY NOT BE USED IN THIS PART OF THE EXAMINATION.

Indicate all of your answers to the multiple-choice questions on the answer sheet. No credit will be given for anything written in this exam booklet, but you may use the booklet for notes or scratch work. After you have decided which of the suggested answers is best, completely fill in the corresponding oval on the answer sheet. Give only one answer to each question. If you change an answer, be sure that the previous mark is erased completely. Here is a sample question and answer.

Sample Question   Sample Answer

Chicago is a
(A) state
(B) city
(C) country
(D) continent

Use your time effectively, working as quickly as you can without losing accuracy. Do not spend too much time on any one question. Go on to other questions and come back to the ones you have not answered if you have time. It is not expected that everyone will know the answers to all the multiple-choice questions.

About Guessing

Many candidates wonder whether or not to guess the answers to questions about which they are not certain. Multiple-choice scores are based on the number of questions answered correctly. Points are not deducted for incorrect answers, and no points are awarded for unanswered questions. Because points are not deducted for incorrect answers, you are encouraged to answer all multiple-choice questions. On any questions you do not know the answer to, you should eliminate as many choices as you can, and then select the best answer among the remaining choices.

CALCULUS AB

SECTION I, Part A

Time—60 Minutes

Number of questions—30

A CALCULATOR MAY NOT BE USED ON THIS PART OF THE EXAMINATION

<u>Directions</u>: Solve each of the following problems, using the available space for scratchwork. After examining the form of the choices, decide which is the best of the choices given and fill in the corresponding oval on the answer sheet. No credit will be given for anything written in the test book. Do not spend too much time on any one problem.

<u>In this test</u>: Unless otherwise specified, the domain of a function f is assumed to be the set of all real numbers x for which $f(x)$ is a real number.

1. If $g(x) = \dfrac{1}{32}x^4 - 5x^2$, find $g'(4)$.

   (A) −72
   (B) −32
   (C)   24
   (D)   32

2. $\displaystyle\lim_{x \to 0} \dfrac{8x^2}{\cos x - 1} =$

   (A) −16
   (B) −1
   (C)   8
   (D)   6

GO ON TO THE NEXT PAGE.

3. $\lim\limits_{x \to 5} \dfrac{x^2 - 25}{x - 5}$ is

(A)   0
(B)   5
(C)   10
(D)   The limit does not exist.

4. If $f(x) = \dfrac{x^5 - x + 2}{x^3 + 7}$, find $f'(x)$.

(A) $\dfrac{(5x^4 - 1)}{(3x^2)}$

(B) $\dfrac{(x^3 + 7)(5x^4 - 1) - (x^5 - x + 2)(3x^2)}{(x^3 + 7)}$

(C) $\dfrac{(x^5 - x + 2)(3x^2) - (x^3 + 7)(5x^4 - 1)}{(x^3 + 7)^2}$

(D) $\dfrac{(x^3 + 7)(5x^4 - 1) - (x^5 - x + 2)(3x^2)}{(x^3 + 7)^2}$

GO ON TO THE NEXT PAGE.

5. Evaluate $\displaystyle\lim_{h\to 0}\frac{\tan\left(\dfrac{\pi}{4}+h\right)-1}{h}$.

(A) 0

(B) 1

(C) 2

(D) The limit does not exist.

6. $\displaystyle\int x\sqrt{3x}\ dx =$

(A) $\dfrac{2\sqrt{3}}{5}x^{\frac{5}{2}}+C$

(B) $\dfrac{5\sqrt{3}}{2}x^{\frac{5}{2}}+C$

(C) $\dfrac{\sqrt{3}}{2}x^{\frac{1}{2}}+C$

(D) $\dfrac{5\sqrt{3}}{2}x^{\frac{3}{2}}+C$

GO ON TO THE NEXT PAGE.

7. For what value of k is f continuous at $x = 1$ if $f(x) = \begin{cases} x^2 - 3kx + 2; x \le 1 \\ 5x - kx^2; x > 1 \end{cases}$.

    (A) -1

    (B) $-\dfrac{1}{2}$

    (C) 2

    (D) 8

---

8. The graph of $f(x)$ is given below. Evaluate $\displaystyle\int_0^7 f(x)\,dx$.

    (A) -11
    (B) -4
    (C) 5
    (D) 11

GO ON TO THE NEXT PAGE.

9. Find $\dfrac{dy}{dx}$ if $y = \sec(\pi x^2)$.

   (A) $\tan(\pi x^2)$
   (B) $(2\pi x)\tan(\pi x^2)$
   (C) $\sec(\pi x^2)\tan(\pi x^2)$
   (D) $\sec(\pi x^2)\tan(\pi x^2)(2\pi x)$

10. Given the curve $y = 5 - (x-2)^{\frac{2}{3}}$, find $\dfrac{dy}{dx}$ at $x = 2$.

   (A) $-\dfrac{2}{3}$

   (B) $-\dfrac{2}{3\sqrt[3]{2}}$

   (C) 5

   (D) The limit does not exist.

11. $\displaystyle\int_0^{\frac{1}{2}} \dfrac{2}{\sqrt{1-x^2}}\, dx =$

   (A) $\dfrac{\pi}{3}$

   (B) $-\dfrac{\pi}{3}$

   (C) $\dfrac{2\pi}{3}$

   (D) $-\dfrac{2\pi}{3}$

GO ON TO THE NEXT PAGE.

12. Let f be the function $f(x) = \begin{cases} x^2 - 3bx + 2 \, ; \, x \le 2 \\ 2bx^2 - 8 \, ; \, x > 2 \end{cases}$. For what value of b is f continuous at $x = 2$?

(A) 0

(B) $\dfrac{1}{4}$

(C) 1

(D) 4

13. Let f be the function defined by $f(x) = xe^{-x}$. What is the absolute maximum value of f ?

(A) 0

(B) $\dfrac{1}{e}$

(C) 1

(D) e

GO ON TO THE NEXT PAGE.

14. Find $\dfrac{dy}{dx}$ at $(1, 2)$ for $y^3 = xy - 2x^2 + 8$.

(A) $-\dfrac{11}{2}$

(B) $-\dfrac{2}{11}$

(C) $\dfrac{2}{11}$

(D) $\dfrac{11}{2}$

15. $\displaystyle\lim_{x \to 0} \dfrac{x \cdot 2^x}{2^x - 1} =$

(A) $\ln 2$

(B) 1

(C) 2

(D) $\dfrac{1}{\ln 2}$

GO ON TO THE NEXT PAGE.

16. $\int x\sec^2(1+x^2)\,dx =$

    (A) $\dfrac{1}{2}\tan(1+x^2)+C$

    (B) $2\tan(1+x^2)+C$

    (C) $\dfrac{x}{2}\tan(1+x^2)+C$

    (D) $2x\tan(1+x^2)+C$

---

17. Find the equation of the tangent line to $9x^2 + 16y^2 = 52$ through $(2, -1)$.

    (A) $-9x + 8y - 26 = 0$
    (B) $9x - 8y - 26 = 0$
    (C) $9x - 8y - 106 = 0$
    (D) $8x + 9y - 17 = 0$

---

18. A particle moves along the x-axis. The velocity of the particle at time t is given by $v(t) = 12t - 3t^2$. What is the total distance traveled by the particle from time $t = 0$ to $t = 6$?

    (A)   0
    (B) 32
    (C) 64
    (D) 96

GO ON TO THE NEXT PAGE.

19. If $f(x) = 3^{\pi x}$, then $f'(x) =$

(A) $\dfrac{3^{\pi x}}{\ln 3}$

(B) $\dfrac{3^{\pi x}}{\pi}$

(C) $\pi(3^{\pi x - 1})$

(D) $\pi \ln 3 (3^{\pi x})$

20. The average value of $f(x) = \dfrac{1}{x}$ from $x = 1$ to $x = e$ is

(A) $\dfrac{1}{e+1}$

(B) $\dfrac{1}{1-e}$

(C) $e - 1$

(D) $\dfrac{1}{e-1}$

GO ON TO THE NEXT PAGE.

21. If $y = \left(x^4 + \sin x\right)^6$, then $\dfrac{dy}{dx} =$

    (A) $6\left(x^4 + \sin x\right)^6 \left(4x^3 - \cos x\right)$

    (B) $6\left(x^4 + \sin x\right)^5 \left(4x^3 + \cos x\right)$

    (C) $6\left(4x^3 + \cos x\right)^5$

    (D) $6\left(4x^3 - \cos x\right)^5$

---

22. Find the slope of the normal line to $y = x + \cos xy$ at $(0, 1)$.

    (A) -1
    (B) 1
    (C) 0
    (D) Undefined

GO ON TO THE NEXT PAGE.

23. $\int \dfrac{\csc^2\left(\sqrt{x}\right)}{\sqrt{x}}\,dx =$

(A) $2\cot\sqrt{x} + C$

(B) $-2\cot\sqrt{x} + C$

(C) $\dfrac{\csc^2\sqrt{x}}{3\sqrt{x}} + C$

(D) $\dfrac{\csc^2\sqrt{x}}{6\sqrt{x}} + C$

24. $\displaystyle\lim_{x \to 0} \dfrac{\tan^3(2x)}{x^3} =$

(A) -8
(B) 2
(C) 8
(D) The limit does not exist.

GO ON TO THE NEXT PAGE.

25. A solid is generated when the region in the first quadrant bounded by the graph of $y = 1 + \sin^2 x$, the line $x = \dfrac{\pi}{2}$, the x-axis, and the y-axis is revolved about the x-axis. Its volume is found by evaluating which of the following integrals?

(A) $\pi \displaystyle\int_0^1 (1 + \sin^4 x)\, dx$

(B) $\pi \displaystyle\int_0^1 (1 + \sin^2 x)^2\, dx$

(C) $\pi \displaystyle\int_0^{\frac{\pi}{2}} (1 + \sin^4 x)\, dx$

(D) $\pi \displaystyle\int_0^{\frac{\pi}{2}} (1 + \sin^2 x)^2\, dx$

26. If $y = \left(\dfrac{x^3 - 2}{2x^5 - 1} \right)^4$, find $\dfrac{dy}{dx}$ at $x = 1$.

(A) −52
(B) −28
(C) 13
(D) 52

GO ON TO THE NEXT PAGE.

27. $\int x\sqrt{5-x}\,dx =$

(A) $\quad -\dfrac{10}{3}(5-x)^{\frac{3}{2}}$

(B) $\quad \dfrac{10}{3}\sqrt{\dfrac{5x^2}{2} - \dfrac{x^3}{3}} + C$

(C) $\quad 10(5-x)^{\frac{1}{2}} + \dfrac{2}{3}(5-x)^{\frac{3}{2}} + C$

(D) $\quad -\dfrac{10}{3}(5-x)^{\frac{3}{2}} + \dfrac{2}{5}(5-x)^{\frac{5}{2}} + C$

---

28. Given the differential equation $\dfrac{dy}{dt} = -2y$, where $y(0) = 100$, find $y(2)$.

(A) $\quad -200$

(B) $\quad -4$

(C) $\quad \dfrac{100}{e^{16}}$

(D) $\quad \dfrac{100}{e^4}$

GO ON TO THE NEXT PAGE.

29. Given $y = x^4 - 6x^3 - 24x^2 + 10$, for which of the following values of x does the graph of y have a point of inflection?

(A) 1
(B) 2
(C) 3
(D) 4

---

30. $\int_0^1 \tan x \, dx =$

(A) 0

(B) $\ln(\cos(1))$

(C) $\ln(\sec(1))$

(D) $\ln(\sec(1)) - 1$

END OF PART A, SECTION I
IF YOU FINISH BEFORE TIME IS CALLED, YOU MAY CHECK YOUR WORK ON PART A ONLY.
DO NOT GO ON TO PART B UNTIL YOU ARE TOLD TO DO SO.

CALCULUS AB

SECTION I, Part B

Time—45 Minutes

Number of questions—15

A GRAPHING CALCULATOR IS REQUIRED FOR SOME QUESTIONS ON THIS PART OF THE EXAMINATION

Directions: Solve each of the following problems, using the available space for scratchwork. After examining the form of the choices, decide which is the best of the choices given and fill in the corresponding oval on the answer sheet. No credit will be given for anything written in the test book. Do not spend too much time on any one problem.

In this test:

1. The **exact** numerical value of the correct answer does not always appear among the choices given. When this happens, select from among the choices the number that best approximates the exact numerical value.

2. Unless otherwise specified, the domain of a function f is assumed to be the set of all real numbers x for which $f(x)$ is a real number.

31. The graph of $y = \frac{1}{2} + \cos x$ has a zero on the interval $[0, \pi]$. What is the slope of the tangent line to the graph at that point?

(A) $\frac{\sqrt{3}}{2}$

(B) $-\frac{\sqrt{3}}{2}$

(C) $-\frac{\sqrt{2}}{2}$

(D) $\frac{\sqrt{2}}{2}$

GO ON TO THE NEXT PAGE.

32. $\dfrac{d}{dx}\displaystyle\int_0^{x^2}\sin^2 t\,dt =$

(A) $x^2\sin^2(x^2)$
(B) $2x\sin^2(x^2)$
(C) $\sin^2(x^2)$
(D) $x^2\cos^2(x^2)$

---

33. Given $y = x^{\cos 4x}$, find $\dfrac{dy}{dx}$.

(A) $\dfrac{dy}{dx} = x^{\cos 4x}\left[-\left(\dfrac{1}{x}\right)(4\sin 4x)\right]$

(B) $\dfrac{dy}{dx} = x^{\cos 4x}\left[(\cos 4x)\left(\dfrac{1}{x}\right) - \ln x(4\sin 4x)\right]$

(C) $\dfrac{dy}{dx} = (\cos 4x)x^{(\cos 4x)-1}$

(D) $\dfrac{dy}{dx} = \left[(\cos 4x)x^{(\cos 4x)-1}\right](-4\sin 4x)$

GO ON TO THE NEXT PAGE.

34. If f is defined by $f(x) = x + e^{-x^2}$ on the interval [0, 10], then f has a point of inflection at which of the following values of x ?

    (A) 0.379
    (B) 0.5
    (C) 0.707
    (D) 0.947

35. Estimate $\int_0^2 3e^x + 1 \, dx$ using a Riemann sum with $n = 4$ right-hand rectangles.

    (A) 26.357
    (B) 33.546
    (C) 52.713
    (D) 56.713

36. The volume generated by revolving about the x-axis the region above the curve $y = x^3$, below the line $y = 1$, and between $x = 0$ and $x = 1$ is

    (A) $\dfrac{\pi}{42}$

    (B) 0.143π

    (C) 0.643π

    (D) $\dfrac{6\pi}{7}$

GO ON TO THE NEXT PAGE.

37. A sphere is increasing in volume at the rate of $20 \frac{in.^3}{s}$. At what rate is the radius of the sphere increasing when the radius is 4 in. ?

(A) $0.025 \frac{in.}{s}$

(B) $0.0995 \frac{in.}{s}$

(C) $0.424 \frac{in.}{s}$

(D) $0.982 \frac{in.}{s}$

38. The function f is given by $f(x) = 2x^3 - 5$ on the interval $[1, 5]$. Which of the following is guaranteed by the Mean Value Theorem for some value c on the interval $(1, 5)$?

(A) 3.136
(B) 3.215
(C) 3.225
(D) 4.160

39. Find two non-negative numbers x and y whose sum is 100 and for which x^2y is a maximum.

(A) $x = 50$ and $y = 50$
(B) $x = 33.333$ and $y = 66.667$
(C) $x = 100$ and $y = 0$
(D) $x = 66.667$ and $y = 33.333$

GO ON TO THE NEXT PAGE.

40. An object is moving along a line with its velocity given by $v(t) = t^2 \sin t$, for time $t \geq 0$. If the object's position at time $t = 0$ is 4, what is its position at time $t = 2$?

(A) 0.469
(B) 2.469
(C) 4.469
(D) 6.469

---

41. $\displaystyle \int \sin^4(\pi x)\cos(\pi x)\,dx =$

(A) $\dfrac{\sin^5(\pi x)}{5\pi} + C$

(B) $\dfrac{\sin^5(\pi x)}{2\pi} + C$

(C) $-\dfrac{\cos^5(\pi x)}{5\pi} + C$

(D) $-\dfrac{\cos^5(\pi x)}{2\pi} + C$

GO ON TO THE NEXT PAGE.

42. A balloon is inflating at the rate $\dfrac{dV}{dt} = 300 - t \ln t \ \dfrac{\text{in.}^3}{\text{s}}$, where t is the number of seconds that the balloon has been inflating. If the initial volume of the balloon is 100 in^3, what is the volume of the balloon after it has been inflating for 8 seconds, to the nearest 10 in.^{3}?

(A) 150 in.^{3}
(B) 280 in.^{3}
(C) 320 in.^{3}
(D) 2450 in.^{3}

---

43. If $x^2 + 3x^2y + y^3 = 13$, find $\dfrac{dy}{dx}$ at $(2, 1)$.

(A) $-\dfrac{16}{15}$

(B) $-\dfrac{4}{15}$

(C) $\dfrac{16}{15}$

(D) $\dfrac{19}{15}$

GO ON TO THE NEXT PAGE.

44. Find the equation of the line tangent to $y = x\tan x$ at $x = 1$.

 (A) $y = 4.983x + 3.426$
 (B) $y = 4.983x - 3.426$
 (C) $y = 4.983x + 6.540$
 (D) $y = 4.983x - 6.540$

---

45. If $f(x)$ is continuous and differentiable and $f(x) = \begin{cases} ax^4 + 5x; & x \le 2 \\ bx^2 - 3x; & x > 2 \end{cases}$, then $b =$

 (A) 0
 (B) 2
 (C) 6
 (D) There is no value of b.

STOP

END OF PART B, SECTION I

IF YOU FINISH BEFORE TIME IS CALLED, YOU MAY CHECK YOUR WORK ON PART B ONLY.

DO NOT GO ON TO SECTION II UNTIL YOU ARE TOLD TO DO SO.

SECTION II
GENERAL INSTRUCTIONS

You may wish to look over the problems before starting to work on them, since it is not expected that everyone will be able to complete all parts of all problems. All problems are given equal weight, but the parts of a particular problem are not necessarily given equal weight.

A GRAPHING CALCULATOR IS REQUIRED FOR SOME PROBLEMS OR PARTS OF PROBLEMS ON THIS SECTION OF THE EXAMINATION.

- You should write all work for each part of each problem in the space provided for that part in the booklet. Be sure to write clearly and legibly. If you make an error, you may save time by crossing it out rather than trying to erase it. Erased or crossed-out work will not be graded.

- Show all your work. You will be graded on the correctness and completeness of your methods as well as your answers. Correct answers without supporting work may not receive credit.

- Justifications require that you give mathematical (noncalculator) reasons and that you clearly identify functions, graphs, tables, or other objects you use.

- You are permitted to use your calculator to solve an equation, find the derivative of a function at a point, or calculate the value of a definite integral. However, you must clearly indicate the setup of your problem, namely the equation, function, or integral you are using. If you use other built-in features or programs, you must show the mathematical steps necessary to produce your results.

- Your work must be expressed in standard mathematical notation rather than calculator syntax. For example, $\int_1^5 x^2\, dx$ may not be written as fnInt (X², X, 1, 5).

- Unless otherwise specified, answers (numeric or algebraic) need not be simplified. If your answer is given as a decimal approximation, it should be correct to three places after the decimal point.

- Unless otherwise specified, the domain of a function f is assumed to be the set of all real numbers x for which $f(x)$ is a real number.

GO ON TO THE NEXT PAGE.

SECTION II, PART A
Time—30 minutes
Number of problems—2

A graphing calculator is required for some problems or parts of problems.

During the timed portion for Part A, you may work only on the problems in Part A.

On Part A, you are permitted to use your calculator to solve an equation, find the derivative of a function at a point, or calculate the value of a definite integral. However, you must clearly indicate the setup of your problem, namely the equation, function, or integral you are using. If you use other built-in features or programs, you must show the mathematical steps necessary to produce your results.

1. The temperature on New Year's Day in Hinterland was given by $T(H) = -A - B \cos\left(\dfrac{\pi H}{12}\right)$, where T is the temperature in

   degrees Fahrenheit and H is the number of hours from midnight ($0 \leq H < 24$).

   (a) The initial temperature at midnight was $-15°$ F and at noon of New Year's Day was $5°$ F. Find A and B.

   (b) Find the average temperature for the first 10 hours.

   (c) Use the Trapezoid Rule with 4 equal subdivisions to estimate $\displaystyle\int_6^8 T(H)\, dH$.

   (d) Find an expression for the rate that the temperature is changing with respect to H.

---

2. Sea grass grows on a lake. The rate of growth of the grass is $\dfrac{dG}{dt} = kG$, where k is a constant.

   (a) Find an expression for G, the amount of grass in the lake (in tons), in terms of t, the number of years, if the amount of grass is 100 tons initially and 120 tons after one year.

   (b) In how many years will the amount of grass available be 300 tons?

   (c) If fish are now introduced into the lake and consume a consistent 80 tons/year of sea grass, how long will it take for the lake to be completely free of sea grass?

GO ON TO THE NEXT PAGE.

SECTION II, PART B
Time—1 hour
Number of problems—4

No calculator is allowed for these problems.

During the timed portion for Part B, you may continue to work on the problems in Part A without the use of any calculator.

---

3. The functions *f* and *g* are twice-differentiable and have the following table of values:

x	f(x)	f'(x)	g(x)	g'(x)
1	3	4	−2	−4
2	2	3	4	−2
3	5	2	−1	3
4	−1	−6	−8	0

(a) Let $h(x) = f(g(x))$. Find the equation of the tangent line to *h* at $x = 2$.

(b) Let $j(x) = f(x)g(x)$. Find $j'(3)$.

(c) Evaluate $\int_1^4 3f''(x)\,dx$.

---

4. Water is being poured into a hemispherical bowl of radius 6 inches at the rate of 4 in.³/sec.

(a) Given that the volume of the water in the spherical segment shown above is $V = \pi h^2\left(R - \dfrac{h}{3}\right)$, where R is the radius of

the *sphere,* find the rate that the water level is rising when the water is 2 inches deep.

(b) Find an expression for *r,* the radius of the *surface of the spherical segment* of water, in terms of *h.*

(c) How fast is the circular area of the surface of the spherical segment of water growing (in in.²/sec) when the water is 2 inches deep?

GO ON TO THE NEXT PAGE.

5. Let R be the region in the first quadrant bounded by $y^2 = x$ and $x^2 = y$.

   (a) Find the area of region R.

   (b) Find the volume of the solid generated when R is revolved about the x-axis.

   (c) The section of a certain solid cut by any plane perpendicular to the x-axis is a circle with the endpoints of its diameter lying on the parabolas $y^2 = x$ and $x^2 = y$. Find the volume of the solid.

---

6. For time $t \geq 0$, a particle moves along the x-axis. The velocity of the particle at time t is given by $v(t) = 1 - 2\cos\left(\dfrac{\pi}{3}t\right)$. The particle's position at time $t = 0$ is $x(0) = 8$.

   (a) Is the particle speeding up or slowing down at time $t = \dfrac{1}{2}$? Justify your answer.

   (b) When does the particle change direction in the interval $0 \leq t \leq 2$? Justify your answer.

   (c) What is the particle's position at time $t = 2$?

   (d) What is the total distance traveled from time $t = 0$ to time $t = 2$?

STOP

END OF EXAM

Practice Test 2:
Answers and
Explanations

ANSWER KEY

Section I

1.	B	11.	A	21.	B	31.	B	41.	A
2.	A	12.	C	22.	A	32.	B	42.	D
3.	C	13.	B	23.	B	33.	B	43.	A
4.	D	14.	B	24.	C	34.	C	44.	B
5.	C	15.	D	25.	D	35.	A	45.	C
6.	A	16.	A	26.	A	36.	D		
7.	A	17.	B	27.	D	37.	B		
8.	B	18.	C	28.	D	38.	B		
9.	D	19.	D	29.	D	39.	D		
10.	D	20.	D	30.	C	40.	D		

ANSWERS AND EXPLANATIONS TO SECTION I

1.   **B**   First, take the derivative.

$$g'(x) = \frac{1}{32}\left(4x^3\right) - 5(2x) = \frac{x^3}{8} - 10x$$

Now, plug in 4 for x.

$$\frac{(4)^3}{8} - 10(4) = 8 - 40 = -32$$

2.   **A**   If we take the limit as x goes to 0, we get an indeterminate form $\frac{0}{0}$, so let's use L'Hôpital's Rule. We

take the derivative of the numerator and the denominator and we get $\lim\limits_{x\to 0}\dfrac{8x^2}{\cos x - 1} = \lim\limits_{x\to 0}\dfrac{16x}{\left(-\sin x\right)}$.

When we take the limit, we again get an indeterminate form $\frac{0}{0}$, so let's use L'Hôpital's Rule a second

time. We take the derivative of the numerator and the denominator and we get

$\lim\limits_{x\to 0}\dfrac{16x}{\left(-\sin x\right)} = \lim\limits_{x\to 0}\dfrac{16}{\left(-\cos x\right)}$. Now, when we take the limit we get: $\lim\limits_{x\to 0}\dfrac{16}{\left(-\cos x\right)} = -16$.

3.   **C**   Notice that if we plug 5 into the expressions in the numerator and the denominator, we

get $\frac{0}{0}$, which is undefined. Before we give up, we need to see if we can simplify the limit so

that it can be evaluated. If we factor the expression in the numerator, we get $\dfrac{(x + 5)(x - 5)}{(x - 5)}$,

which can be simplified to $x + 5$. Now, if we take the limit (by plugging in 5 for x), we get 10.

4.   **D**   We need to use the Quotient Rule, which is

Given $f(x) = \dfrac{g(x)}{h(x)}$, then $f'(x) = \dfrac{h(x)g'(x) - g(x)h'(x)}{\left[h(x)\right]^2}$.

Here we have

$$f'(x) = \frac{(x^3 + 7)(5x^4 - 1) - (x^5 - x + 2)(3x^2)}{(x^3 + 7)^2}$$

5.　**C**　There are two ways we could evaluate this limit. First, we could recognize that this limit is in the

form of the Definition of the Derivative $\lim\limits_{h \to 0} \dfrac{f(x+h) - f(x)}{h}$. Then the limit is the derivative

of $f(x) = \tan x$ at $x = \dfrac{\pi}{4}$. The derivative of $f(x) = \tan x$ is $f'(x) = \sec^2 x$, and at $x = \dfrac{\pi}{4}$, we get

$f'\left(\dfrac{\pi}{4}\right) = \sec^2\left(\dfrac{\pi}{4}\right) = 2$. Second, we have a limit of the indeterminate form $\dfrac{0}{0}$, so we can use

L'Hôpital's Rule to find the limit. Take the derivative of the numerator and the denominator:

$\lim\limits_{h \to 0} \dfrac{\tan\left(\dfrac{\pi}{4} + h\right) - 1}{h} = \lim\limits_{h \to 0} \dfrac{\sec^2\left(\dfrac{\pi}{4} + h\right)}{1}$. Now we take the limit to get $\sec^2\left(\dfrac{\pi}{4}\right) = 2$.

6.　**A**　First, rewrite the integral as $\int x\left(\sqrt{3}\right)x^{\frac{1}{2}}\,dx$.

Now, we can simplify the integral to $\sqrt{3}\int x^{\frac{3}{2}}\,dx$.

Next, use the power rule for integrals, which is $\int x^n\,dx = \dfrac{x^{n+1}}{n+1} + C$.

Then, we get $\sqrt{3}\int x^{\frac{3}{2}}\,dx = \sqrt{3}\,\dfrac{x^{\frac{5}{2}}}{\dfrac{5}{2}} + C = \dfrac{2\sqrt{3}}{5}x^{\frac{5}{2}} + C$.

7.　**A**　The simplest thing to do here is to find $\lim\limits_{x \to 1}$ of both pieces of the function and set them equal to
each other. We can do this by plugging $x = 1$ into both pieces: $1 - 3k + 2 = 5 - k$. If we solve for k,
we get $k = -1$.

8.　**B**　We have to find the areas of the regions between the curve and the x-axis on the interval from $x = 0$

to $x = 7$. The regions are all triangles, so we can easily find the areas and add them up. Note that the

region from $x = 2$ to $x = 6$ is negative. We get $\dfrac{1}{2}(2)(2) - \dfrac{1}{2}(4)(4) + \dfrac{1}{2}(1)(2) = -4$.

The answer is (B).

9.　**D**　Use the Chain Rule: $\dfrac{dy}{dx} = \sec(\pi x^2)\tan(\pi x^2)(2\pi x)$.

10.　**D**　First, take the derivative: $\dfrac{dy}{dx} = -\dfrac{2}{3}(x-2)^{-\frac{1}{3}}$, which can be rewritten as $\dfrac{dy}{dx} = -\dfrac{2}{3\sqrt[3]{x-2}}$. If we

plug in $x = 2$, we get zero in the denominator, so the derivative does not exist at $x = 2$.

11.　**A**　This integral is of the form $\int \dfrac{dx}{\sqrt{a^2 - x^2}} = \sin^{-1}\left(\dfrac{x}{a}\right) + C$, where $a = 1$.

Thus, we get

$$\int_0^{\frac{1}{2}} \frac{2\,dx}{\sqrt{1 - x^2}} = 2\sin^{-1}(x)\Big|_0^{\frac{1}{2}} = 2\left[\sin^{-1}\left(\frac{1}{2}\right) - \sin^{-1}(0)\right] = 2\left(\frac{\pi}{6} - 0\right) = \frac{\pi}{3}$$

12.　**C**　In order for f to be continuous, three conditions must be met. First, $f(2)$ must exist: $f(2) = 4 - 6b + 2 = 6 - 6b$. Next, $\lim\limits_{x \to 2} f(x)$ must exist. The limit from the left is $\lim\limits_{x \to 2^-} f(x) = 6 - 6b$, and the limit from the right is $\lim\limits_{x \to 2^+} f(x) = 8b - 8$. We set the two limits equal to each other and solve for b: $6 - 6b = 8b - 8$. So $b = 1$. The final condition is that $\lim\limits_{x \to 2} f(x) = f(2)$. At $b = 1$, $\lim\limits_{x \to 2} f(x) = f(2) = 0$. Therefore, b equals 1, and the answer is (C), 1.

13.　**B**　First, we need to take the derivative of f using the Product Rule: $f'(x) = e^{-x} - xe^{-x}$.

Next, we set it equal to zero and solve: $e^{-x} - xe^{-x} = 0$. Factor: $e^{-x}(1 - x) = 0$. Because e^{-x} is always positive, the only solution is $x = 1$. Next, we need to verify that $x = 1$ is a maximum. Plug in values less than and greater than 1 to make sure that the derivative changes sign there: $f'(0) = e^0 - (0)e^0 > 0$ and $f'(2) = e^{-2} - (2)e^{-2} < 0$. So, $x = 1$ is the x-coordinate of the maximum. The maximum value occurs at the y-coordinate, which we find by plugging $x = 1$ into the original equation: $f(1) = (1)e^{-1} = \dfrac{1}{e}$.

14.　**B**　We can use implicit differentiation to find $\dfrac{dy}{dx}$. First, differentiate with respect to x:

$3y^2\dfrac{dy}{dx} = x\dfrac{dy}{dx} + y - 4x$. Next, plug in (1, 2) for x and y: $3(2)^2\dfrac{dy}{dx} = (1)\dfrac{dy}{dx} + (2) - 4(1)$. Simplify:

$12\dfrac{dy}{dx} = \dfrac{dy}{dx} - 2$. And solve for $\dfrac{dy}{dx}$ on the left and the terms without $\dfrac{dy}{dx}$ on the right: $\dfrac{dy}{dx} = -\dfrac{2}{11}$.

15.　**D**　If we take the limit as x goes to 0, we get an indeterminate form $\dfrac{0}{0}$, so let's use L'Hôpital's Rule. We

take the derivative of the numerator and the denominator, and we get $\lim\limits_{x \to 0} \dfrac{x \cdot 2^x}{2^x - 1} = \lim\limits_{x \to 0} \dfrac{x \cdot 2^x \ln 2 + 2^x}{2^x \ln 2}$.

Now, when we take the limit, we get $\lim\limits_{x \to 0} \dfrac{x \cdot 2^x \ln 2 + 2^x}{2^x \ln 2} = \dfrac{1}{\ln 2}$.

16.　**A**　We can evaluate this integral using u-substitution. Let $u = 1 + x^2$ and $du = 2x\,dx$, so $\dfrac{1}{2}du = x\,dx$. Sub-

stitute into the integrand: $\int x\sec^2(1 + x^2)dx = \dfrac{1}{2}\int \sec^2 u\,du$. Integrate: $\dfrac{1}{2}\int \sec^2 u\,du = \dfrac{1}{2}\tan u + C$

and substitute back: $\dfrac{1}{2}\tan u + C = \dfrac{1}{2}\tan(1 + x^2) + C$.

17.   **B**   First, we need to find $\dfrac{dy}{dx}$. It's simplest to find it implicitly.

$$18x + 32y\,\dfrac{dy}{dx} = 0$$

Now, solve for $\dfrac{dy}{dx}$.

$$\dfrac{dy}{dx} = -\dfrac{18x}{32y} = -\dfrac{9x}{16y}$$

Next, plug in $x = 2$ and $y = -1$ to get the slope of the tangent line at the point.

$$\dfrac{dy}{dx} = \dfrac{-18}{-16} = \dfrac{9}{8}$$

Now, use the point-slope formula to find the equation of the tangent line.

$$(y + 1) = \dfrac{9}{8}(x - 2)$$

If we multiply through by 8, we get $8y + 8 = 9x - 18$ or $9x - 8y - 26 = 0$.

18.   **C**   First, we need to check the sign of the particle's velocity in the time interval. $v(t) = 12t - 3t^2$ $= 3t(4 - t) = 0$ at $t = 0$ and $t = 4$. Let's pick a number between 0 and 4, like 1: $v(1) = 12(1) - 3(1)^2 > 0$. Next, we pick a number greater than 4, like 5: $v(5) = 12(5) - 3(5)^2 < 0$. This means that the particle is moving to the right for the first four seconds and to the left for the last two seconds. We get the distance traveled in the first interval with $\int_0^4 12x - 3x^2\,dx = \left(6x^2 - x^3\right)\Big|_0^4 = 32 - 0 = 32$. We get the distance traveled in the second interval with $-\int_4^6 12x - 3x^2\,dx = -\left(6x^2 - x^3\right)\Big|_4^6 = -(0 - 32) = 32$. Therefore, the distance that the particle travels is $32 + 32 = 64$.

19.   **D**   The derivative of an expression of the form a^u, where u is a function of x, is

$$\dfrac{d}{dx}\,a^u = a^u\,(\ln a)\,\dfrac{du}{dx}$$

Here we get

$$\dfrac{d}{dx}\,3^{\pi x} = 3^{\pi x}\,(\ln 3)\pi$$

20.　**D**　In order to find the average value, we use the Mean Value Theorem for Integrals, which says that the average value of $f(x)$ on the interval $[a, b]$ is $\dfrac{1}{b-a}\displaystyle\int_a^b f(x)\,dx$.

Here we have $\dfrac{1}{e-1}\displaystyle\int_1^e \dfrac{1}{x}\,dx$.

Evaluating the integral, we get $\ln x\big|_1^e = \ln e - \ln 1 = 1$. Therefore, the answer is $\dfrac{1}{e-1}$.

21.　**B**　Use the Chain Rule to find the derivative: $\dfrac{dy}{dx} = 6\left(x^4 + \sin x\right)^5\left(4x^3 + \cos x\right)$.

22.　**A**　First, we need to find $\dfrac{dy}{dx}$ using implicit differentiation.

$$\frac{dy}{dx} = 1 - \left(x\frac{dy}{dx} + y\right)\sin xy$$

Rather than simplifying this, simply plug in $(0, 1)$ to find $\dfrac{dy}{dx}$.

We get $\dfrac{dy}{dx} = 1$.

This means that the slope of the tangent line at $(0, 1)$ is 1, so the slope of the normal line at $(0, 1)$ is the negative reciprocal, which is -1.

23.　**B**　We can evaluate this integral using u-substitution. Let $u = \sqrt{x}$ and $du = \dfrac{1}{2\sqrt{x}}\,dx$, so $2\,du = \dfrac{1}{\sqrt{x}}\,dx$.

Substituting into the integrand, we get $2\displaystyle\int \csc^2 u\,du = -2\cot u + C$. Now we substitute back:

$$-2\cot u + C = -2\cot\sqrt{x} + C.$$

24.　**C**　We will need to use the fact that $\displaystyle\lim_{x\to 0}\dfrac{\sin x}{x} = 1$ to find the limit.

First, rewrite the limit as

$$\lim_{x\to 0}\frac{\sin^3(2x)}{x^3\cos^3(2x)}$$

Next, break the fraction into

$$\lim_{x\to 0}\left(\frac{\sin^3(2x)}{x^3}\,\frac{1}{\cos^3(2x)}\right)$$

Now, if we multiply the top and bottom of the first fraction by 8, we get

$$\lim_{x\to 0}\frac{8\sin^3(2x)}{(2x)^3}\,\frac{1}{\cos^3(2x)}$$

Now, we can take the limit, which gives us $8(1)(1) = 8$.

25.   **D**   First, let's graph the curve.

We can find the volume by taking a vertical slice of the region. The formula for the volume of a solid of revolution around the x-axis, using a vertical slice bounded from above by the curve $f(x)$ and from below by $g(x)$, on the interval $[a, b]$ is

$$\pi \int_a^b \left[f(x)^2 - g(x)^2 \right] dx$$

Here we get

$$\pi \int_0^{\frac{\pi}{2}} (1 + \sin^2 x)^2 \, dx$$

26.   **A**   We use the Chain Rule and the Quotient Rule.

$$\frac{dy}{dx} = 4 \left(\frac{x^3 - 2}{2x^5 - 1} \right)^3 \left[\frac{(2x^5 - 1)(3x^2) - (x^3 - 2)(10x^4)}{(2x^5 - 1)^2} \right]$$

If we plug in 1 for x, we get

$$\frac{dy}{dx} = 4(-1)^3 \left[\frac{3 + 10}{1^2} \right] = -52$$

27.   **D**   We can evaluate this integral using u-substitution. Let $u = 5 - x$ and $5 - u = x$. Then $-du = dx$.

Substituting, we get

$$-\int (5 - u) u^{\frac{1}{2}} \, du$$

The integral can be rewritten as

$$-\int \left(5u^{\frac{1}{2}} - u^{\frac{3}{2}} \right) du$$

Evaluating the integral, we get

$$-5\dfrac{u^{\frac{3}{2}}}{\dfrac{3}{2}}+\dfrac{u^{\frac{5}{2}}}{\dfrac{5}{2}}+C$$

This can be simplified to

$$-\dfrac{10}{3}u^{\frac{3}{2}}+\dfrac{2}{5}u^{\frac{5}{2}}+C$$

Finally, substituting back, we get

$$-\dfrac{10}{3}(5-x)^{\frac{3}{2}}+\dfrac{2}{5}(5-x)^{\frac{5}{2}}+C$$

28. **D** First, separate the variables: $\dfrac{dy}{y}=-2dt$. Integrate both sides: $\int\dfrac{dy}{y}=-2\int dt$, $\ln y=-2t+C$. We can isolate y by exponentiating both sides: $y=e^{-2t+C}=Ae^{-2t}$. Now, plug in the initial condition to solve for the constant: $100=Ae^0=A$. We get $y=100e^{-2t}$. Therefore, $y(2)=100e^{-2(2)}=100e^{-4}=\dfrac{100}{e^4}$.

29. **D** In order to find possible points of inflection, we first need to find the second derivative:

$$\dfrac{dy}{dx}=4x^3-18x^2-48x$$
$$\dfrac{d^2y}{dx^2}=12x^2-36x-48$$

Next, we set the second derivative equal to zero and solve: $12x^2-36x-48=0$. Divide through by 12: $x^2-3x-4=0$, which factors to $(x-4)(x+1)=0$. We get $x=4$ and $x=-1$. In order to be sure that $x=4$ is a point of inflection, we need to check that the second derivative changes sign there. Plug in values less than and greater than 4 and check the sign of the second derivative: $12(3)^2-36(3)-48<0$ and $12(5)^2-36(5)-48>0$, so $x=4$ is a point of inflection.

30. **C** First, rewrite the integral as $\int_0^1\dfrac{\sin x}{\cos x}\,dx$.

Now, we can use u-substitution to evaluate the integral. Let $u=\cos x$. Then $du=-\sin x$. We can also change the limits of integration. The lower limit becomes $\cos 0=1$ and the upper limit becomes $\cos 1$, which we leave alone. Now we perform the substitution, and we get

$$-\int_1^{\cos 1}\dfrac{du}{u}$$

Evaluating the integral, we get $-\ln u\Big|_1^{\cos 1}=-\ln(\cos 1)+\ln 1=-\ln(\cos 1)$. This log is also equal to $\ln(\sec 1)$.

31. **B** We can find the slope of the tangent line by taking the derivative: $\frac{dy}{dx} = -\sin x$. Next, we need to find where the graph has a zero: $\frac{1}{2} + \cos x = 0$. $\cos x = -\frac{1}{2}$ in the interval at $x = \frac{2\pi}{3}$. Plug this into the derivative: $\frac{dy}{dx} = -\sin\left(\frac{2\pi}{3}\right) = -\frac{\sqrt{3}}{2}$.

32. **B** The second fundamental theorem of calculus tells us how to find the derivative of an integral. It says that $\frac{d}{dx}\int_c^u f(t)\,dt = f(u)\frac{du}{dx}$, where c is a constant and u is a function of x.

    Here we can use the theorem to get

    $$\frac{d}{dx}\int_0^{x^2} \sin^2 t\,dt = (\sin^2(x^2))(2x), \text{ or } 2x\sin^2(x^2)$$

33. **B** Here we have a function raised to another function. The best way to find the derivative is with logarithmic differentiation. First, take the log of both sides: $\ln y = \ln x^{\cos 4x}$. Next, use the log laws to rewrite the right-hand side: $\ln y = (\cos 4x)\ln x$. Now take the derivative of both sides: $\frac{1}{y}\frac{dy}{dx} = (\cos 4x)\left(\frac{1}{x}\right) - \ln x(4\sin 4x)$. Multiply both sides by y: $\frac{dy}{dx} = y\left[(\cos 4x)\left(\frac{1}{x}\right) - \ln x(4\sin 4x)\right]$. Remember that $y = x^{\cos 4x}$, so the derivative is $\frac{dy}{dx} = x^{\cos 4x}\left[(\cos 4x)\left(\frac{1}{x}\right) - \ln x(4\sin 4x)\right]$.

34. **C** We need to take the second derivative of f: $f'(x) = 1 - 2xe^{-x^2}$ and $f''(x) = -2x\left(-2xe^{-x^2}\right) - 2e^{-x^2} = 4x^2e^{-x^2} - 2e^{-x^2} = e^{-x^2}\left(4x^2 - 2\right)$. Now we set the second derivative equal to zero. Note that e^{-x^2} is always positive, so we get $4x^2 - 2 = 0$ and $x = \pm 0.707$.

35. **A** The width of each rectangle of the Riemann sum will be $\frac{2-0}{4} = \frac{1}{2}$, so the sum is $\frac{1}{2}\left[f\left(\frac{1}{2}\right) + f(1) + f\left(\frac{3}{2}\right) + f(2)\right]$. Evaluate this with a calculator and we get 26.357.

36. **D** First, make a quick sketch of the region.

We can find the volume by taking a vertical slice of the region. The formula for the volume of a solid of revolution around the x-axis, using a vertical slice bounded from above by the curve $f(x)$ and from below by $g(x)$, on the interval $[a, b]$, is

$$\pi \int_a^b \left[f(x)^2 - g(x)^2 \right] dx$$

Here we get

$$\pi \int_0^1 \left[(1)^2 - (x^3)^2 \right] dx$$

Now we have to evaluate the integral. First, expand the integrand to get

$$\pi \int_0^1 (1 - x^6) \, dx$$

Next integrate to get

$$\pi \left(x - \frac{x^7}{7} \right) \Big|_0^1 = \pi \left(1 - \frac{1}{7} \right) = \frac{6\pi}{7}$$

37. **B** We need an equation that relates the volume of a sphere to its radius, namely $V = \frac{4}{3}\pi r^3$. If we differentiate both sides with respect to t, we get $\frac{dV}{dt} = \frac{4}{3}\pi(3r^2)\frac{dr}{dt} = 4\pi r^2 \frac{dr}{dt}$. Next plug in $\frac{dV}{dt} = 20$ and $r = 4$: $20 = 4\pi(4)^2 \frac{dr}{dt} = 64\pi \frac{dr}{dt}$, so $\frac{dr}{dt} = \frac{20}{64\pi} \approx 0.0995 \frac{\text{in.}}{\text{s}}$.

38. **B** The Mean Value Theorem states that if a function is continuous on the interval $[a, b]$, then there exists a value c in the interval (a, b) such that $\frac{f(b) - f(a)}{b - a} = f'(c)$. Here, f is a polynomial, so it is continuous everywhere. First, let's find $\frac{f(b) - f(a)}{b - a}$. We get $\frac{f(5) - f(1)}{5 - 1} = \frac{245 - (-3)}{4} = 62$. Next, we need $f'(c)$. We get $f'(x) = 6x^2$ so $f'(c) = 6c^2$. Now we set them equal to each other and solve: $6c^2 = 62$, so $c = \pm 3.215$. We don't use the negative value because it is not in the interval.

39.   **D**   Let's set $P = x^2y$. We want to maximize P, so we need to eliminate one of the variables. We are also given that $x + y = 100$, so we can solve this for y and substitute: $y = 100 - x$, so $P = x^2(100 - x) = 100x^2 - x^3$.

Now we can take the derivative.

$$\frac{dP}{dx} = 200x - 3x^2$$

Set the derivative equal to zero and solve for x.

$$200x - 3x^2 = 0$$
$$x(200 - 3x) = 0$$

$$x = 0 \text{ or } x = \frac{200}{3} \approx 66.667$$

Now we can use the second derivative to find the maximum: $\frac{d^2P}{dx^2} = 200 - 6x$. If we plug in $x = 66.667$, the second derivative is negative, so P is a maximum at $x = 66.667$. Solving for y, we get $y \approx 33.333$.

40.   **D**   First, we need to see if the velocity changes sign in the interval from $t = 0$ to $t = 2$. If you graph the velocity equation on your calculator, you will see that all of the y values are positive, so the object will always be moving to the right. Next, let's see how far the object travels in those 2 seconds. We can find this by evaluating $\int_0^2 t^2 \sin t \, dt$. If you plug this in your calculator, you get 2.469. (If you are unsure how to evaluate the integral on your calculator, check the online Appendix for some tips on using a TI-84 calculator.) Because the object's initial position is 4, the position at time $t = 2$ will be 6.469.

41.   **A**   We can evaluate this integral using u-substitution. Let $u = \sin(\pi x)$ and $du = \pi\cos(\pi x)dx$, so $\frac{1}{\pi}du = \cos(\pi x)dx$. Substituting into the integrand, we get $\int \sin^4(\pi x)\cos(\pi x)dx = \frac{1}{\pi}\int u^4\,du$. Evaluate: $\frac{1}{\pi}\int u^4\,du = \frac{1}{\pi}\frac{u^5}{5} + C = \frac{u^5}{5\pi} + C$. Now we substitute back: $\frac{u^5}{5\pi} + C = \frac{\sin^5(\pi x)}{5\pi} + C$.

42.   **D**   If we want to find the amount that the balloon has inflated in the first 8 seconds, we need to evaluate $\int_0^8 300 - t\ln t \, dt$. Use your calculator to get 2,349.458. (If you have a TI-84 series calculator, you can access the integration function by pressing MATH and then 9.) Add the original 100 in.³ to get 2,449.458, which rounds to 2,450 in.³.

43. **A** We can find $\dfrac{dy}{dx}$ using implicit differentiation. We get $2x + 3x^2 \dfrac{dy}{dx} + 6xy + 3y^2 \dfrac{dy}{dx} = 0$. Plug in the

point (2, 1) and solve for $\dfrac{dy}{dx}$:

$$2(2) + 3(2)^2 \dfrac{dy}{dx} + 6(2)(1) + 3(1)^2 \dfrac{dy}{dx} = 0$$

$$4 + 12\dfrac{dy}{dx} + 12 + 3\dfrac{dy}{dx} = 0$$

$$\dfrac{dy}{dx} = -\dfrac{16}{15}$$

44. **B** First, we will need to find the y-coordinate that corresponds to $x = 1$: $y = \tan 1 \approx 1.557$. Next, we need to find the derivative: $\dfrac{dy}{dx} = \tan x + x\sec^2 x$. We plug in $x = 1$ to get the slope of the tangent line: $\dfrac{dy}{dx} = \tan 1 + \sec^2 1 \approx 4.983$. Now we can plug this into the equation of a line: $y - 1.557 = 4.983(x - 1)$ or $y = 4.983x - 3.426$.

45. **C** In order to solve this for b, we need $f(x)$ to be continuous at $x = 2$. If we plug $x = 2$ into both pieces of this piecewise function, we get

$$f(x) = \begin{cases} 16a + 10; & x \le 2 \\ 4b - 6; & x > 2 \end{cases}$$

So, we need $16a + 10 = 4b - 6$. Now, if we take the derivative of both pieces of this function and plug in $x = 2$, we get

$$f'(x) = \begin{cases} 32a + 5; & x \le 2 \\ 4b - 3; & x > 2 \end{cases}, \text{ so we need } 32a + 5 = 4b - 3$$

Solving the simultaneous equations, we get $a = \dfrac{1}{2}$ and $b = 6$.

ANSWERS AND EXPLANATIONS TO SECTION II

1. The temperature on New Year's Day in Hinterland was given by $T(H) = -A - B\cos\left(\dfrac{\pi H}{12}\right)$, where T is the temperature in degrees Fahrenheit and H is the number of hours from midnight $(0 \leq H < 24)$.

(a) The initial temperature at midnight was $-15°$ F, and at noon of New Year's Day was $5°$ F. Find A and B.

Simply plug in the temperature, -15, for T and the time, midnight $(H = 0)$, for H into the equation. We get $-15 = -A - B\cos 0$, which simplifies to $-15 = -A - B$.

Now plug the temperature, 5, for T and the time, noon $(H = 12)$, for H into the equation. We get $5 = -A - B\cos(\pi)$, which simplifies to $5 = -A + B$.

Now we can solve the pair of simultaneous equations for A and B, and we get $A = 5°$ F and $B = 10°$ F.

(b) Find the average temperature for the first 10 hours.

In order to find the average value, we use the Mean Value Theorem for Integrals, which says that the average value of $f(x)$ on the interval $[a, b]$ is

$$\frac{1}{b-a}\int_a^b f(x)\,dx$$

Here, we have $\dfrac{1}{10-0}\displaystyle\int_0^{10}\left(-5 - 10\cos\left(\dfrac{\pi H}{12}\right)\right)dH$.

Evaluating the integral, we get

$$\frac{1}{10}\left[\left(-5H - \frac{120}{\pi}\sin\left(\frac{\pi H}{12}\right)\right)\right]_0^{10} = \frac{1}{10}\left[\left(-50 - \frac{120}{\pi}\sin\left(\frac{5\pi}{6}\right)\right)\right] = \frac{1}{10}\left[\left(-50 - \frac{60}{\pi}\right)\right] \approx -6.910°\text{F}$$

(c) Use the Trapezoid Rule with 4 equal subdivisions to estimate $\displaystyle\int_6^8 T(H)\,dH$.

The Trapezoid Rule enables us to approximate the area under a curve with a fair degree of accuracy. The rule says that the area between the x-axis and the curve $y = f(x)$ on the interval $[a, b]$, with n trapezoids, is

$$\frac{1}{2}\frac{b-a}{n}\left[y_0 + 2y_1 + 2y_2 + 2y_3 + \ldots + 2y_{n-1} + y_n\right]$$

Using the rule here, with $n = 4$, $a = 6$, and $b = 8$, we get

$$\frac{1}{2} \cdot \frac{1}{2}\left[\left(-5 - 10\cos\frac{6\pi}{12}\right) + 2\left(-5 - 10\cos\frac{13\pi}{24}\right) + 2\left(-5 - 10\cos\frac{7\pi}{12}\right) + 2\left(-5 - 10\cos\frac{15\pi}{24}\right) + \left(-5 - 10\cos\frac{8\pi}{12}\right)\right]$$

This is approximately $-4.890°F$.

(d) Find an expression for the rate that the temperature is changing with respect to H.

We simply take the derivative with respect to H.

$$\frac{dT}{dH} = -10\left(\frac{\pi}{12}\right)\left(-\sin\frac{\pi H}{12}\right) = \frac{5\pi}{6}\sin\frac{\pi H}{12}$$

2.   Sea grass grows on a lake. The rate of growth of the grass is $\frac{dG}{dt} = kG$, where k is a constant.

(a) Find an expression for G, the amount of grass in the lake (in tons), in terms of t, the number of years, if the amount of grass is 100 tons initially and 120 tons after one year.

We solve this differential equation using separation of variables.

First, move the G to the left side and the dt to the right side, to get $\frac{dG}{G} = k\,dt$.

Now, integrate both sides.

$$\int\frac{dG}{G} = k\int dt$$

$$\ln G = kt + C$$

Next, solve for G by exponentiating both sides to the base e. We get $G = e^{kt+C}$.

Using the rules of exponents, we can rewrite this as $G = e^{kt}\,e^{C}$. Finally, because e^{C} is a constant, we can rewrite the equation as $G = Ce^{kt}$.

Now, we use the initial condition that $G = 100$ at time $t = 0$ to solve for C.

$$100 = Ce^{0} = C(1) = C$$

This gives us $G = 100e^{kt}$.

Next, we use the condition that $G = 120$ at time $t = 1$ to solve for k.

$$120 = 100e^k$$
$$1.2 = e^k$$
$$\ln 1.2 = k \approx 0.1823$$

This gives us $G = 100e^{0.1823t}$.

(b)   In how many years will the amount of grass available be 300 tons?

All we need to do is set G equal to 300 and solve for t.

$$300 = 100e^{0.1823t}$$
$$3 = e^{0.1823t}$$
$$\ln 3 = 0.1823t$$
$$t \approx 6.026 \text{ years}$$

(c)   If fish are now introduced into the lake and consume a consistent 80 tons/year of sea grass, how long will it take for the lake to be completely free of sea grass?

Now we have to account for the fish's consumption of the sea grass. So we have to evaluate the differential equation $\dfrac{dG}{dt} = kG - 80$.

First, separate the variables, to get

$$\frac{dG}{kG - 80} = dt$$

Now, integrate both sides.

$$\int \frac{dG}{kG - 80} = \int dt \text{ or } \int \frac{dG}{G - \dfrac{80}{k}} = k \int dt$$

$$\ln\left(G - \frac{80}{k}\right) = kt + C$$

Next, exponentiate both sides to the base e. We get

$$G - \frac{80}{k} = Ce^{kt}$$

Solving for G, we get

$$G = \left(G_0 - \frac{80}{k}\right)e^{kt} + \frac{80}{k}$$

Now, set $G = 0$. We get

$$0 = \left(G_0 - \frac{80}{k}\right)e^{kt} + \frac{80}{k}$$

Now, set $G_0 = 300$, and solve for e^{kt}.

$$e^{kt} = \frac{\dfrac{-80}{k}}{300 - \dfrac{80}{k}} = \frac{80}{80 - 300k}$$

Take the log of both sides.

$$kt = \ln\left(\frac{80}{80 - 300k}\right)$$

$$\text{and } t = \frac{1}{k}\ln\left(\frac{80}{80 - 300k}\right)$$

Now, we plug in the value for k that we got in part (a) above to get $t \approx 6.313$ years.

3.   The functions f and g are twice-differentiable and have the following table of values:

x	$f(x)$	$f'(x)$	$g(x)$	$g'(x)$
1	3	4	-2	-4
2	2	3	4	-2
3	5	2	-1	3
4	-1	-6	-8	0

(a)   Let $h(x) = f(g(x))$. Find the equation of the tangent line to h at $x = 2$.

We can find the derivative of $h(x)$ using the Chain Rule: $h'(x) = f'(g(x))g'(x)$. At $x = 2$, we get $h'(2) = f'(g(2))g'(2) = f'(4)g'(2) = (-6)(4) = -24$.

And, at $x = 2$, $h(2) = f(g(2)) = f(4) = -1$.

Therefore, the equation of the tangent line is $y + 1 = -24(x - 2)$.

(b)   Let $j(x) = f(x)g(x)$. Find $j'(3)$.

We can find the derivative of $j(x)$ using the Product Rule: $j'(x) = f(x)g'(x) + f'(x)g(x)$. At $x = 3$, we get: $j'(3) = f(3)g'(3) + f'(3)g(3) = (5)(3) + (2)(-1) = 13$.

(c)   Evaluate $\int_1^4 3f''(x)\,dx$.

We simply evaluate $\int_1^4 3f''(x)\,dx = 3f'(x)\big|_1^4 = 3f'(4) - 3f'(1) = 3(-6) - 3(4) = -30$.

4.   Water is being poured into a hemispherical bowl of radius 6 inches at the rate of 4 in.³/sec.

(a)   Given that the volume of the water in the spherical segment shown above is $V = \pi h^2 \left(R - \dfrac{h}{3} \right)$, where R is the radius of the *sphere*, find the rate that the water level is rising when the water is 2 inches deep.

First, rewrite the equation as

$$V = \pi R h^2 - \frac{\pi}{3} h^3$$

Now, take the derivative of the equation with respect to t.

$$\frac{dV}{dt} = 2\pi R h \frac{dh}{dt} - \pi h^2 \frac{dh}{dt}$$

If we plug in $\dfrac{dV}{dt} = 4$, $R = 6$, and $h = 2$, we get

$$4 = 20\pi \frac{dh}{dt} \quad \text{or} \quad \frac{dh}{dt} = \frac{4}{20\pi} = \frac{1}{5\pi}$$

(b)   Find an expression for r, the radius of the *surface of the spherical segment* of water, in terms of h.

Notice that we can construct a right triangle using the radius of the sphere and the radius of the surface of the water.

Notice that the distance from the center of the sphere to the surface of the water is $R - h$. Now, we can use the Pythagorean Theorem to find r.

$$R^2 = (R - h)^2 + r^2$$

We can rearrange this to get

$$r = \sqrt{R^2 - (R - h)^2} = \sqrt{2Rh - h^2}$$

Because $R = 6$, we get

$$r = \sqrt{12h - h^2}$$

(c) How fast is the circular area of the surface of the spherical segment of water growing (in in.²/sec) when the water is 2 inches deep?

The area of the surface of the water is $A = \pi r^2$, where $r = \sqrt{12h - h^2}$. Thus, $A = \pi(12h - h^2)$.

Taking the derivative of the equation with respect to t, we get

$$\frac{dA}{dt} = \pi\left(12\frac{dh}{dt} - 2h\frac{dh}{dt}\right)$$

We found in part (a) above that

$$\frac{dh}{dt} = \frac{1}{5\pi}, \text{ so } \frac{dA}{dt} = \pi\left(\frac{12}{5\pi} - \frac{4}{5\pi}\right) = \frac{8}{5} \text{ in.}^2/\sec$$

5.   Let R be the region in the first quadrant bounded by $y^2 = x$ and $x^2 = y$.

(a) Find the area of region R.

First, let's sketch the region.

In order to find the area, we "slice" the region vertically and add up all of the slices. Now, we use the formula for the area of the region between $y = f(x)$ and $y = g(x)$, from $x = a$ to $x = b$.

$$\int_a^b \left[f(x) - g(x) \right] dx$$

We need to rewrite the equation $y^2 = x$ as $y = \sqrt{x}$ so that we can integrate with respect to x. Our integral for the area is

$$\int_0^1 \left(\sqrt{x} - x^2 \right) dx$$

Evaluating the integral, we get

$$\left(\frac{2x^{\frac{3}{2}}}{3} - \frac{x^3}{3} \right)\Bigg|_0^1 = \frac{2}{3} - \frac{1}{3} = \frac{1}{3}$$

(b)   Find the volume of the solid generated when R is revolved about the x-axis.

In order to find the volume of a region between $y = f(x)$ and $y = g(x)$, from $x = a$ to $x = b$, when it is revolved around the x-axis, we use the following formula:

$$\pi \int_a^b \left[f(x)^2 - g(x)^2 \right] dx$$

Here, our integral for the area is

$$\pi \int_0^1 \left(x - x^4 \right) dx$$

Evaluating the integral, we get

$$\pi \left(\frac{x^2}{2} - \frac{x^5}{5} \right)\Bigg|_0^1 = \pi \left(\frac{1}{2} - \frac{1}{5} \right) = \frac{3\pi}{10}$$

(c)   The section of a certain solid cut by any plane perpendicular to the x-axis is a circle with the endpoints of its diameter lying on the parabolas $y^2 = x$ and $x^2 = y$. Find the volume of the solid.

Whenever we want to find the volume of a solid, formed by the region between $y = f(x)$ and $y = g(x)$, with a known cross-section, from $x = a$ to $x = b$, when it is revolved around the x-axis, we use the following formula:

$$\int_a^b A(x)\, dx$$

(Note: $A(x)$ is the area of the cross-section.) We find the area of the cross-section by using the vertical slice formed by $f(x) - g(x)$ and then plugging it into the appropriate area formula. In the case of a circle, $f(x) - g(x)$ gives us the length of the diameter and we use the following formula:

$$A(x) = \frac{\pi(diameter)^2}{4}$$

This gives us the integral,

$$\int_0^1 \frac{\pi}{4}\left(\sqrt{x} - x^2\right)^2 dx$$

Expand the integrand.

$$\int_0^1 \frac{\pi}{4}\left(\sqrt{x} - x^2\right)^2 dx = \frac{\pi}{4}\int_0^1 \left(x - 2x^{\frac{5}{2}} + x^4\right) dx =$$

Evaluate the integral.

$$\frac{\pi}{4}\int_0^1 \left(x - 2x^{\frac{5}{2}} + x^4\right) dx = \frac{\pi}{4}\left(\frac{x^2}{2} - \frac{4x^{\frac{7}{2}}}{7} + \frac{x^5}{5}\right)\Bigg|_0^1 = \frac{\pi}{4}\left(\frac{1}{2} - \frac{4}{7} + \frac{1}{5}\right) = \frac{9\pi}{280}$$

6. For time $t \geq 0$, a particle moves along the x-axis. The velocity of the particle at time t is given by $v(t) = 1 - 2\cos\left(\frac{\pi}{3}t\right)$. The particle's position at time $t = 0$ is $x(0) = 8$.

(a) Is the particle speeding up or slowing down at time $t = \frac{1}{2}$? Justify your answer.

An object is speeding up if its velocity and acceleration have the same sign at a given time, and it is slowing down if they have opposite signs at a given time. We can find the acceleration by taking the derivative of the velocity: $a(t) = \frac{2}{3}\sin\left(\frac{\pi}{3}t\right)$. Now we can check the signs of the velocity and acceleration at time $t = \frac{1}{2}$: $v = 1 - 2\cos\frac{\pi}{6} = 1 - 2\left(\frac{\sqrt{3}}{2}\right) = 1 - \sqrt{3}$ and $a = \frac{2}{3}\sin\frac{\pi}{6} = \frac{1}{3}$. The acceleration and velocity have opposite signs, so the particle is slowing down.

(b) When does the particle change direction in the interval $0 \leq t \leq 2$? Justify your answer.

The particle changes direction at times when the velocity is zero and changes sign. Find when the velocity is zero.

$$1 - 2\cos\left(\frac{\pi}{3}t\right) = 0$$

$$\cos\left(\frac{\pi}{3}t\right) = \frac{1}{2}$$

$$\frac{\pi}{3}t = \frac{\pi}{3}$$

$$t = 1$$

The next solution $t = 5$ is outside the interval. Now pick a value less than 1, such as $t = \frac{1}{2}$ and note that the velocity is negative (see the explanation in part a). Pick another value greater than 1, such as $t = 2$. The velocity is $1 - 2\cos\frac{2\pi}{3} = 2$, which is positive. Therefore, the particle is changing direction at time $t = 1$.

(c) What is the particle's position at time $t = 2$?

We can find the change in the object's position in the first 6 seconds by evaluating the integral

We can find the change in the particle's position in the first 2 seconds by evaluating the integral

$\int_0^2 1 - \cos\left(\frac{\pi}{3}t\right)dt$. We get $\int_0^2 1 - \cos\left(\frac{\pi}{3}t\right)dt = t - \dfrac{\sin\left(\frac{\pi}{3}t\right)}{\frac{\pi}{3}}\Bigg|_0^2 = t - \frac{3}{\pi}\sin\left(\frac{\pi}{3}t\right)\Bigg|_0^2 = 2 - \frac{3\sqrt{3}}{2\pi}$.

(d) What is the total distance traveled from time $t = 0$ to time $t = 2$?

If we want to find the total distance traveled in the first 2 seconds, we do something similar to what we did in part (c), but in the time between $t = 0$ and $t = 1$, the particle is moving to the left (in part (b) we found that its velocity is negative in that interval), so we need to make the integral negative so that we don't get a negative distance. Then we add it to the integral for when the particle is moving to the right between $t = 1$ and $t = 2$.

We get $-\int_0^1 1 - \cos\left(\frac{\pi}{3}t\right)dt + \int_1^2 1 - \cos\left(\frac{\pi}{3}t\right)dt$.

Evaluate the first integral: $-\int_0^1 1 - \cos\left(\frac{\pi}{3}t\right)dt = -\left[t - \frac{3}{\pi}\sin\left(\frac{\pi}{3}t\right)\right]\Bigg|_0^1 = -\left(1 - \frac{3\sqrt{3}}{2\pi}\right) - 0 = -1 + \frac{3\sqrt{3}}{2\pi}$

Evaluate the second integral: $\int_1^2 1 - \cos\left(\frac{\pi}{3}t\right)dt = \left[t - \frac{3}{\pi}\sin\left(\frac{\pi}{3}t\right)\right]\Bigg|_1^2 = \left(2 - \frac{3\sqrt{3}}{2\pi}\right) - \left(1 - \frac{3\sqrt{3}}{2\pi}\right) = 1$

Combine the two values for the total distance: $-1 + \frac{3\sqrt{3}}{2\pi} + 1 = \frac{3\sqrt{3}}{2\pi}$

Practice Test 3

AP® Calculus AB Exam

SECTION I: Multiple-Choice Questions

DO NOT OPEN THIS BOOKLET UNTIL YOU ARE TOLD TO DO SO.

At a Glance

Total Time
1 hour and 45 minutes
Number of Questions
45
Percent of Total Grade
50%
Writing Instrument
Pencil required

Instructions

Section I of this examination contains 45 multiple-choice questions. Fill in only the ovals for numbers 1 through 45 on your answer sheet.

CALCULATORS MAY NOT BE USED IN THIS PART OF THE EXAMINATION.

Indicate all of your answers to the multiple-choice questions on the answer sheet. No credit will be given for anything written in this exam booklet, but you may use the booklet for notes or scratch work. After you have decided which of the suggested answers is best, completely fill in the corresponding oval on the answer sheet. Give only one answer to each question. If you change an answer, be sure that the previous mark is erased completely. Here is a sample question and answer.

Sample Question   Sample Answer

Chicago is a
(A) state
(B) city
(C) country
(D) continent

Use your time effectively, working as quickly as you can without losing accuracy. Do not spend too much time on any one question. Go on to other questions and come back to the ones you have not answered if you have time. It is not expected that everyone will know the answers to all the multiple-choice questions.

About Guessing

Many candidates wonder whether or not to guess the answers to questions about which they are not certain. Multiple-choice scores are based on the number of questions answered correctly. Points are not deducted for incorrect answers, and no points are awarded for unanswered questions. Because points are not deducted for incorrect answers, you are encouraged to answer all multiple-choice questions. On any questions you do not know the answer to, you should eliminate as many choices as you can, and then select the best answer among the remaining choices.

CALCULUS AB

SECTION I, Part A

Time—60 Minutes

Number of questions—30

A CALCULATOR MAY NOT BE USED ON THIS PART OF THE EXAMINATION

Directions: Solve each of the following problems, using the available space for scratchwork. After examining the form of the choices, decide which is the best of the choices given and fill in the corresponding oval on the answer sheet. No credit will be given for anything written in the test book. Do not spend too much time on any one problem.

In this test: Unless otherwise specified, the domain of a function f is assumed to be the set of all real numbers x for which $f(x)$ is a real number.

1. $\int_{\frac{\pi}{4}}^{x} \cos(2t)\, dt =$

  (A) $\cos(2x)$

  (B) $\dfrac{\sin(2x) - 1}{2}$

  (C) $\cos(2x) - 1$

  (D) $\dfrac{\sin 2(x)}{2}$

2. Let f be the function given by $f(x) = 4x^3 - 48x + 12$. On which of the following intervals is the function $f(x)$ increasing?

  (A) $\left(-4\sqrt{3}, 4\sqrt{3}\right)$

  (B) $(-2, 2)$

  (C) $\left(-\infty, -4\sqrt{3}\right)$ and $\left(4\sqrt{3}, \infty\right)$

  (D) $(-\infty, -2)$ and $(2, \infty)$

GO ON TO THE NEXT PAGE.

3. If $3x^2 - 2xy + 3y = 1$, then when $x = 2$, $\dfrac{dy}{dx} =$

  (A) -12

  (B) -10

  (C) $-\dfrac{10}{7}$

  (D) 12

---

4. $\displaystyle\int \dfrac{2x^3 + 4x^2 - 2x}{x^2}\, dx =$

  (A) $x^2 + 4x - 2\ln x + C$

  (B) $2x + 4 - \dfrac{2}{x} + C$

  (C) $x^2 + 4x - \dfrac{2}{x^2} + C$

  (D) $\dfrac{\dfrac{x^4}{2} + \dfrac{4x^3}{3} - x^2}{x^3} + C$

GO ON TO THE NEXT PAGE.

5.

The graph of a piecewise linear function f, for $0 \le x \le 8$, is shown above. What is the value of $\int_0^8 f(x)\,dx$?

(A) 1
(B) 4
(C) 8
(D) 10

6. $\lim\limits_{x \to 0} \dfrac{x - \sin x}{x^2} =$

(A) 0
(B) 1
(C) 2
(D) The limit does not exist.

GO ON TO THE NEXT PAGE.

7. If $y = 2x \cos x$, then $\dfrac{dy}{dx} =$

   (A) $2 + \cos x$
   (B) $2 - \sin x$
   (C) $2\cos x - 2x \sin x$
   (D) $2\cos x + 2x \sin x$

---

8. Given $f(x) = \dfrac{3x-2}{2x-3}$, find $f'(x)$.

   (A) $\dfrac{3}{2}$

   (B) $\dfrac{5}{(2x-3)^2}$

   (C) $-\dfrac{5}{(2x-3)^2}$

   (D) $\dfrac{12x}{(2x-3)^2}$

GO ON TO THE NEXT PAGE.

9. $\int_{2}^{e+1}\left(\dfrac{4}{x-1}\right)dx =$

   (A) 4
   (B) 4e
   (C) 0
   (D) −4

10.

A car's velocity is shown on the graph above. Which of the following gives the total distance traveled from $t = 0$ to $t = 16$ (in kilometers)?

   (A) 360
   (B) 390
   (C) 780
   (D) 1,000

GO ON TO THE NEXT PAGE.

11. If $f(x) = \sqrt{\tan(5x)}$, then $f'\left(\dfrac{\pi}{4}\right) =$

(A) -5
(B) 0
(C) 5
(D) The derivative does not exist at $x = \dfrac{\pi}{4}$

---

12. What is the equation of the line tangent to the graph of $y = \sin^2 x$ at $x = \dfrac{\pi}{4}$?

(A) $y - \dfrac{1}{2} = \left(x - \dfrac{\pi}{4}\right)$

(B) $y - \dfrac{1}{\sqrt{2}} = \left(x - \dfrac{\pi}{4}\right)$

(C) $y - \dfrac{1}{\sqrt{2}} = \dfrac{1}{2}\left(x - \dfrac{\pi}{4}\right)$

(D) $y - \dfrac{1}{2} = \dfrac{1}{2}\left(x - \dfrac{\pi}{4}\right)$

GO ON TO THE NEXT PAGE.

13. Given $f(x) = \begin{cases} 2x+3\,;\, x<3 \\ 2x^2-3x\,;\, x \geq 3 \end{cases}$, which of the following statements is true?

    (A) $f(x)$ is continuous at $x = 3$.

    (B) $f(x)$ is differentiable at $x = 3$.

    (C) $f(x)$ is not continuous at $x = 3$.

    (D) $\lim\limits_{x \to 3} f(x)$ does not exist.

14. The graph of $y = x^4 + 8x^3 - 72x^2 + 4$ is concave down for

    (A) $-6 < x < 2$

    (B) $x > 2$

    (C) $x < -6$

    (D) $-3 - 3\sqrt{5} < x < -3 + 3\sqrt{5}$

15. $\lim\limits_{x \to \infty} \dfrac{\ln(x+1)}{\log_2 x} =$

    (A) $\dfrac{1}{\ln 2}$

    (B) 0

    (C) 1

    (D) $\ln 2$

GO ON TO THE NEXT PAGE.

16.

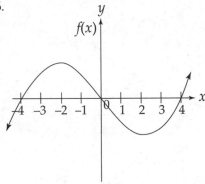

The graph of $f(x)$ is shown in the figure above. Which of the following could be the graph of $f'(x)$?

(A)

(C)

(B)

(D)

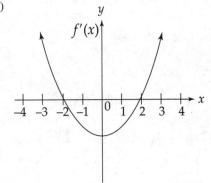

GO ON TO THE NEXT PAGE.

17. If $f(x) = \ln(\cos(3x))$, then $f'(x) =$

    (A) $3\sec(3x)$

    (B) $3\tan(3x)$

    (C) $-3\tan(3x)$

    (D) $-3\cot(3x)$

18. $\dfrac{d}{dx}\displaystyle\int_{3}^{4x}\cos^2 t \, dt =$

    (A) $\sin^2(4x)$

    (B) $\cos^2(4x)$

    (C) $4\sin^2(4x)$

    (D) $4\cos^2(4x)$

19. A particle moves along the x-axis with its position at time t, where $t \geq 0$, given by $x(t) = 2t^3 - 15t^2 + 36t + 4$. For which of the following values of t is the particle at rest?

    (A) $t = 0$ and $t = 6$

    (B) $t = 3$

    (C) $t = 0$

    (D) $t = 2$ and $t = 3$

GO ON TO THE NEXT PAGE.

20. $\displaystyle\int_0^{\frac{\pi}{2}} \sin(2x)e^{\sin^2 x}\,dx =$

   (A) $e-1$
   (B) $1-e$
   (C) $e+1$
   (D) 1

21. The average value of $\sec^2 x$ on the interval $\left[\dfrac{\pi}{6}, \dfrac{\pi}{4}\right]$ is

   (A) $\dfrac{12\sqrt{3}-12}{\pi}$

   (B) $\dfrac{12-4\sqrt{3}}{\pi}$

   (C) $\dfrac{6\sqrt{2}-6}{\pi}$

   (D) $\dfrac{6-6\sqrt{2}}{\pi}$

GO ON TO THE NEXT PAGE.

22. What is the region in the first quadrant bounded from above by the graph of the line $y = \dfrac{1}{2}$ and from below by $y = \sin x$ on the interval $x = 0$ to $x = \dfrac{\pi}{6}$?

(A) $\dfrac{\pi}{12} + \dfrac{\sqrt{3}}{2} - 2$

(B) $\dfrac{\pi}{12} + \dfrac{\sqrt{3}}{2} - 1$

(C) $\dfrac{\pi}{12} + \dfrac{\sqrt{3}}{2} - \dfrac{1}{2}$

(D) $\dfrac{\pi}{12} + \dfrac{\sqrt{3}}{2} + \dfrac{1}{2}$

23. The function f is given by $f(x) = x^4 + 4x^3$. On which of the following intervals is f decreasing?

(A) $(-3, 0)$

(B) $(0, \infty)$

(C) $(-3, \infty)$

(D) $(-\infty, -3)$

GO ON TO THE NEXT PAGE.

24. $\lim\limits_{x \to 0} \dfrac{\tan(3x) + 3x}{\sin(5x)} =$

    (A) 0

    (B) $\dfrac{3}{5}$

    (C) $\dfrac{6}{5}$

    (D) The limit does not exist.

25. If the region enclosed by the y-axis, the curve $y = 4\sqrt{x}$, and the line $y = 8$ is revolved about the x-axis, the volume of the solid generated is

    (A) $\dfrac{32\pi}{3}$

    (B) 128π

    (C) $\dfrac{128}{3}$

    (D) $\dfrac{128\pi}{3}$

GO ON TO THE NEXT PAGE.

26. The maximum velocity attained on the interval $0 \leq t \leq 5$, by the particle whose displacement is given by $s(t) = 2t^3 - 12t^2 + 16t + 2$, is

    (A) 286
    (B) 46
    (C) 16
    (D) 0

27. If $f(x) = x^2 + 2x$ on the interval [2, 6], find the value of c guaranteed by the conclusion of the Mean Value Theorem on the interval (2, 6).

    (A) 3
    (B) 3.5
    (C) 4
    (D) 4.5

GO ON TO THE NEXT PAGE.

28. If $f(x) = \sin^{-1}(4x)$, then $f'(x) =$

    (A) $\dfrac{1}{\sqrt{1-4x^2}}$

    (B) $\dfrac{4}{\sqrt{1-4x^2}}$

    (C) $\dfrac{1}{\sqrt{1-16x^2}}$

    (D) $\dfrac{4}{\sqrt{1-16x^2}}$

---

29. $\dfrac{d}{dx}\displaystyle\int_{2x}^{5x} \cos t \; dt =$

    (A) $5\cos 5x - 2\cos 2x$

    (B) $5\sin 5x - 2\sin 2x$

    (C) $\cos 5x - \cos 2x$

    (D) $\sin 5x - \sin 2x$

GO ON TO THE NEXT PAGE.

30.

Let $g(x) = \int_0^x f(t)\, dt$, where $f(t)$ has the graph shown above. Which of the following could be the graph of g?

(A)

(C)

(B)

(D)

END OF PART A, SECTION I
IF YOU FINISH BEFORE TIME IS CALLED, YOU MAY CHECK YOUR WORK ON PART A ONLY.
DO NOT GO ON TO PART B UNTIL YOU ARE TOLD TO DO SO.

CALCULUS AB

SECTION I, Part B

Time—45 Minutes

Number of questions—15

A GRAPHING CALCULATOR IS REQUIRED FOR SOME QUESTIONS ON THIS PART OF THE EXAMINATION

<u>Directions:</u> Solve each of the following problems, using the available space for scratchwork. After examining the form of the choices, decide which is the best of the choices given and fill in the corresponding oval on the answer sheet. No credit will be given for anything written in the test book. Do not spend too much time on any one problem.

<u>In this test:</u>

1. The **exact** numerical value of the correct answer does not always appear among the choices given. When this happens, select from among the choices the number that best approximates the exact numerical value.

2. Unless otherwise specified, the domain of a function f is assumed to be the set of all real numbers x for which $f(x)$ is a real number.

31. If $f(x)$ is the function given by $f(x) = e^{3x} + 1$, at what value of x is the slope of the tangent line to $f(x)$ equal to 2 ?

    (A) −0.366
    (B) −0.135
    (C)   0
    (D)   0.693

GO ON TO THE NEXT PAGE.

32. The graph of the function $y = x^3 + 12x^2 + 15x + 3$ has a relative maximum at $x =$

    (A) −10.613
    (B)   −7.317
    (C)   −1.138
    (D)   −0.248

33. The side of a square is increasing at a constant rate of 0.4 cm/s. In terms of the perimeter, P, what is the rate of change of the area of the square, in cm²/s?

    (A) $0.05P$
    (B) $0.2P$
    (C) $0.4P$
    (D) $51.2P$

34. Let f be the function given by $f(x) = 3^x$. For what value of x is the slope of the line tangent to the curve at $(x, f(x))$ equal to 1 ?

    (A)   0.086
    (B)   0
    (C) −0.086
    (D) −1.099

GO ON TO THE NEXT PAGE.

35. The table below gives values for the functions f and g and its derivatives for certain values of x. If the function $h(x) = f(g(x))$, what is the value of $h'(2)$?

x	$f(x)$	$g(x)$	$f'(x)$	$g'(x)$
0	3	2	0	1
1	−4	6	4	0
2	5	1	−2	6

(A) −12
(B) 12
(C) 24
(D) 30

36. Let R be the region in the first quadrant bounded by $f(x) = e^x + 3$ and $g(x) = \ln(x + 1)$, from $x = 0$ to $x = 2$. What is the area of R?

(A) 5.547
(B) 11.093
(C) 13.685
(D) 127.032

37. What is the trapezoidal approximation of $\int_0^3 e^x \, dx$ using $n = 4$ subintervals?

(A) 6.407
(B) 13.565
(C) 19.972
(D) 27.879

GO ON TO THE NEXT PAGE.

38. If $f(x) = x^4 - 3x^2 + 5$, on which of the following intervals is f concave up?

(A) $\left(-\infty, -\sqrt{\dfrac{3}{2}}\right)$ and $\left(\sqrt{\dfrac{3}{2}}, \infty\right)$

(B) $\left(-\sqrt{\dfrac{3}{2}}, \sqrt{\dfrac{3}{2}}\right)$

(C) $\left(-\sqrt{\dfrac{1}{2}}, \sqrt{\dfrac{1}{2}}\right)$

(D) $\left(-\infty, -\sqrt{\dfrac{1}{2}}\right)$ and $\left(\sqrt{\dfrac{1}{2}}, \infty\right)$

39. Evaluate $\displaystyle\lim_{x \to \frac{\pi}{2}} \dfrac{1 - \sin x}{\cos x}$.

(A) −1
(B) 0
(C) 1
(D) The limit does not exist.

GO ON TO THE NEXT PAGE.

40. The base of a solid S is the region enclosed by the graph of $4x + 5y = 20$, the x-axis, and the y-axis. If the cross-sections of S perpendicular to the x-axis are semicircles, then the volume of S is

(A) $\dfrac{5\pi}{3}$

(B) $\dfrac{10\pi}{3}$

(C) $\dfrac{50\pi}{3}$

(D) $\dfrac{225\pi}{3}$

41. Which of the following is an equation of the line tangent to the graph of $y = x^3 + x^2$ at $y = 3$?

(A) $y = 33x - 63$
(B) $y = 33x - 13$
(C) $y = 6.488x - 4.620$
(D) $y = 6.488x - 10.620$

42. If $f'(x) = \ln x - x + 2$, at which of the following values of x does f have a relative minimum value?

(A) 5.146
(B) 3.146
(C) 0.159
(D) 0

GO ON TO THE NEXT PAGE.

43. Find the total area of the region between the curve $y = \cos x$ and the x-axis from $x = 1$ to $x = 2$ in radians.

(A) 0
(B) 0.068
(C) 0.249
(D) 1.751

---

44. Let $f(x) = \int \cot x \, dx; \, 0 < x < \pi$. If $f\left(\dfrac{\pi}{6}\right) = 1$, then $f(1) =$

(A) −1.861
(B) −0.480
(C)   0.524
(D)   1.521

---

45. Given $\dfrac{dy}{dx} = 2xy - y$ and $y(0) = 4$, which of the following is equal to y?

(A) $e^{x^2 - x - 4}$

(B) $e^{x^2 - x + 4}$

(C) $4 + e^{x^2 - x}$

(D) $4e^{x^2 - x}$

STOP
END OF PART B, SECTION I
IF YOU FINISH BEFORE TIME IS CALLED, YOU MAY CHECK YOUR WORK ON PART B ONLY.
DO NOT GO ON TO SECTION II UNTIL YOU ARE TOLD TO DO SO.

SECTION II
GENERAL INSTRUCTIONS

You may wish to look over the problems before starting to work on them, since it is not expected that everyone will be able to complete all parts of all problems. All problems are given equal weight, but the parts of a particular problem are not necessarily given equal weight.

A GRAPHING CALCULATOR IS REQUIRED FOR SOME PROBLEMS OR PARTS OF PROBLEMS ON THIS SECTION OF THE EXAMINATION.

- You should write all work for each part of each problem in the space provided for that part in the booklet. Be sure to write clearly and legibly. If you make an error, you may save time by crossing it out rather than trying to erase it. Erased or crossed-out work will not be graded.

- Show all your work. You will be graded on the correctness and completeness of your methods as well as your answers. Correct answers without supporting work may not receive credit.

- Justifications require that you give mathematical (noncalculator) reasons and that you clearly identify functions, graphs, tables, or other objects you use.

- You are permitted to use your calculator to solve an equation, find the derivative of a function at a point, or calculate the value of a definite integral. However, you must clearly indicate the setup of your problem, namely the equation, function, or integral you are using. If you use other built-in features or programs, you must show the mathematical steps necessary to produce your results.

- Your work must be expressed in standard mathematical notation rather than calculator syntax. For example, $\int_{1}^{5} x^2 \, dx$ may not be written as fnInt (X², X, 1, 5).

- Unless otherwise specified, answers (numeric or algebraic) need not be simplified. If your answer is given as a decimal approximation, it should be correct to three places after the decimal point.

- Unless otherwise specified, the domain of a function f is assumed to be the set of all real numbers x for which $f(x)$ is a real number.

GO ON TO THE NEXT PAGE.

SECTION II, PART A
Time—30 minutes
Number of problems—2

A graphing calculator is required for some problems or parts of problems.

During the timed portion for Part A, you may work only on the problems in Part A.

On Part A, you are permitted to use your calculator to solve an equation, find the derivative of a function at a point, or calculate the value of a definite integral. However, you must clearly indicate the setup of your problem, namely the equation, function, or integral you are using. If you use other built-in features or programs, you must show the mathematical steps necessary to produce your results.

1.

Let R be the region in the first quadrant shown in the figure above.

(a) Find the area of R.

(b) Find the volume of the solid generated when R is revolved about the x-axis.

(c) Find the volume of the solid generated when R is revolved about the line $x = -1$.

GO ON TO THE NEXT PAGE.

2. A body is coasting to a stop and the only force acting on it is a resistance proportional to its speed, according to the equation

$$\frac{ds}{dt} = v_f = v_0 e^{-\left(\frac{k}{m}\right)t}; \quad s(0) = 0,$$ where v_0 is the body's initial velocity (in m/s), v_f is its final velocity, m is its mass, k is a constant, and t is time.

(a) If a body with mass $m = 50$ kg and $k = 1.5$ kg/s initially has a velocity of 30 m/s, how long, to the nearest second, will it take to slow to 1 m/s?

(b) How far, to the 10 nearest meters, will the body coast during the time it takes to slow from 30 m/s to 1 m/s?

(c) If the body coasts from 30 m/s to a stop, how far will it coast?

GO ON TO THE NEXT PAGE.

SECTION II, PART B
Time—1 hour
Number of problems—4

No calculator is allowed for these problems.

During the timed portion for Part B, you may continue to work on the problems in Part A without the use of any calculator.

---

3. An object moves with velocity $v(t) = 9t^2 + 18t - 7$ for $t \geq 0$ from an initial position of $s(0) = 3$.

    (a) Write an equation for the position of the particle.

    (b) When is the particle changing direction?

    (c) What is the total distance covered from $t = 2$ to $t = 5$?

GO ON TO THE NEXT PAGE.

4.

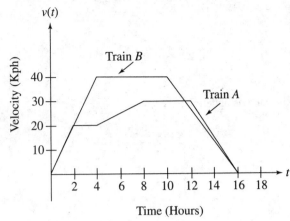

Three trains, A, B, and C each travel on a straight track for $0 \leq t \leq 16$ hours. The graphs above, which consist of line segments, show the velocities, in kilometers per hour, of trains A and B. The velocity of C is given by

$$v(t) = 8t - 0.25t^2$$

(Indicate units of measure for all answers.)

(a) Find the velocities of A and C at time $t = 6$ hours.

(b) Find the accelerations of B and C at time $t = 6$ hours.

(c) Find the positive difference between the total distance that A traveled and the total distance that B traveled in 16 hours.

(d) Find the total distance that C traveled in 16 hours.

GO ON TO THE NEXT PAGE.

5. Consider the curve $y^2 + 3xy = 4$.

   (a) Find an equation of the tangent line to the curve at the point (1, 1).

   (b) Find all x-coordinates where the slope of the tangent line to the curve is undefined.

   (c) Evaluate $\dfrac{d^2y}{dx^2}$ at the point (1, 1).

GO ON TO THE NEXT PAGE.

6. Consider the differential equation $\dfrac{dy}{dx} = \dfrac{xy}{2}$.

(a) Sketch the slope field at the 9 points indicated below.

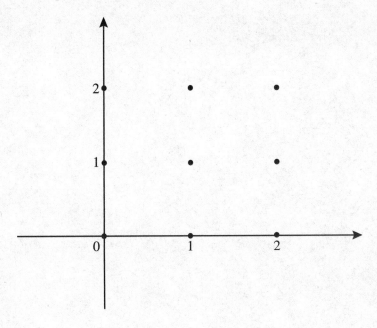

(b) Find the particular solution to the differential equation $y = f(x)$ with initial condition $f(0) = 5$.

(c) Write an equation for the tangent line to the graph at $(2, 1)$. Use the tangent line to estimate y at $x = 2.1$.

STOP

END OF EXAM

Practice Test 3:
Answers and
Explanations

ANSWER KEY

Section I

1.	B	11.	C	21.	B	31.	B	41.	C
2.	D	12.	A	22.	B	32.	B	42.	C
3.	B	13.	A	23.	D	33.	B	43.	C
4.	A	14.	A	24.	C	34.	C	44.	D
5.	B	15.	D	25.	B	35.	C	45.	D
6.	A	16.	D	26.	B	36.	B		
7.	C	17.	C	27.	C	37.	C		
8.	C	18.	D	28.	D	38.	D		
9.	A	19.	D	29.	A	39.	B		
10.	C	20.	A	30.	B	40.	B		

ANSWERS AND EXPLANATIONS TO SECTION I

1. **B** First, take the antiderivative: $\int \cos(2t)\, dt = \dfrac{1}{2}\sin(2t)$. Next, plug in x and $\dfrac{\pi}{4}$ for t and take the difference: $\dfrac{1}{2}\sin(2x) - \dfrac{1}{2}\sin\left(2\left(\dfrac{\pi}{4}\right)\right)$. This can be simplified to $\dfrac{\sin(2x) - 1}{2}$.

2. **D** First, take the derivative: $f'(x) = 12x^2 - 48$. Next, set the derivative equal to zero and solve: $12x^2 - 48 = 0$. We get $x = \pm 2$. Now we sign test to find where the function is increasing or decreasing. Let's pick a number less than -2, like -3. Plug it into the derivative and we get $f'(-3) = 12(-3)^2 - 48 > 0$. Next we pick a number between -2 and 2, like 0: $f'(0) = 12(0)^2 - 48 < 0$. Finally, we pick a number greater than 2, like 3: $f'(3) = 12(3)^2 - 48 > 0$. Therefore, $f(x)$ is increasing when x is less than -2 or greater than 2. In interval notation, we get $\left(-\infty, -2\right)$ and $\left(2, \infty\right)$.

3. **B** We need to use implicit differentiation to find $\dfrac{dy}{dx}$.

$$6x - 2\left(x\,\dfrac{dy}{dx} + y\right) + 3\dfrac{dy}{dx} = 0$$

$$6x - 2x\dfrac{dy}{dx} - 2y + 3\dfrac{dy}{dx} = 0$$

Now, if we wanted to solve for $\dfrac{dy}{dx}$ in terms of x and y, we would have to do some algebra to isolate $\dfrac{dy}{dx}$. But because we are asked to solve for $\dfrac{dy}{dx}$ at a specific value of x, we don't need to simplify.

We need to find the y-coordinate that corresponds to the x-coordinate $x = 2$. We plug $x = 2$ into the original equation and solve for y.

$$3(2)^2 - 2(2)y + 3y = 12 - y = 1$$

$$y = 11$$

Finally, we plug $x = 2$ and $y = 11$ into the derivative, and we get

$$6(2) - 2(2)\dfrac{dy}{dx} - 2(11) + 3\dfrac{dy}{dx} = 0$$

$$12 - 4\dfrac{dy}{dx} - 22 + 3\dfrac{dy}{dx} = 0$$

$$\dfrac{dy}{dx} = -10$$

4. **A** Simplify the integrand by separating it into three fractions:

$$\int \dfrac{2x^3 + 4x^2 - 2x}{x^2}\, dx = \int \dfrac{2x^3}{x^2} + \dfrac{4x^2}{x^2} - \dfrac{2x}{x^2}\, dx$$

We can simplify this to $\int \dfrac{2x^3}{x^2} + \dfrac{4x^2}{x^2} - \dfrac{2x}{x^2}\, dx = \int 2x + 4 - \dfrac{2}{x}\, dx$. And integrate:

$$\int 2x + 4 - \dfrac{2}{x}\, dx = x^2 + 4x - 2\ln x + C$$

5. **B** We need to add the areas of the regions between the graph and the x-axis. Note that the area of the region between 0 and 5 has a positive value and the area of the region between 5 and 8 has a negative value. The area of the former region can be found by calculating the area of a trapezoid with bases of 2 and 5 and a height of 2. The area is $\dfrac{1}{2}(2 + 5)(2) = 7$. The area of the latter region can be found by calculating the area of a triangle with a base of 3 and a height of 2. The area is $\dfrac{1}{2}(3)(2) = 3$. Thus, the value of the integral is $7 - 3 = 4$.

6. **A** If we take the limit as x goes to 0, we get an indeterminate form $\dfrac{0}{0}$, so let's use L'Hôpital's Rule. We take the derivative of the numerator and the denominator to get $\lim\limits_{x \to 0} \dfrac{x - \sin x}{x^2} = \lim\limits_{x \to 0} \dfrac{1 - \cos x}{2x}$. When we take the limit, we again get an indeterminate form $\dfrac{0}{0}$, so let's use L'Hôpital's Rule a second time. We take the derivative of the numerator and the denominator and to $\lim\limits_{x \to 0} \dfrac{1 - \cos x}{2x} = \lim\limits_{x \to 0} \dfrac{\sin x}{2}$. Now, when we take the limit, we get $\lim\limits_{x \to 0} \dfrac{\sin x}{2} = 0$.

7. **C** Use the Product Rule to find the derivative: $\dfrac{dy}{dx} = 2x(-\sin x) + 2\cos x = -2x\sin x + 2\cos x$.

8. **C** We can find the derivative with the Quotient Rule: $f'(x) = \dfrac{(2x - 3)(3) - (3x - 2)(2)}{(2x - 3)^2} = \dfrac{-5}{(2x - 3)^2}$.

9. **A** You should know that $\int \dfrac{dx}{x} = \ln|x| + C$. We take the antiderivative and we get $\int \left(\dfrac{4}{x - 1} \right) dx = 4\ln|x - 1| + C$. Next, plug in $e + 1$ and 2 for x, and take the difference: $4\ln(e) - 4\ln(1)$. You should know that $\ln e = 1$ and $\ln 1 = 0$. Thus, we get $4\ln(e) - 4\ln(1) = 4$.

10. **C** We find the total distance traveled by finding the area of the region between the curve and the x-axis. Normally, we would have to integrate but here we can find the area of the region easily because it consists of geometric objects whose areas are simple to calculate. The area of the region between

$t = 0$ and $t = 4$ can be found by calculating the area of a triangle with a base of 4 and a height of 60. The area is $\frac{1}{2}(4)(60) = 120$. The area of the region between $t = 4$ and $t = 8$ can be found by calculating the area of a rectangle with a base of 4 and a height of 30. The area is $(4)(30) = 120$. The area of the region between $t = 8$ and $t = 16$ can be found by calculating the area of a trapezoid with bases of 4 and 8, and a height of 90 (or you could break it up into a rectangle and a triangle). The area is $\frac{1}{2}(4 + 8)(90) = 540$. Thus, the total distance traveled is $120 + 120 + 540 = 780$ kilometers.

11. **C** Use the Chain Rule: $f'(x) = \frac{1}{2}\left(\tan(5x)\right)^{-\frac{1}{2}}\left(\sec^2(5x)\right)(5)$. Next, plug in $x = \frac{\pi}{4}$:

$$f'\left(\frac{\pi}{4}\right) = \frac{1}{2}\left(\tan\left(\frac{5\pi}{4}\right)\right)^{-\frac{1}{2}}\left(\sec^2\left(\frac{5\pi}{4}\right)\right)(5) = \frac{1}{2}(1)^{-\frac{1}{2}}(2)(5) = 5$$

12. **A** If we want to find the equation of the tangent line, first we need to find the y-coordinate that cor-responds to $x = \frac{\pi}{4}$. It is $y = \sin^2\left(\frac{\pi}{4}\right) = \left(\frac{1}{\sqrt{2}}\right)^2 = \frac{1}{2}$.

Next, we need to find the derivative of the curve at $x = \frac{\pi}{4}$, using the Chain Rule.

We get $\frac{dy}{dx} = 2\sin x \cos x$. At $x = \frac{\pi}{4}$, $\frac{dy}{dx}\Big|_{x=\frac{\pi}{4}} = 2\sin\left(\frac{\pi}{4}\right)\cos\left(\frac{\pi}{4}\right) = 2\left(\frac{1}{\sqrt{2}}\right)\left(\frac{1}{\sqrt{2}}\right) = 1$.

Now we have the slope of the tangent line and a point that it goes through. We can use the point-slope formula for the equation of a line, $(y - y_1) = m(x - x_1)$, and plug in what we have just found.

We get $\left(y - \frac{1}{2}\right) = (1)\left(x - \frac{\pi}{4}\right)$.

13. **A** Let's test whether f is continuous. First, $f(3) = 9$ exists. Next, $\lim\limits_{x \to 3^-} f(x) = 9$ and $\lim\limits_{x \to 3^+} f(x) = 9$, so $\lim\limits_{x \to 3} f(x) = 9$. Finally, because $\lim\limits_{x \to 3} f(x) = f(3) = 9$, $f(x)$ is continuous at $x = 3$.

14. **A** A graph is concave down where the second derivative is negative.

First, we find the first and second derivative.

$$\frac{dy}{dx} = 4x^3 + 24x^2 - 144x$$

$$\frac{d^2y}{dx^2} = 12x^2 + 48x - 144$$

Next, we want to determine on which intervals the second derivative of the function is positive and on which it is negative. We do this by finding where the second derivative is zero.

$$12x^2 + 48x - 144 = 0$$
$$x^2 + 4x - 12 = 0$$
$$(x + 6)(x - 2) = 0$$
$$x = -6 \text{ or } x = 2$$

We can test where the second derivative is positive and negative by picking a point in each of the three regions $-\infty < x < -6, -6 < x < 2$, and $2 < x < \infty$, plugging the point into the second derivative, and seeing what the sign of the answer is. You should find that the second derivative is negative on the interval $-6 < x < 2$.

15. **D**   If we take the limit as x goes to ∞, we get an indeterminate form $\dfrac{\infty}{\infty}$, so let's use L'Hôpital's Rule. We

take the derivative of the numerator and the denominator to get $\lim\limits_{x \to \infty} \dfrac{\ln(x+1)}{\log_2 x} = \lim\limits_{x \to \infty} \dfrac{\dfrac{1}{x+1}}{\dfrac{1}{x\ln 2}} = \lim\limits_{x \to \infty} \dfrac{x\ln 2}{x+1}$.

When we take the limit, we again get an indeterminate form, $\dfrac{\infty}{\infty}$, so let's use L'Hôpital's Rule a

second time. We take the derivative of the numerator and the denominator to get

$\lim\limits_{x \to \infty} \dfrac{x\ln 2}{x+1} = \lim\limits_{x \to \infty} \dfrac{\ln 2}{1} = \ln 2$.

16. **D**   Here we want to examine the slopes of various pieces of the graph of $f(x)$. Notice that the graph has a positive slope from $x = -\infty$ to $x = -2$, where the slope is zero. Thus, we are looking for a graph of $f'(x)$ that is positive from $x = -\infty$ to $x = -2$ and zero at $x = -2$. Notice that the graph of $f(x)$ has a negative slope from $x = -2$ to $x = 2$, where the slope is zero. Thus, we are looking for a graph of $f'(x)$ that is negative from $x = -2$ to $x = 2$ and zero at $x = 2$. Finally, notice that the graph of $f(x)$ has a positive slope from $x = 2$ to $x = \infty$. Thus, we are looking for a graph of $f'(x)$ that is positive from $x = 2$ to $x = \infty$. The graph in (D) satisfies all of these requirements.

17. **C**   Remember that $\dfrac{d}{dx}\ln(u(x)) = \dfrac{u'(x)}{u(x)}$.

We will need to use the Chain Rule to find the derivative.

$$f'(x) = \left(\dfrac{-\sin(3x)}{\cos(3x)}\right)(3) = -3\tan(3x)$$

18. **D** Use the Second Fundamental Theorem of Calculus: $\dfrac{d}{dx}\displaystyle\int_{3}^{4x}\cos^2 t\, dt = \cos^2\left(4x\right)(4)$.

19. **D** The particle is at rest where its velocity is zero. We can obtain the velocity from the derivative of the position function. We get $v(t) = x'(t) = 6t^2 - 30t + 36$. Setting the velocity equal to zero we get $6t^2 - 30t + 36 = 0$. Divide through by 6: $t^2 - 5t + 6 = 0$. $(t - 3)(t - 2) = 0$ So the particle is at rest at $t = 2$ and $t = 3$.

20. **A** We can use u-substitution to evaluate the integral. Let $u = \sin^2 x$ and $du = 2\sin x \cos x\, dx$. Next, recall from trigonometry that $2\sin x \cos x = \sin(2x)$. Now we can substitute into the integral $\displaystyle\int e^u\, du$, leaving out the limits of integration for the moment. Evaluate the integral to get $\displaystyle\int e^u\, du = e^u$.

Now, we substitute back to get $e^{\sin^2 x}$. Finally, we evaluate at the limits of integration to get

$$e^{\sin^2 x}\Big|_{0}^{\frac{\pi}{2}} = e^{\sin^2 \frac{\pi}{2}} - e^{\sin^2 0} = e - 1$$

21. **B** In order to find the average value, we use the Mean Value Theorem for Integrals, which says that the average value of $f(x)$ on the interval $[a, b]$ is $\dfrac{1}{b - a}\displaystyle\int_{a}^{b} f(x)\, dx$.

Here we have $\dfrac{1}{\dfrac{\pi}{4} - \dfrac{\pi}{6}}\displaystyle\int_{\frac{\pi}{6}}^{\frac{\pi}{4}}\sec^2 x\, dx$. Next, recall that $\dfrac{d}{dx}\tan x = \sec^2 x$.

We evaluate the integral.

$$\frac{1}{\dfrac{\pi}{4} - \dfrac{\pi}{6}}(\tan x)\Big|_{\frac{\pi}{6}}^{\frac{\pi}{4}} = \frac{1}{\dfrac{\pi}{4} - \dfrac{\pi}{6}}\left[\tan\frac{\pi}{4} - \tan\frac{\pi}{6}\right] = \frac{1}{\dfrac{\pi}{4} - \dfrac{\pi}{6}}\left(1 - \frac{\sqrt{3}}{3}\right)$$

Get a common denominator for each of the two expressions.

$$\frac{1}{\dfrac{\pi}{4} - \dfrac{\pi}{6}}\left(1 - \frac{\sqrt{3}}{3}\right) = \frac{1}{\dfrac{6\pi}{24} - \dfrac{4\pi}{24}}\left(\frac{3}{3} - \frac{\sqrt{3}}{3}\right)$$

We can simplify this to $\dfrac{1}{\dfrac{2\pi}{24}}\left(\dfrac{3 - \sqrt{3}}{3}\right) = \dfrac{12}{\pi}\left(\dfrac{3 - \sqrt{3}}{3}\right) = \dfrac{12 - 4\sqrt{3}}{\pi}$.

22.　**B**　First, let's make a sketch of the region.

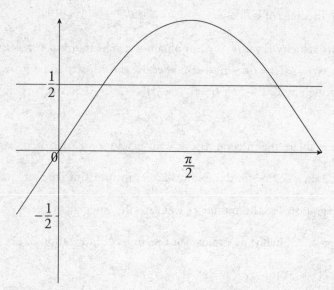

Note that the line intersects the curve at $x = \dfrac{\pi}{6}$, and that the line is above the curve for the whole

region. So in order to find the area, we need to evaluate the integral

$$\int_0^{\frac{\pi}{6}} \left(\frac{1}{2} - \sin x \right) dx = \left(\frac{1}{2} x + \cos x \right) \Big|_0^{\frac{\pi}{6}} = \left(\frac{1}{2}\left(\frac{\pi}{6} \right) + \cos\left(\frac{\pi}{6} \right) \right) - \left(\frac{1}{2}(0) + \cos(0) \right) = \frac{\pi}{12} + \frac{\sqrt{3}}{2} - 1$$

23.　**D**　A function is decreasing on an interval where the derivative is negative. The derivative is $f'(x) = 4x^3 + 12x^2$. We want to determine on which intervals the derivative of the function is positive and on which it is negative. We do this by finding where the derivative is zero.

$$4x^3 + 12x^2 = 0$$

$$4x^2(x + 3) = 0$$

$$x = -3 \text{ or } x = 0$$

We can test where the derivative is positive and negative by picking a point in each of the three regions $-\infty < x < -3$, $-3 < x < 0$, and $0 < x < \infty$, plugging the point into the derivative, and seeing what the sign of the answer is. Because x^2 is never negative, you should find that the derivative is negative on the interval $-\infty < x < -3$.

24.　**C**　We will need to use the fact that $\displaystyle\lim_{x \to 0} \frac{\sin x}{x} = 1$ to find the limit.

First, rewrite the limit as $\displaystyle\lim_{x \to 0} \dfrac{\dfrac{\sin(3x)}{\cos(3x)} + 3x}{\sin(5x)} =$

Next, break the expression into two rational expressions.

$$\lim_{x \to 0} \frac{\sin(3x)}{\sin(5x)\cos(3x)} + \frac{3x}{\sin(5x)} =$$

This can be broken up further into

$$\lim_{x \to 0} \frac{\sin(3x)}{\sin(5x)} \frac{1}{\cos(3x)} + \frac{3x}{\sin(5x)} =$$

We will evaluate the limit of each separately.

<u>First expression</u>

Divide the top and bottom by x: $\lim_{x \to 0} \dfrac{\dfrac{\sin(3x)}{x}}{\dfrac{\sin(5x)}{x}}$.

Then, multiply the top and bottom of the upper expression by 3, and the top and bottom of the

lower expression by 5: $\lim_{x \to 0} \dfrac{\dfrac{3\sin(3x)}{3x}}{\dfrac{5\sin(5x)}{5x}}$.

Now, if we take the limit, we get $\lim_{x \to 0} \dfrac{\dfrac{3\sin(3x)}{3x}}{\dfrac{5\sin(5x)}{5x}} = \dfrac{3(1)}{5(1)} = \dfrac{3}{5}$.

<u>Second expression</u>

This limit is straightforward: $\lim_{x \to 0} \dfrac{1}{\cos(3x)} = \dfrac{1}{\cos(0)} = 1$.

<u>Third expression</u>

First, pull the constant, 3, out of the limit: $\lim_{x \to 0} \dfrac{3x}{\sin(5x)} = 3\lim_{x \to 0} \dfrac{x}{\sin(5x)}$.

Now, if we multiply the top and bottom of the expression by 5, we get $3\lim_{x \to 0} \dfrac{5x}{5\sin(5x)}$.

Now, if we take the limit, we get $3\lim_{x \to 0} \dfrac{5x}{5\sin(5x)} = 3\left(\dfrac{1}{5}\right) = \dfrac{3}{5}$.

Combine the three numbers to get $\dfrac{3}{5}(1) + \dfrac{3}{5} = \dfrac{6}{5}$.

25.   **B**   First, we graph the curves.

We can find the volume by taking a vertical slice of the region. The formula for the volume of a solid of revolution around the x-axis, using a vertical slice bounded from above by the curve $f(x)$ and from below by $g(x)$, on the interval $[a, b]$, is

$$\pi \int_a^b \left[f(x)^2 - g(x)^2 \right] dx$$

The upper curve is $y = 8$, and the lower curve is $y = 4\sqrt{x}$.

Next, we need to find the point(s) of intersection of the two curves, which we do by setting them equal to each other and solving for x.

$$8 = 4\sqrt{x}$$

$$2 = \sqrt{x}$$

$$x = 4$$

Thus, the limits of integration are $x = 0$ and $x = 4$.

Now, we evaluate the integral.

$$\pi \int_0^4 \left[(8)^2 - \left[4\sqrt{x} \right]^2 \right] dx = \pi \int_0^4 \left(64 - 16x \right) dx = \pi \left(64x - 8x^2 \right) \Big|_0^4 = 128\pi$$

26.   **B**   Velocity is the first derivative of position with respect to time.

The first derivative is

$$v(t) = 6t^2 - 24t + 16$$

If we want to find the maximum velocity, we take the derivative of velocity (which is acceleration) and find where the derivative is zero.

$$v'(t) = 12t - 24$$

Next, we set the derivative equal to zero and solve for t, in order to find the critical value.

$$12t - 24 = 0$$

$$t = 2$$

Note that the second derivative of velocity is 12, which is positive. Remember the second derivative test: If the sign of the second derivative at a critical value is positive, then the curve has a local minimum there. If the sign of the second derivative is negative, then the curve has a local maximum there.

Thus, the velocity is a *minimum* at $t = 2$. In order to find where it has an absolute *maximum*, we plug the endpoints of the interval into the original equation for velocity, and the larger value will be the answer. At $t = 0$, the velocity is 16. At $t = 5$, the velocity is 46.

27. **C** First, we need to find $\dfrac{f(6) - f(2)}{(6-2)} = \dfrac{48-8}{4} = 10$. Next, we take the derivative at $x = c$. We get $f'(x) = 2x + 2$ and $f'(c) = 2c + 2$. Setting them equal to each other, we get $2c + 2 = 10$ and $c = 4$.

28. **D** Remember that $\dfrac{d}{dx} \sin^{-1} u = \dfrac{1}{\sqrt{1-u^2}} \dfrac{du}{dx}$, so here we get $\dfrac{d}{dx} \sin^{-1}(4x) = \dfrac{1}{\sqrt{1-(4x)^2}}(4) = \dfrac{4}{\sqrt{1-16x^2}}$.

29. **A** The Second Fundamental Theorem of Calculus tells us how to find the derivative of an integral: $\dfrac{dF}{dx} = \dfrac{d}{dx} \displaystyle\int_a^x f(t)\,dt$. Here we can use the theorem to get $\dfrac{d}{dx} \displaystyle\int_{2x}^{5x} \cos t\,dt = 5\cos 5x - 2\cos 2x$.

30. **B** The function $g(x) = \displaystyle\int_0^x f(t)\,dt$ is called an accumulation function and stands for the area between the curve and the x-axis to the point x. At $x = 0$, the area is 0, so $g(0) = 0$. From $x = 0$ to $x = 2$, the area grows, so $g(x)$ has a positive slope. Then, from $x = 2$ to $x = 4$, the area shrinks (because we subtract the area of the region under the x-axis from the area of the region above it), so $g(x)$ has a negative slope. Finally, from $x = 4$ to $x = 6$, the area again grows, so $g(x)$ has a positive slope. The curve that best represents this is shown in (B).

31. **B** The slope of the tangent line is the derivative of the function. We get $f'(x) = 3e^{3x}$. Now we set the derivative equal to 2 and solve for x.

$$3e^{3x} = 2$$

$$e^{3x} = \frac{2}{3}$$

$$3x = \ln\frac{2}{3}$$

$$x = \frac{1}{3}\ln\frac{2}{3} \approx -0.135$$

Remember to round all answers to three decimal places on the AP Exam.

32. **B** First, let's find the derivative: $\dfrac{dy}{dx} = 3x^2 + 24x + 15$.

Next, set the derivative equal to zero and solve for x.

$$3x^2 + 24x + 15 = 0$$
$$x^2 + 8x + 5 = 0$$

Using the quadratic formula (or a calculator), we get

$$x = \frac{-8 \pm \sqrt{64 - 20}}{2} \approx -0.683, -7.317$$

Let's use the second derivative test to determine which is the maximum. We take the second derivative and then plug in the critical values that we found when we set the first derivative equal to zero. If the sign of the second derivative at a critical value is positive, then the curve has a local minimum there. If the sign of the second derivative is negative, then the curve has a local maximum there.

The second derivative is $\dfrac{d^2 y}{dx^2} = 6x + 24$. The second derivative is negative at $x = -7.317$, so the curve has a local maximum there.

33. **B** The formula for the perimeter of a square is $P = 4s$, where s is the length of a side of the square.

If we differentiate this with respect to t, we get $\dfrac{dP}{dt} = 4\dfrac{ds}{dt}$. We plug in $\dfrac{ds}{dt} = 0.4$ to get $\dfrac{dP}{dt} = 4(0.4) = 1.6$.

The formula for the area of a square is $A = s^2$. If we solve the perimeter equation for s in terms of P and substitute it into the area equation, we get

$$s = \frac{P}{4}, \text{ so } A = \left(\frac{P}{4}\right)^2 = \frac{P^2}{16}$$

If we differentiate this with respect to t, we get $\dfrac{dA}{dt} = \dfrac{P}{8}\dfrac{dP}{dt}$.

Now we plug in $\dfrac{dP}{dt} = 1.6$ to get $\dfrac{dA}{dt} = \dfrac{P}{8}(1.6) = 0.2P$.

34. **C** The slope of the tangent line is the derivative of the function. Recall that $\dfrac{d}{dx}a^x = a^x \ln a$. Here we get $f'(x) = 3^x \ln 3$. Now we set the derivative equal to 1 and solve for x. Using the calculator, we get $3^x \ln 3 = 1$, so $x \approx -0.086$.

35. **C** First, use the Chain Rule to find $h'(x) = f'(g(x))g'(x)$. Next, evaluate the derivative at $x = 2$. We get $h'(2) = f'(g(2))g'(2)$. Use the table to find $g(2) = 1$ and $g'(2) = 6$. Next, because $g(2) = 1$, we need to find $f'(1) = 4$ (from the table). Therefore, $h'(2) = f'(g(2))g'(2) = f'(1)g'(2) = (4)(6) = 24$.

36. **B** First, graph the equations to see which one is on top and which is on the bottom.

Note that $f(x) = e^x + 3$ is always above $g(x) = \ln(x + 1)$, so in order to find the area, we simply have to integrate $\int_0^2 (e^x + 3) - \ln(x+1)\,dx$. If you plug this in your calculator, you get 11.093. (If you are unsure how to evaluate the integral on your calculator, check the online Appendix for some tips on using a TI-84 calculator.)

37. **C** The Trapezoid Rule enables us to approximate the area under a curve with a fair degree of accuracy. The rule says that the area between the x-axis and the curve $y = f(x)$, on the interval $[a, b]$, with n trapezoids, is

$$\frac{1}{2}\frac{b-a}{n}\,[y_0 + 2y_1 + 2y_2 + 2y_3 + \ldots + 2y_{n-1} + y_n]$$

Using the rule here, with $n = 4$, $a = 0$, and $b = 3$, we get

$$\frac{1}{2}\left(\frac{3}{4}\right)\left[e^0 + 2e^{\frac{3}{4}} + 2e^{\frac{6}{4}} + 2e^{\frac{9}{4}} + e^3\right] \approx 19.972$$

38. **D** The function will be concave up where the second derivative is positive. We need to differentiate f twice: $f'(x) = 4x^3 - 6x$ $f''(x) = 12x^2 - 6$. Now we set the second derivative equal to zero and solve. We get $x = \pm\sqrt{\frac{1}{2}}$. Now we just have to sign test the different intervals. Pick a number less than $-\sqrt{\frac{1}{2}}$, say -1, and plug it into the second derivative: $f''(-1) = 12(-1)^2 - 6 > 0$, so f is concave up for x

less than $-\sqrt{\dfrac{1}{2}}$. Next try a number between $-\sqrt{\dfrac{1}{2}}$ and $\sqrt{\dfrac{1}{2}}$, say 0: $f''(0) = 12(0)^2 - 6 < 0$, so f is concave down on that interval. Finally, pick a number greater than $\sqrt{\dfrac{1}{2}}$, say 1: $f''(1) = 12(1)^2 - 6 > 0$, so f is concave up for x greater than $\sqrt{\dfrac{1}{2}}$. Therefore, f is concave up on $\left(-\infty, -\sqrt{\dfrac{1}{2}}\right)$ and $\left(\sqrt{\dfrac{1}{2}}, \infty\right)$.

39. **B** If we plug $x = \dfrac{\pi}{2}$ into the limit, we get $\dfrac{1 - \sin\dfrac{\pi}{2}}{\cos\dfrac{\pi}{2}} = \dfrac{0}{0}$, which is an indeterminate form. This means that we can use L'Hôpital's Rule. Take the derivative of the numerator and denominator:

$\lim\limits_{x \to \frac{\pi}{2}} \dfrac{-\cos x}{-\sin x}$. Now if we evaluate the limit, we get $\dfrac{-\cos\dfrac{\pi}{2}}{-\sin\dfrac{\pi}{2}} = 0$.

40. **B** First, sketch the region.

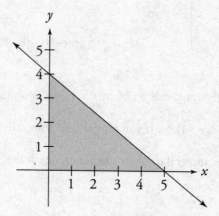

The rule for finding the volume of a solid with known cross-sections is $V = \displaystyle\int_a^b A(x)\, dx$, where A is the formula for the area of the cross-section. So x represents the diameter of a semi-circular cross-section.

The area of a semi-circle in terms of its diameter is $A = \pi \dfrac{d^2}{8}$. We find the length of the diameter by solving the equation $4x + 5y = 20$ for y: $y = \dfrac{20 - 4x}{5}$. Next, we need to find where the graph intersects the x-axis. You should get $x = 5$. Thus, we find the volume by evaluating the integral.

$$\int_0^5 \pi \dfrac{\left(\dfrac{20 - 4x}{5}\right)^2}{8}\, dx$$

This integral can be simplified to

$$\frac{\pi}{200} \int_0^5 \left(20 - 4x\right)^2 dx = \frac{\pi}{200} \int_0^5 \left(400 - 160x + 16x^2\right) dx$$

You can evaluate the integral by hand or with a calculator. You should get

$$\frac{\pi}{200} \int_0^5 \left(400 - 160x + 16x^2\right) dx = \frac{10\pi}{3}$$

41. **C** If we want to find the equation of the tangent line, first we need to find the x-coordinate that corresponds to $y = 3$. If you use your calculator to solve $x^3 + x^2 = 3$, you should get $x = 1.1746$.

Next, we need to find the derivative of the curve at $x = 1.1746$. We get

$$\frac{dy}{dx} = 3x^2 + 2x. \text{ At } x = 1.1746, \left.\frac{dy}{dx}\right|_{x=1.1746} = 3(1.1746)^2 + 2(1.1746) = 6.488$$

(It is rounded to three decimal places.)

Now we have the slope of the tangent line and a point that it goes through. We can use the point-slope formula for the equation of a line, $(y - y_1) = m(x - x_1)$, and plug in what we have just found. We get $(y - 3) = (6.488)(x - 1.1746)$. This simplifies to $y = 6.488x - 4.620$.

42. **C** Set the derivative equal to zero and solve for x. Using your calculator, you should get $\ln x - x + 2 = 0$.

$$x = 3.146 \text{ or } x = 0.159 \text{ (rounded to three decimal places)}$$

Let's use the second derivative test to determine which is the minimum. We take the second derivative and then plug in the critical values that we found when we set the first derivative equal to zero. If the sign of the second derivative at a critical value is positive, then the curve has a local minimum there. If the sign of the second derivative is negative, then the curve has a local maximum there.

The second derivative is $f''(x) = \frac{1}{x} - 1$. The second derivative is positive at $x = 0.159$, so the curve has a local minimum there.

43.   **C**   First, we should graph the curve.

Note that the curve is above the x-axis from $x = 1$ to $x = \dfrac{\pi}{2}$ and below the x-axis from $x = \dfrac{\pi}{2}$ to

$x = 2$. Thus, we need to evaluate two integrals to find the area.

$$\int_{1}^{\frac{\pi}{2}} \cos x \, dx + \int_{\frac{\pi}{2}}^{2} \left(-\cos x \right) dx$$

We will need a calculator to evaluate these integrals.

$$\int_{1}^{\frac{\pi}{2}} \cos x \, dx + \int_{\frac{\pi}{2}}^{2} \left(-\cos x \right) dx \approx 0.249$$

44.   **D**   First, we find $\int \cot x \, dx$ by rewriting the integral as $\int \dfrac{\cos x}{\sin x} \, dx$. Then, we use u-substitution. Let

$u = \sin x$ and $du = \cos x$. Substituting, we can get $\int \dfrac{\cos x}{\sin x} \, dx = \int \dfrac{du}{u} = \ln |u| + C$. Then substitut-

ing back, we get $\ln(\sin x) + C$. (We can get rid of the absolute value bars because sine is always posi-

tive on the interval.)

Next, we use $f\left(\dfrac{\pi}{6} \right) = 1$ to solve for C. We get $1 = \ln\left(\sin \dfrac{\pi}{6} \right) + C$.

$$1 = \ln\left(\dfrac{1}{2} \right) + C$$

$$1 = \ln\left(\dfrac{1}{2} \right) + C = 1.693147$$

Thus, $f(x) = \ln(\sin x) + 1.693147$.

At $x = 1$, we get $f(1) = \ln(\sin 1) + 1.693147 = 1.521$ (rounded to three decimal places).

45. **D** We can solve this differential equation by first separating the variables: $\dfrac{dy}{dx} = y(2x-1)$

$\dfrac{dy}{y} = (2x-1)dx$. Next, we integrate both sides: $\displaystyle\int \dfrac{dy}{y} = \int (2x-1)dx$ $\ln|y| = x^2 - x + C$. Now we

exponentiate both sides: $y = e^{x^2-x+C} = Ce^{x^2-x}$. Finally, plug in the initial condition to solve for C:

$4 = Ce^0 = C$, so we get $y = 4e^{x^2-x}$

ANSWERS AND EXPLANATIONS TO SECTION II

1.

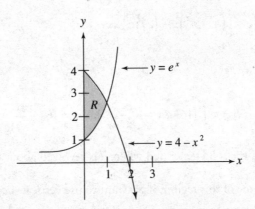

Let R be the region in the first quadrant shown in the figure above.

(a)   Find the area of R.

In order to find the area, we "slice" the region vertically and add up all of the slices. We use the formula for the area of the region between $y = f(x)$ and $y = g(x)$, from $x = a$ to $x = b$,

$$\int_a^b \left[f(x) - g(x) \right] dx$$

We have

$$f(x) = 4 - x^2 \text{ and } g(x) = e^x$$

Next, we need to find the point of intersection in the first quadrant. Use your calculator to find that the point of intersection is $x = 1.058$ (rounded to three decimal places). Plugging into the formula, we get

$$\int_0^{1.058} \left[(4 - x^2) - e^x \right] dx$$

Evaluating the integral, we get $\int_0^{1.058} \left[(4 - x^2) - e^x \right] dx = \left(4x - \dfrac{x^3}{3} - e^x \right)\Bigg|_0^{1.058} = 1.957.$

(b)   Find the volume of the solid generated when R is revolved about the x-axis.

In order to find the volume of a region between $y = f(x)$ and $y = g(x)$, from $x = a$ to $x = b$, when it is revolved about the x-axis, we use the following formula:

$$\pi \int_a^b \left[f(x)^2 - g(x)^2 \right] dx$$

Here our integral is $\pi \int_0^{1.058} \left[\left(4 - x^2\right)^2 - \left(e^x\right)^2 \right] dx.$

Evaluating the integral, we get

$$\pi \int_0^{1.058} \left[\left(4 - x^2\right) \right]^2 - \left[e^x \right]^2 dx = \pi \int_0^{1.058} \left(16 - 8x^2 + x^4 - e^{2x} \right) dx = \pi \left[16x - \dfrac{8x^3}{3} + \dfrac{x^5}{5} - \dfrac{e^{2x}}{2} \right]_0^{1.058} = 32.629$$

(c)   Find the volume of the solid generated when R is revolved about the line $x = -1$.

In order to find the volume of this region, if we want to use vertical slices, we will use the method of cylindrical shells. Also, because we are revolving about the line $x = -1$, we will need to add 1 to the radius of the cylindrical shell. We will use the formula

$$2\pi \int_a^b (x + 1)\left[f(x) - g(x) \right] dx$$

We get

$$2\pi \int_0^{1.058} (x + 1)\left[(4 - x^2) - e^x \right] dx$$

We suggest that you use your calculator to evaluate the integral.

$$2\pi \int_0^{1.058} (x + 1)\left[(4 - x^2) - e^x \right] dx = 2\pi \int_0^{1.058} \left[4x - x^3 - xe^x + 4 - x^2 - e^x \right] dx = 17.059$$

2.   A body is coasting to a stop and the only force acting on it is a resistance proportional to its speed, according to the equation $\frac{ds}{dt} = v_f = v_0 \, e^{-\left(\frac{k}{m}\right)t}$; $s(0) = 0$, where v_0 is the body's initial velocity (in m/s), v_f is its final velocity, m is its mass, k is a constant, and t is time.

(a) If a body, with mass $m = 50$ kg and $k = 1.5$ kg/s, initially has a velocity of 30 m/s, how long, to the nearest second, will it take to slow to 1 m/s?

We simply plug into the formula and solve for t.

$$v_f = v_0 \, e^{-\left(\frac{k}{m}\right)t}$$

$$1 = 30 \, e^{-\left(\frac{1.5}{50}\right)t}$$

Divide both sides by 30: $\frac{1}{30} = e^{-\left(\frac{1.5}{50}\right)t}$.

Take the log of both sides: $\ln\frac{1}{30} = -\left(\frac{1.5}{50}\right)t$.

Multiply both sides by $-\frac{50}{1.5}$: $-\frac{50}{1.5}\ln\frac{1}{30} = t \approx 113$ seconds.

(b) How far, to the nearest 10 meters, will the body coast during the time it takes to slow from 30 m/s to 1 m/s?

Now we need to solve the differential equation $\frac{ds}{dt} = v_0 \, e^{-\left(\frac{k}{m}\right)t}$, which we can do with separation of variables.

First, multiply both sides by dt: $ds = v_0 \, e^{-\left(\frac{k}{m}\right)t} \, dt$.

Integrate both sides: $\int ds = \int v_0 \, e^{-\left(\frac{k}{m}\right)t} \, dt$.

Evaluate the integrals: $s = -\frac{mv_0}{k} e^{-\left(\frac{k}{m}\right)t} + C$.

Now, plug in the initial conditions to solve for C: $0 = -\frac{(50)(30)}{1.5} e^{-\left(\frac{1.5}{50}\right)(0)} + C$.

$$C = \frac{(30)(50)}{1.5} = 1,000$$

Therefore, $s = -\dfrac{mv_0}{k} e^{-\left(\frac{k}{m}\right)t} + 1,000$. Now, we plug in the time $t = 113$ that we found in part (a) as well as the initial conditions to solve for s, which yields the following:

$$s = -\dfrac{(50)(30)}{1.5} e^{-\left(\frac{1.5}{50}\right)113} + 1,000 \approx 970 \text{ meters}$$

(c) If the body coasts from 30 m/s to a stop, how far will it coast?

Here, because the braking force is an exponential function, the object will coast to a stop after an infinite amount of time. In other words, we need to find

$$\lim_{t \to \infty} s(t) = \lim_{t \to \infty}\left[1,000 - 1,000e^{-\left(\frac{k}{m}\right)t}\right] = 1,000 \text{ meters}$$

3. An object moves with velocity $v(t) = 9t^2 + 18t - 7$ for $t \geq 0$ from an initial position of $s(0) = 3$.

(a) Write an equation for the position of the particle.

The position function of the particle can be determined by integrating the velocity with respect to time, thus $s(t) = \int v(t)dt$. For this problem, $s(t) = \int (9t^2 + 18t - 7)dt = 3t^3 + 9t^2 - 7t + C$. Because we are given the initial position, $s(0) = 3$, plug that in to solve for C. Thus, $C = 3$ and the equation for the position of the particle is $s(t) = 3t^3 + 9t^2 - 7t + 3$.

(b) When is the particle changing direction?

The particle changes direction when the velocity is zero, but the acceleration is not. In order to determine when those times are, set the velocity equal to zero and solve for t: $v(t) = 9t^2 + 18t - 7 = 0$ when $t = \dfrac{1}{3}$ and $t = -\dfrac{7}{3}$. Because the time range in question is $t \geq 0$, we can ignore $t = -\dfrac{7}{3}$. Then, take the derivative of the velocity function to find the acceleration function, as $\dfrac{d}{dt}(v(t)) = a(t)$. For the given $v(t)$, $a(t) = 18t + 18$. Check that the acceleration at time $t = \dfrac{1}{3}$ is not zero by plugging into the acceleration function: $a(t) = 24$. Therefore, the particle is changing direction at $t = \dfrac{1}{3}$ because $v(t) = 0$ and $a(t) \neq 0$.

(c) What is the total distance covered from $t = 2$ to $t = 5$?

The distance covered is found by using the position function found in part (a). Determine the position at $t = 2$ and subtract it from the position at $t = 5$. From part (b), we know that the object does not change direction over this time interval, so we do not need to find the time piecewise. Thus, $s(5) - s(2) = 568 - 49 = 519$.

4.

Three trains, *A, B,* and *C,* each travel on a straight track for $0 \leq t \leq 16$ hours. The graphs above, which consist of line segments, show the velocities, in kilometers per hour, of trains *A* and *B.* The velocity of *C* is given by $v(t) = 8t - 0.25t^2$.

(Indicate units of measure for all answers.)

(a)   Find the velocities of *A* and *C* at time $t = 6$ hours.

We can find the velocity of train *A* at time $t = 6$ simply by reading the graph. We get $v_A(6) = 25$ kph. We find the velocity of train *C* at time $t = 6$ by plugging $t = 6$ into the formula. We get $v_C(6) = 8(6) - 0.25(6^2) = 39$ kilometers per hour.

(b)   Find the accelerations of *B* and *C* at time $t = 6$ hours.

Acceleration is the derivative of velocity with respect to time. For train *B,* we look at the *slope* of the graph at time $t = 6$. We get $a_B(6) = 0$ km/hr². For train *C,* we take the derivative of *v.* We get $a(t) = 8 - 0.5t$. At time $t = 6$, we get $a_C(6) = 5$ km/hr².

(c)   Find the positive difference between the total distance that *A* traveled and the total distance that *B* traveled in 16 hours.

In order to find the total distance that train *A* traveled in 16 hours, we need to find the area under the graph. We can find this area by adding up the areas of the different geometric objects that are under the graph. From time $t = 0$ to $t = 2$, we need to find the area of a triangle with a base of 2 and a height of 20. The area is 20. Next, from time $t = 2$ to $t = 4$, we need to find the area of a rectangle with a base of 2 and a height of 20. The area is 40. Next, from time $t = 4$ to $t = 8$, we need to find the area of a trapezoid with bases of 20 and 30 and a height of 4. The area is 100. Next, from time $t = 8$ to $t = 12$, we need to find the area of a rectangle with a base of 4 and a height of 30. The area is 120. Finally, from time $t = 12$ to $t = 16$, we need to find the area of a triangle with a base of 4 and a height of 30. The area is 60. Thus, the total distance that train *A* traveled is 340 km.

Let's repeat the process for train B. From time $t = 0$ to $t = 4$, we need to find the area of a triangle with a base of 4 and a height of 40. The area is 80. Next, from time $t = 4$ to $t = 10$, we need to find the area of a rectangle with a base of 6 and a height of 40. The area is 240. Finally, from time $t = 10$ to $t = 16$, we need to find the area of a triangle with a base of 6 and a height of 40. The area is 120. Thus, the total distance that train B traveled is 440 km.

Therefore, the positive difference between their distances is 100 km.

(d)   Find the total distance that C traveled in 16 hours.

First, note that the graph of train C's velocity, $v(t) = 8t - 0.25t^2$, is above the x-axis on the entire interval. Therefore, in order to find the total distance traveled, we integrate $v(t)$ over the interval. We get

$$\int_0^{16} (8t - 0.25t^2)\, dt$$

Evaluate the integral: $\int_0^{16} (8t - 0.25t^2)\, dt = \left(4t^2 - \dfrac{t^3}{12} \right)_0^{16} = \dfrac{2,048}{3}$ km.

5.   Consider the curve $y^2 + 3xy = 4$.

(a)   Find an equation of the tangent line to the curve at the point $(1, 1)$.

First, we need to use implicit differentiation to find $\dfrac{dy}{dx}$. We get $2y\dfrac{dy}{dx} + 3\left(x\dfrac{dy}{dx} + y \right) = 0$. Let's isolate $\dfrac{dy}{dx}$ because we will need it in part (c). Otherwise, we would plug in $(1, 1)$ here to solve for the slope.

$$2y\dfrac{dy}{dx} + 3x\dfrac{dy}{dx} + 3y = 0$$

$$\dfrac{dy}{dx}(2y + 3x) = -3y$$

$$\dfrac{dy}{dx} = \dfrac{-3y}{2y + 3x}$$

Now plug in $(1, 1)$ to solve for the slope: $\dfrac{dy}{dx} = \dfrac{-3}{2 + 3} = -\dfrac{3}{5}$.

Therefore, the equation of the tangent line is $y - 1 = -\dfrac{3}{5}(x - 1)$.

(b)   Find all *x*-coordinates where the slope of the tangent line to the curve is undefined.

The tangent line will be undefined where the denominator is zero. That is,

$$2y + 3x = 0$$

$$2y = -3x$$

$$y = -\frac{3x}{2}$$

Plug $y = -\frac{3x}{2}$ into the original equation: $\left(-\frac{3x}{2}\right)^2 - 3x\left(-\frac{3x}{2}\right) = 4$. And solve:

$$\frac{9x^2}{4} + \frac{9x^2}{2} = 4$$

$$\frac{27x^2}{4} = 4$$

$$x^2 = \frac{16}{27}$$

(c)   Evaluate $\dfrac{d^2y}{dx^2}$ at the point (1, 1).

We need to take the derivative of $\dfrac{dy}{dx}$ that we found in part (a). Using the Quotient Rule and implicit differentiation, we get $\dfrac{d^2y}{dx^2} = \dfrac{(2y+3x)\left(-3\dfrac{dy}{dx}\right) - 3y\left(2\dfrac{dy}{dx}+3\right)}{(2y+3x)^2}$.

Note that at (1, 1), $\dfrac{dy}{dx} = -\dfrac{3}{5}$. Therefore, at (1, 1),

$$\frac{d^2y}{dx^2} = \frac{(2+3)\left(-3\left(\frac{-3}{5}\right)\right) - 3\left(2\left(\frac{-3}{5}\right)+3\right)}{(2+3)^2} = \frac{(5)\left(\frac{9}{5}\right) - (3)\left(\frac{9}{5}\right)}{25} = \frac{8}{125}$$

6. Consider the differential equation $\frac{dy}{dx} = \frac{xy}{2}$.

    (a) Sketch the slope field at the 9 points indicated below.

Let's make a table of the slopes at each of the points:

x	y	$\frac{dy}{dx} = \frac{xy}{2}$
0	0	0
1	0	0
2	0	0
0	1	0
1	1	$\frac{1}{2}$
2	1	1
0	2	0
1	2	1
2	2	2

Now we will graph a line segment with the slope at each of the points.

(b) Find the particular solution to the differential equation $y = f(x)$ with initial condition $f(0) = 5$.

We can solve this differential equation using Separation of Variables. Divide both

sides by y and multiply both sides by dx: $\dfrac{dy}{y} = \dfrac{x\,dx}{2}$. Integrate both sides: $\int \dfrac{dy}{y} = \int \dfrac{x\,dx}{2}$.

$\ln|y| = \dfrac{x^2}{4} + C.$ Exponentiate both sides: $y = e^{\frac{x^2}{4}+C} = e^{\frac{x^2}{4}} e^{C} = Ce^{\frac{x^2}{4}}.$ Now solve for C

using the initial condition: $5 = Ce^{0} = C.$ Therefore, the solution is $y = 5e^{\frac{x^2}{4}}.$

(c) Write an equation for the tangent line to the graph at $(2, 1)$. Use the tangent line to estimate
y at $x = 2.1$.

The slope of the tangent line is $\dfrac{dy}{dx} = \dfrac{(2)(1)}{2} = 1,$ so the equation of the tangent line is
$y - 1 = 1(x - 2)$, or $y = x - 1$. Therefore, at $x = 2.1, y = 2.1 - 1 = 1.1.$

About the Author

David S. Kahn studied applied mathematics and physics at the University of Wisconsin and has taught courses in calculus, precalculus, algebra, trigonometry, and geometry at the college and high school levels. He has worked as an educational consultant for many years and tutored more students in mathematics than he can count! He has worked for The Princeton Review since 1989, and, in addition to AP Calculus, he has taught math and verbal courses for the SAT, SAT Subject Tests, LSAT, GMAT, and the GRE, trained other teachers, and written several other math books.

NOTES

NOTES

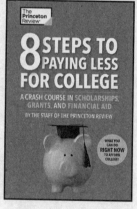